Molecular Rotation Spectra

H. W. KROTO

The School of Chemistry and Molecular Sciences
University of Sussex, Brighton, England

Dover Publications, Inc.
New York

Published in Canada by General Publishing Company, Ltd., 30 Lesmill Road, Don Mills, Toronto, Ontario.
Published in the United Kingdom by Constable and Company, Ltd., 3 The Lanchesters, 162–164 Fulham Palace Road, London W6 9ER.

This Dover edition, first published in 1992, is an unabridged, slightly corrected republication of the work first published by John Wiley & Sons, London, 1975, as "A Wiley-Interscience Publication." The author has written a new Preface for this edition.

Manufactured in the United States of America
Dover Publications, Inc., 31 East 2nd Street, Mineola, N.Y. 11501

Library of Congress Cataloging-in-Publication Data

Kroto, H. W.
Molecular rotation spectra / H. W. Kroto.
p. cm.
Originally published: New York : Wiley, c1975. With corrections.
Includes bibliographical references (p.) and indexes.
ISBN 0-486-67259-X
1. Molecular spectroscopy. 2. Molecular rotation. I. Title.
[QC454.M6K76 1992]
539′.6—dc20 92-5893
CIP

Preface to the Dover Edition

P.1 Introduction

Since 1975, when this monograph was first published, all aspects of spectroscopy have advanced dramatically. Laser techniques in particular have enabled molecular properties to be probed in the optical range with resolving powers that rival those which were only previously available to microwave spectroscopists. Microwave techniques have also been revolutionised. The Hewlett–Packard A8460 spectrometer presented microwave specialists with an instrument that freed them almost entirely from the technological problems involved in scanning the microwave range and determining the radiation frequency. This advance cannot be overemphasised, as it allowed researchers to concentrate on producing new molecules rather than on the technical problems involved in detecting them.[1]* Another advance that has taken us on further into major new fields has been the development of microwave–Fourier transform spectroscopy by Flygare and coworkers.[2] Molecular beam methods have made the study of van der Waals and hydrogen-bonded species accessible[3,4] and the marriage of the microwave Fourier transform technique[2] with pulsed supersonic nozzle systems has made such fascinating species even more amenable to detailed study.[5] These complexes are often highly non-rigid and very difficult to handle theoretically; however, the molecules of intermediate flexibility—such as the quasi-linear molecules—have in many cases been analysed with a most satisfying degree of success.[6] In addition, radioastronomy has enabled us to discover many known and also unknown molecules and ions in interstellar and circumstellar space.[7,8] Indeed "space" can be considered to be a vast sample cell with a vast range of novel experimental conditions. These range from cold dark clouds to warm stars that are bright—at least in the infrared.

Although spectroscopy has been transformed, it appears that the demand for the present monograph has diminished little over the years since it was published in 1975, and I am delighted that Dover Publications has decided to republish it. It has been a source of some satisfaction to discover so many photocopies (even bound) in so many laboratories in all parts of the world. Autographing a photocopy of one's own book has a certain curious appeal, especially as in recent years the book has been unavailable. During the period just after it was published I and my coworkers, here at Sussex, aimed at publishing spectra that displayed not only textbook examples but also the beautiful patterns of microwave spectroscopy.[1,7]

*Note: All citations of references in this new preface are to the list at the end of the preface, *not* to the main list beginning on page 278.

Some of these are presented in this special Preface to the Dover Edition. I am most grateful to my coworkers (see the References) for their contributions to this new preface.

P.2 The Spectra: Linear Molecules

In many ways the most beautiful spectra are those of linear polyatomic molecules—partly because they are the simplest to analyse. In Figure P.1 is shown the spectrum of $J=3\leftarrow2$ microwave transition of FC≡P,[9] which was produced by passing CF_3PH_2 vapour over KOH. It nicely shows l-type doubling pattern in the bending vibration satellites. The pseudo first-order Stark effect discussed in Section 7.6 is depicted in Figure P.2.

The spectrum of the first cyanopolyyne to be studied, cyanobutadiyne, HC≡C−C≡C−C≡N,[10] is shown in Figure P.3. This particular spectrum led to the detection of the polyynes in space[1,7] and ultimately to the discovery of C_{60} Buckminsterfullerene.[11,12] The molecule was originally synthesised and its rotational spectrum analysed for two reasons: one was to study how electron delocalisation along a conjugated linear chain affects the bond lengths and to probe the way in which long chains of atoms flex and rotate simultaneously. The r_s structure calculation may be readily carried out[1] from the spectrum of the $J=14\leftarrow13$ transition shown in Figure P.4. The transitions of the various singly substituted ^{13}C and ^{15}N isotopic modifications are also observed; being heavier than the most abundant species, the lines lie to low frequency (in fact they can be seen just barely in Figure P.3). The calculation is summarised in Table P.1. Note that the calculation detailed here has been simplified by neglecting centrifugal

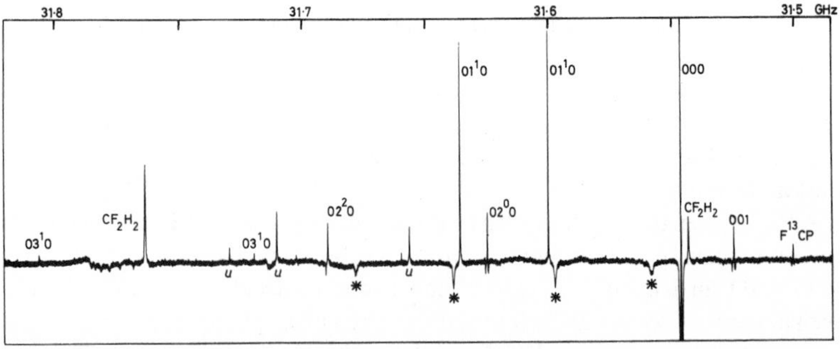

Figure P.1 The $J = 3\leftarrow2$ transition of FC≡P.[9] The ground-vibrational-state line is assigned (000), i.e. ($v_1 = 0$, $v_2 = 0$, $v_3 = 0$). The bending vibration satellite is split into two by l-type doubling. The effect can be considered as a Coriolis splitting *or* to be related to an asymmetry splitting of $K_A = 1$ line in an asymmetric top due to the fact that on bending the molecule is effectively no longer linear and B and C are no longer equal. The doublet components are labelled (01^10). The superscript indicates that the quantum number for vibrational angular momentum, $l = 1$. The $v_2 = 2$ state gives rise to a component with $l = 2$, labelled (02^20), and a state with $l = 0$ (02^00). The (02^00) state is shifted by a Fermi resonance with (001). The asterisks indicate Stark lobes belonging to the two (01^10) lines. The u-lines belong to an unidentified species. Note also that CF_2H_2 impurity lines are also detected

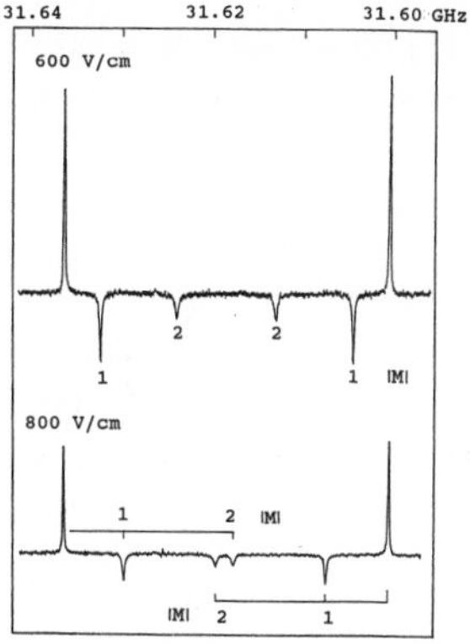

Figure P.2 The Stark effect on the (010) *l*-doublet lines is shown at two voltages, 600 V/cm *above* and 800 V/cm *below*. This voltage is too low to modulate the (02°0) satellite (Figure P.1). Note that in Figure P.1 the voltage is sufficient that the Stark lobes lie *outside* the *l*-doublets, whereas here they still lie between them

Figure P.3 The microwave spectrum of HC≡C–C≡C–C≡N.[10] Each J transition consists of a bunch of closely spaced lines; the ground-state line lies at the low-frequency end and an entourage of vibrational satellites lie to high frequency. The pattern within a bunch is shown in more detail for the $J = 14 \leftarrow 13$ transition in Figure P.4

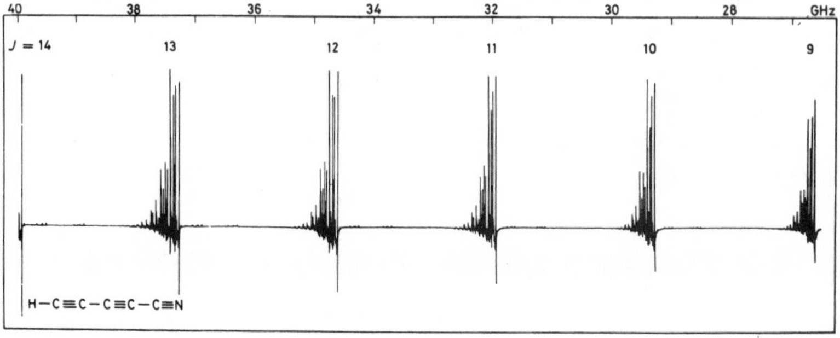

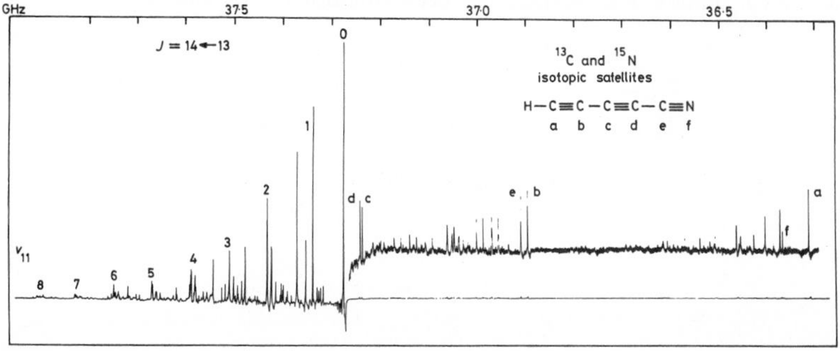

Figure P.4 The J = 14←13 transition of HC_5N under moderate resolution. The bending vibrational satellites march out with exponentially decreasing intensity to high frequency. Transitions of the various singly substituted ^{13}C and ^{15}N isotopic modifications are also observed; being heavier than the most abundant species, the lines lie to low frequency. The ground-state lines for the isotopically substituted species are identified. As the change in the moment of inertia is essentially proportional to the square of the distance of the substituted atom from the centre of mass a neat and simple semiquantitative analysis is possible assuming that all the bonds are roughly the same length and that the c. of m. is in the middle of the central triple bond. The distances of atoms d and c, e and b, and f(N) and a from the c. of m. are then roughly in the ratio 1:3:5 respectively. This should result in isotope shifts approximately in the ratio 1:9:25. From the above spectrum one can see that this is roughly correct, as the ratios are approximately 1:10.3:26.6. An accurate analysis is detailed in Table P.1

distortion, so the B values in this table are not quite correct. However, because the structure calculation utilises moment-of-inertia *differences* (cf. equation 6.143) the discrepancies cancel out and a very good structure is obtained.[1,10] As the conjugated chain lengthens, the electron density does indeed delocalise as the single bonds shorten and the triple bonds lengthen. The vibration–rotation dynamics in this molecule have also been probed[13] and the spectrum found to fit, almost perfectly, standard vibration–rotation theory (Chapter 6), as is indicated by the way in which the simulated and observed patterns match, Figure P.5. The wavefunctions associated with the lines identified in Figure P.5 have the form $\{ | l_{10} l_{11} \rangle \pm | l_{10} l_{11} \rangle \}$; the notation used[13] is simplified to $\sigma l_{10} l_{11}{}^{\sigma}$, where the prefix sign σ indicates whether the + or − combination is involved and the superscript indicates the relative sign of l_{10} and l_{11}.

In Figure P.6 the rotational spectrum of HC_7N in the range 36–39 GHz is presented.[14] This spectrum led to the detection of this species in space.[1,7] In Figure

Table P.1 Determination of the Substitution Bond Lengths (r_s) of Cyanobutadiyne H−C≡C−C≡C−C≡N[a]

Species[b]	$\Delta E(13)$[c]/ MHz	B[d]/ MHz	I[e]/ amu Å²	ΔI[f]/ amu Å²	μ[g]	r_s[h]/Å
H−C≡C−C≡C−C≡N	37 276.99	1331.321	379.6162	—	—	—
D−C≡C−C≡C−C≡N	35 589.32	1271.047	397.6179	18.0017	0.992956	4.2579
H−C*≡C−C≡C−C≡N	36 306.63	1296.665	389.7622	10.1460	0.990111	3.2011
H−C≡C*−C≡C−C≡N	36 894.99	1317.678	383.5467	3.9305	0.990111	1.9924
H−C≡C−C*≡C−C≡N	36 238.39	1329.943	380.0095	0.3933	0.990111	0.63026
H−C≡C−C≡C*−C≡N	36 242.92	1330.104	379.9635	0.3473	0.990111	0.59226
H−C≡C−C≡C−C*≡N	36 908.73	1318.169	383.4038	3.7876	0.990111	1.9559
H−C≡C−C≡C−C≡N*	36 361.62	1298.629	389.1727	9.5565	0.983955	3.1165

1.0568 1.2087 1.3621 1.2225 1.3636 1.1606
H——C≡C——C≡C——C≡N Å

1.058 1.205 1.378 1.159
H——C≡C——C≡N Å

[a] This calculation[1] has been simplified by neglecting centrifugal distortion (see text of Preface and Section 6.11).
[b] An asterisk (*) indicates the nucleus substituted.
[c] Measured frequency of $J = 14 \leftarrow 13$ transition depicted in Figure P.4.
[d] Calculated B value neglecting centrifugal distortion—i.e., $B = \{E(J+1) - E(J)\}/2(J+1)$.
[e] $I(\text{amu\AA}^2) = 505391/B(\text{MHz})$.
[f] $\Delta I = I^* - I_n$ where $I_n = I(H^{12}C_5{}^{14}N)$.
[g] $\mu = M_n \Delta m/(M_n + \Delta m)$; masses for H, D, ^{12}C, ^{13}C, ^{14}N and ^{15}N are 1.007825, 2.014102, 12.0, 13.00335, 14.00307 and 15.00011 amu respectively.
[h] $r_s = \{(I^* - I_n)/\mu\}^{1/2}$.

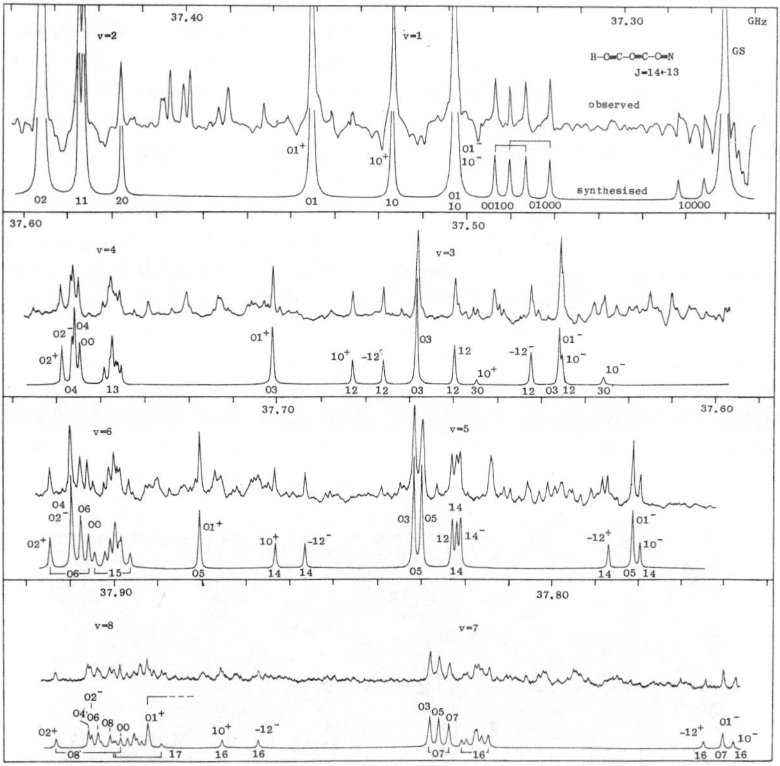

Figure P.5 A comparison[13] of the observed J = 14←13 satellite pattern of HC_5N (*above*) with the theoretically predicted pattern (*below*) for states with up to 8 quanta in the bending vibrations excited. The spectrum was synthesised for up to 8 quanta of υ_{10} and υ_{11}

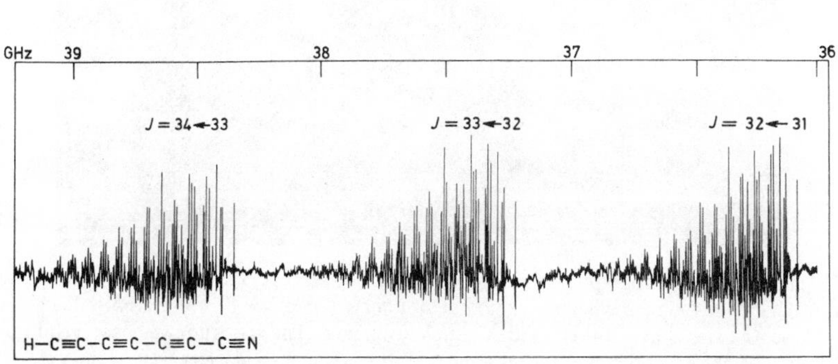

Figure P.6 The spectrum of HC_7N in the range 36–39 GHz.[14] Each J transition consists of a bunch of closely spaced lines. The ground-state transition lies at the low-frequency end of each bunch and the rest, the bending vibrational satellites, lie to higher frequency

P.7 a photograph of the computer display shows the raw data for the original detection by radioastronomy.[15,1,7] Each dot represents a signal in a 10-kHz-wide frequency channel. The spectrum was observed from a cold molecular cloud in the constellation of Taurus which yields very narrow lines no more than 10 kHz wide. The range was centered so that the line should lie in the central channels. The high signals in the central two channels indicate that radiation at the expected frequency has been detected.

The sulphido-boron molecules are isoelectronic with the phosphaalkynes, and the family (X−B=S with X = F, Cl and Br) have been studied[1]—the iodine compound was too unstable to be detected. In Figures P.8 and P.9 the spectra of ClBS[1,16] are shown. Lines of all 12 possible isotopic variants have been measured allowing substitution and double substitution, as well as first and second moment structure determinations (see Section 6.11) to be carried out. These methods have been compared for general reliability in accurate structural analysis.[16] Even though the boron atom is very close to the center of mass (ca. 0.1Å),[16] very reliable bond lengths r(Cl−B) = 1.681 ± 0.001Å and r(B=S) = 1.606 ± 0.001Å have been determined by combining r_s and first moment data.[16]

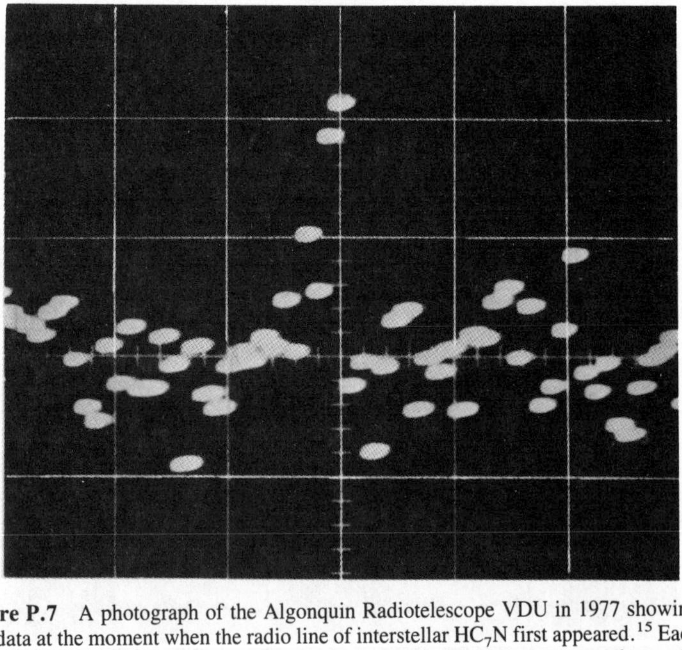

Figure P.7 A photograph of the Algonquin Radiotelescope VDU in 1977 showing the raw data at the moment when the radio line of interstellar HC_7N first appeared.[15] Each dot represents a signal in a 10-kHz-wide frequency channel. The spectrum was observed from a cold molecular cloud in the constellation of Taurus, which yields very narrow lines ca. 10 kHz wide. The range was centred so that the line should lie in the central channels

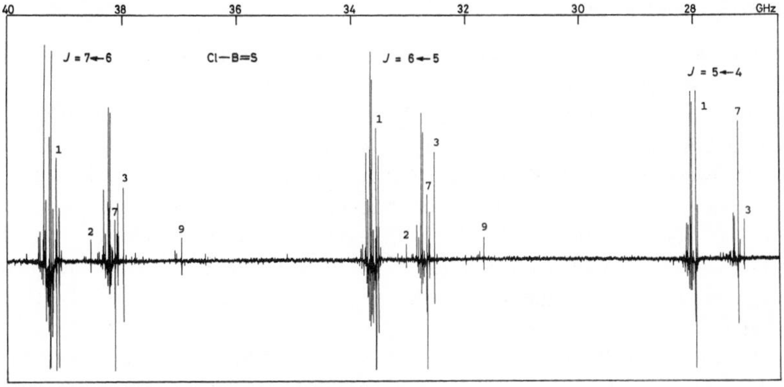

Figure P.8 The spectrum of ClBS between 26.5 and 40 GHz.[16] The more prominent ground-state lines are identified by an isotopic variant number: [1] $^{35}Cl^{11}B^{32}S$, [2] $^{35}Cl^{11}B^{33}S$, [3] $^{35}Cl^{11}B^{34}S$, [4] $^{35}Cl^{10}B^{32}S$, [7] $^{37}Cl^{11}B^{32}S$, [9] $^{37}Cl^{11}B^{34}S$, [10] $^{37}Cl^{10}B^{32}S$. As $^{35}Cl^{11}B^{32}S$ and $^{37}Cl^{11}B^{32}S$ are the most abundant variants, each J transition appears to be split into two main bunches—the higher-frequency bunch associated mainly with ^{35}Cl variants and the lower-frequency bunch with mainly ^{37}Cl variants. There are in fact 12 possible isotopic variants involving ^{35}Cl, ^{37}Cl, ^{10}B, ^{11}B, ^{32}S, ^{33}S and ^{34}S, and lines of all variants have been measured. This has made possible an exhaustively accurate structural analysis, even though the boron atom is very close to the centre of mass (ca. 0.1Å).[16] The resulting bond lengths are r(Cl−B) = 1.681 ± 0.001Å and r(B=S) = 1.606 ± 0.001Å. Note that the heights of the lines in this scan do not correspond well with the correct intensity, mainly as a result of varying Stark modulation characteristics

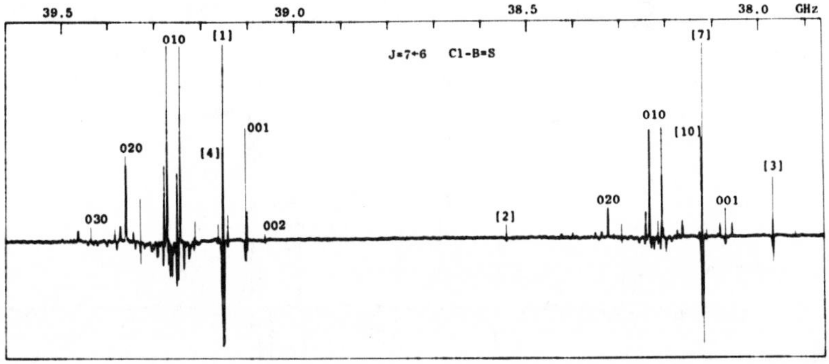

Figure P.9 The $J = 7 \leftarrow 6$ transition of ClBS. The ground-state transitions are identified by their isotopic variant number (see caption to Figure P.8). The more prominent vibrational satellites of [1] $^{35}Cl^{11}B^{32}S$ and [7] $^{37}Cl^{11}B^{32}S$ are also assigned on the LHS and RHS, respectively

P.3. Symmetric Tops

Figure P.10 shows a textbook example of a symmetric rotor transition (Section 6.10) in this case for the J = 4←3 transition of CH_3CP.[17] If the methyl group is partly substituted, as in the case of CD_2HCP shown in Figure P.11, the effect of introducing asymmetry is dramatically displayed. The spectrum[18] of CH_3BS is very similar to that of CH_3CP except that there are two abundant isotopic variants and there are quadrupole splittings to further complicate the issue. In Figure P.12 the $J = 4 \leftarrow 3$ transitions of $^{12}CH_3{}^{11}B^{32}S$ and $^{12}CH_3{}^{10}B^{32}S$ are depicted[18] and in Figure P.13 a comparison between the observed and theoretical quadrupole hyperfine splittings patterns are shown for $^{12}CH_3{}^{11}B^{32}S$ and $^{12}CH_3{}^{10}B^{32}S$ ($^{11}B(I = 3/2)$ and $^{10}B(I = 3)$).

Figures P.14–P.16 show the set of spectra for the methyl cyanopolyyne $CH_3C{\equiv}C{-}C{\equiv}C{-}C{\equiv}N$.[19,1] The way in which the bending vibrational satellites of the $J = 23 \leftarrow 22$ transition march out to high frequency is depicted in Figure P.15. Note that they tend to lie in groups with the same number of vibrational quanta excited. For the ground state, part of the *K* structure of the $J = 24 \leftarrow 23$ transition is depicted in Figure P.16. This spectrum shows the effects of spin statistics, which cause the lines with $|K|$ a multiple of three to be strong (Section 3.11c). In the case $Me_3SiC{\equiv}C{-}C{\equiv}C{-}CN$[1,20] shown in Figure P.17 the spectrum is so congested that the individual rotational lines are unresolved.

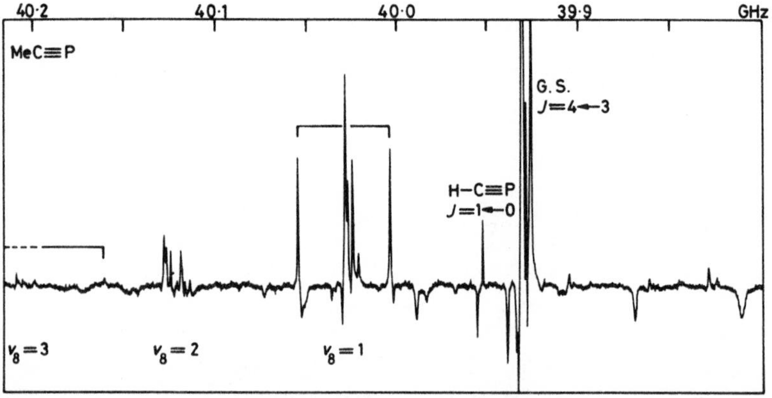

Figure P.10 The $J = 4 \leftarrow 3$ transition of CH_3CP,[17] produced by pyrolysing $CH_3CH_2PCl_2$ and eliminating two molecules of HCl. The ground-vibrational-state lines with $K = 0{-}3$ bunch together at ca. 39930 MHz. The $\upsilon_2 = 1$ satellite shows a pattern somewhat similar to that of a slightly asymmetric rotor due to *l*-doubling. The $J = 1$ line of HCP is also detected

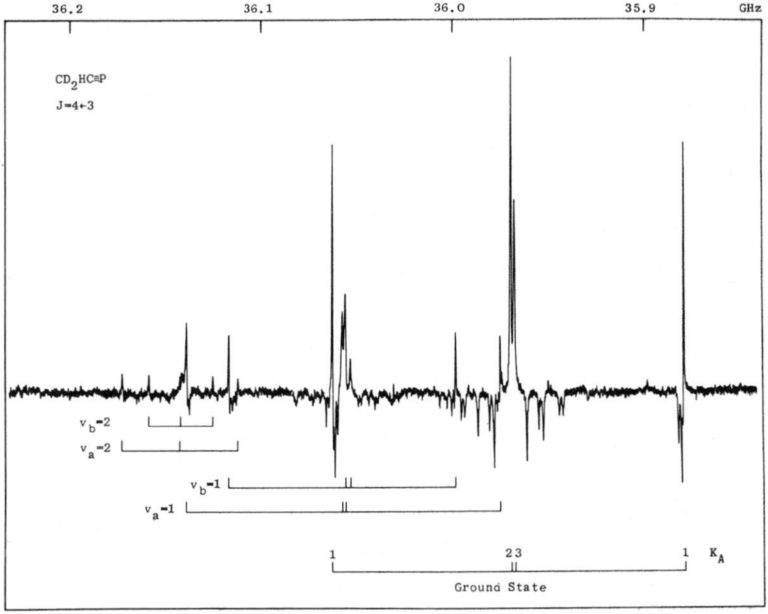

Figure P.11 The $J = 4 \leftarrow 3$ transition of CD_2HCP.[17] Isotopic substitution has produced a slightly asymmetric rotor. Note that the bending vibration is no longer degenerate and the associated pattern splits into two sets. The Stark modulation field is only 200 V/cm—at this strength the $K_A = 0$ lines are not modulated, and do not appear in the spectrum

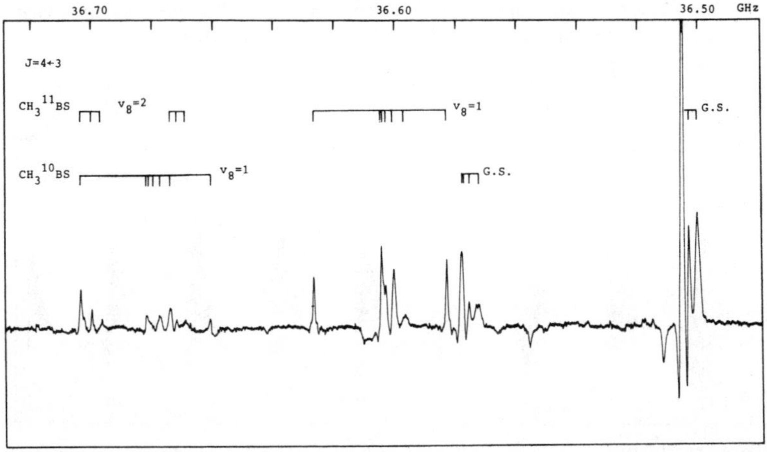

Figure P.12 Medium-resolution scan of the $J = 4 \leftarrow 3$ transition of $^{12}CH_3{}^{11}B^{32}S$ and $^{12}CH_3{}^{10}B^{32}S$.[18] The spectrum is similar to that of CH_3CP except for the fact that there are two sets of prominent isotopic variants and the lines exhibit quadrupole hyperfine splittings (Figure P.13)

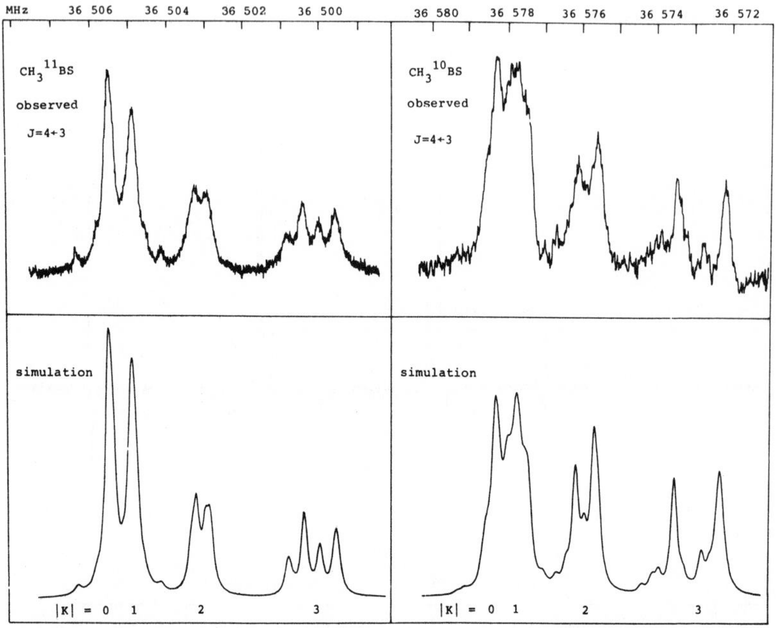

Figure P.13 A comparison of the observed ground-state $J = 4 \leftarrow 3$ transition (*above*) with theoretical simulations of the quadrupole structures (*below*).[18] The spectrum of $^{12}CH_3{}^{11}B^{32}S$ is on the left and that of $^{12}CH_3{}^{10}B^{32}S$ is on the right. The simulations use Lorentzian lineshape functions with a width at half peak height of 0.2 MHz. The spectra nicely show how quadrupole splitting increases with K and also the difference between the patterns for the $^{11}B(I = 3/2)$ and $^{10}B(I = 3)$ species

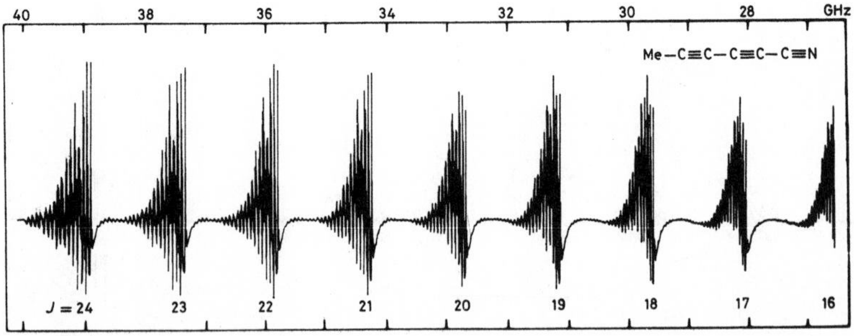

Figure P.14 Wide-band scan of the microwave spectrum of the cyanopentadiyne $CH_3(C{\equiv}C)_2N$.[19] As in the case of HC_5N the first feature at the *RH* end of each J group belongs to the vibrational ground state and the rest to bending vibrational satellites (see Figure P.15). The ground-state lines can be resolved into K multiplets under high resolution as shown in Figure P.16

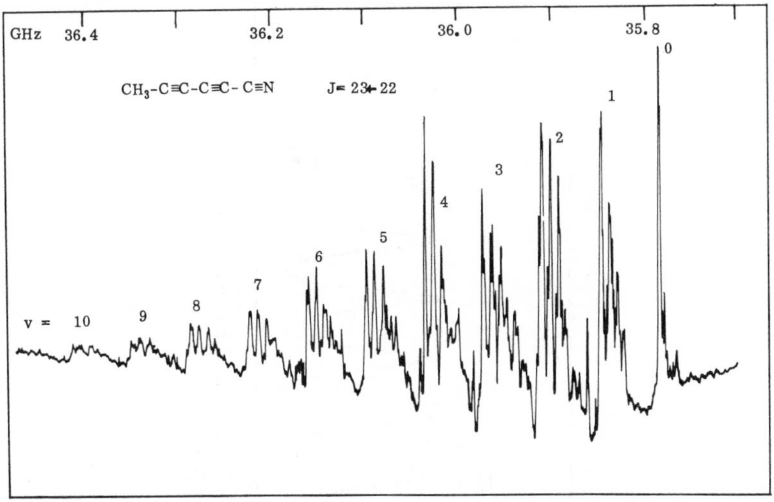

Figure P.15 The structure associated with the $J = 23 \leftarrow 22$ transition for $CH_3(C{\equiv}C)_2CN$. The bending vibrational satellites march out to high frequency and lie in groups with given numbers of vibrational quanta excited.

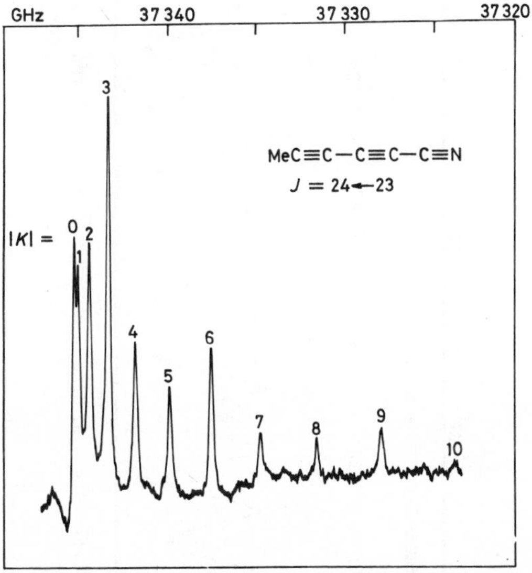

Figure P.16 Part of the K structure of the $J = 24 \leftarrow 23$ transition of $CH_3(C{\equiv}C)_2CN$. The higher K levels are not sufficiently populated so the structure peters out at $|K|>10$. Note the increased intensity of lines when $|K|$ is a multiple of 3, in agreement with C_{3v} statistical weights (see Section 3.11c)

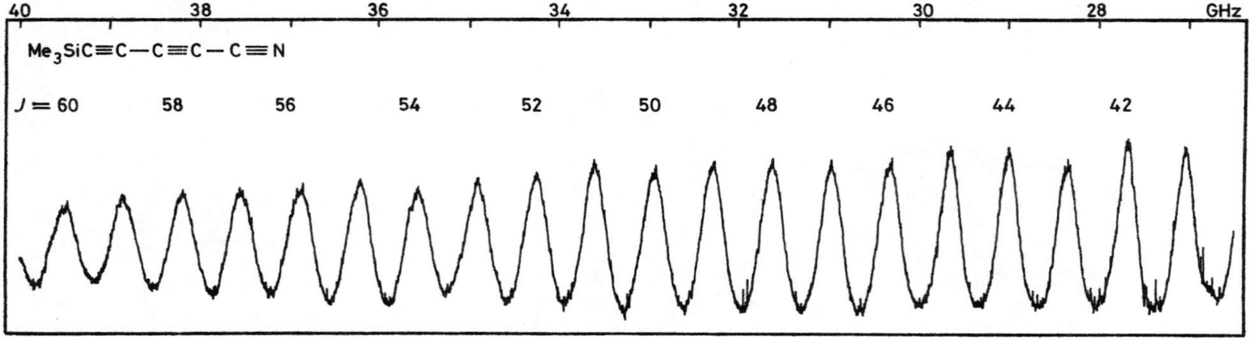

Figure P.17 The wide band of $Me_3SiC{\equiv}C{-}C{\equiv}C{-}CN$ (the precursor of HC_5N).[20] Each band for a given J consists of a multitude of unresolved bending and torsional satellites

P.4 Asymmetric Tops

The $J = 4 \leftarrow 3$ transition of $CH_2{=}PCl$ shown in Figure P.18 is an ideal example of a near prolate asymmetric rotor.[21] The widely spaced $K_A = 1$ lines stand out and the other K_A lines bunch together in the center, more or less. The moderate resolution scan in Figure P.19 depicts the K_A structure more clearly and in Figure P.20 the observed and simulated quadrupole hyperfine structures of these multiplets are compared and shown to be perfectly in accord with theory (compare with Figure 8.3 for a symmetric top).

P.5 Flexible Molecules

The microwave spectrum of NCNCS[1,22] (Figures P.21–P.23) is a beautiful example of a flexible molecule that shows the effects of quasi-linearity.[6] The molecule is wing shaped with angles at the carbon atoms close to 180° and an estimated angle of 150° at the central N atom. The low-frequency bending vibration, which involves mainly the angle at the central nitrogen atom, has levels well above the hump in the double minimum potential excited even at room temperature.[22] In such states the hump has limited influence on the rotation-vibration motion and the molecule behaves more like a linear molecule than an asymmetric rotor. The gradual change from asymmetric rotor to pseudo-linear molecule, which takes place as the bending vibrational quantum number increases, is beautifully depicted in the vibrational satellite structure of the $J = 12 \leftarrow 11$ rotational transition depicted in Figure P.22. The schematic diagram shown in Figure P.23 shows how the pattern which for the ground vibrational state is characteristic of an asymmetric top gradually metamorphoses so that for $\upsilon = 4$ it is much more characteristic of a linear molecule.

The microwave spectrum of CH_3CH_2NO is depicted in Figure P.24.[1,23] It shows the presence of a gauche isomer (two forms) in which the NO bond eclipses the methylene C—H bonds and a cis isomer in which it eclipses the C—C bond. The gauche conformer is somewhat elongated and the spectrum is compact—typical of a near prolate rotor. On the other hand the cis conformer is structurally more compact and its spectrum is more spread out and characteristic of a much more asymmetric top. Lines belonging to a $Q_{1,-1}$ branch are indicated by the series of doublet markers just below the spectrum. The doublet splitting appears to be due to tunneling between the two gauche conformers.

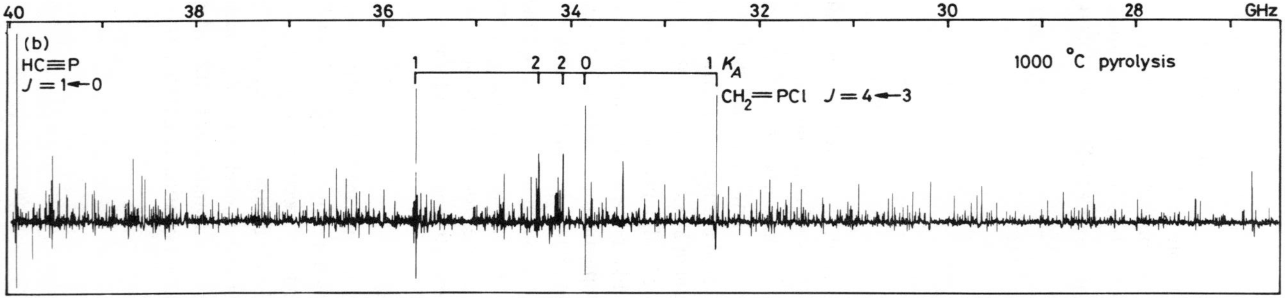

Figure P.18 On pyrolysis of CH_3PCl_2, one molecule of HCl can be eliminated to produce $CH_2{=}PCl$. The $J = 4 \leftarrow 3$ transition is depicted here[21]—only the lines of the more abundant ^{35}Cl isotopic variant are identified. The $J = 1 \leftarrow 0$ line of HCP at 39.952 GHz is also detected, indicating that two HCl fragments have been eliminated from the parent on pyrolysis

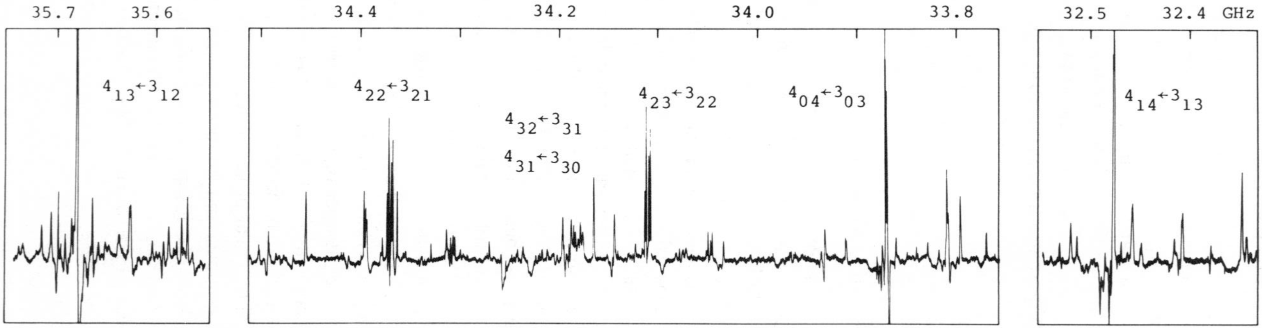

Figure P.19 The $J = 4 \leftarrow 3$ transition of $CH_2{=}P^{35}Cl$ is shown here in more detail. Some of the less intense lines belong to $CH_2{=}P^{37}Cl$ and the parent species, CH_3PCl_2

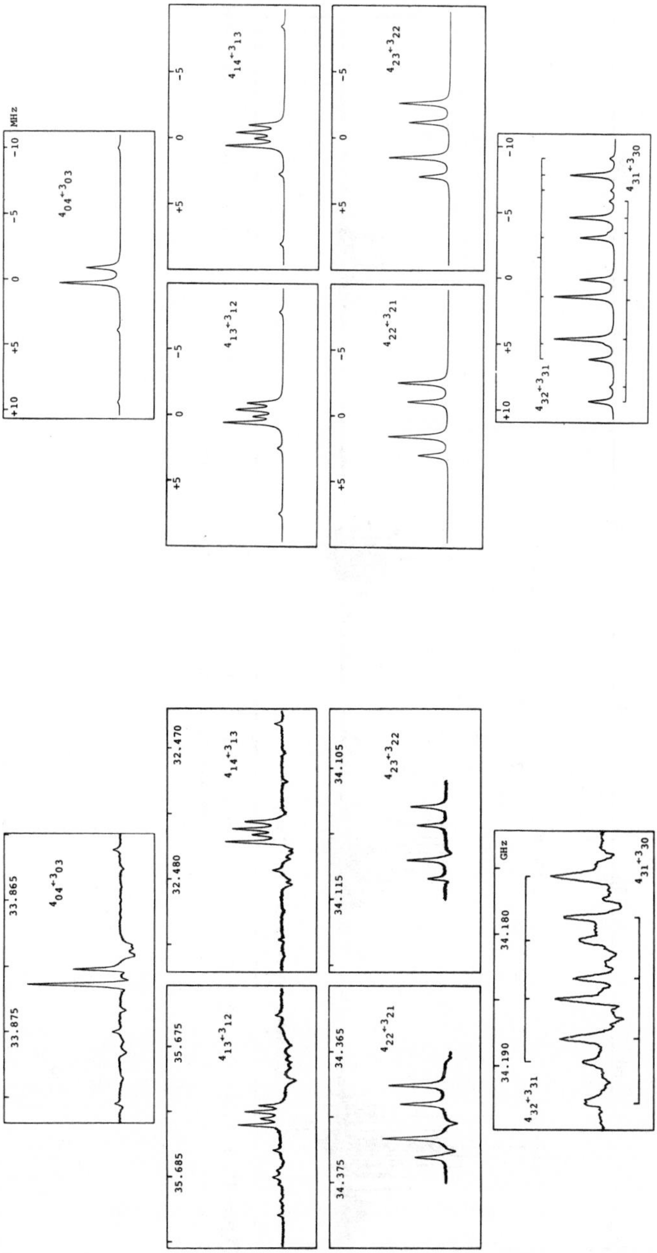

Figure P.20 The individual $J = 4 \leftarrow 3$ quadrupole multiplets for a given K_A of $CH_2{=}P^{35}Cl$: *left*, observed; *right*, computer simulation.[21] Compare these patterns with Figure 8.3

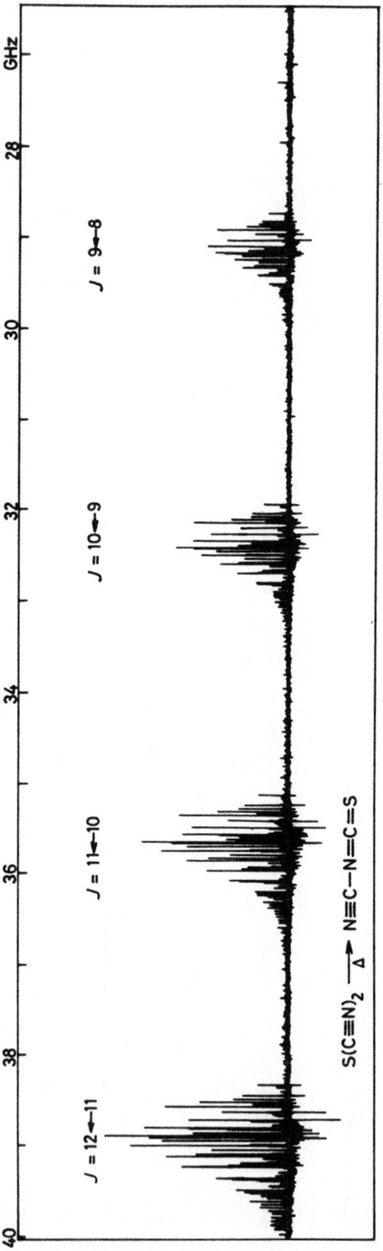

Figure P.21 The microwave spectrum of NCNCS obtained by thermolysing $S(CN)_2$.[22] The bending force constant is, however, very weak and the molecule exhibits large amplitude dynamics

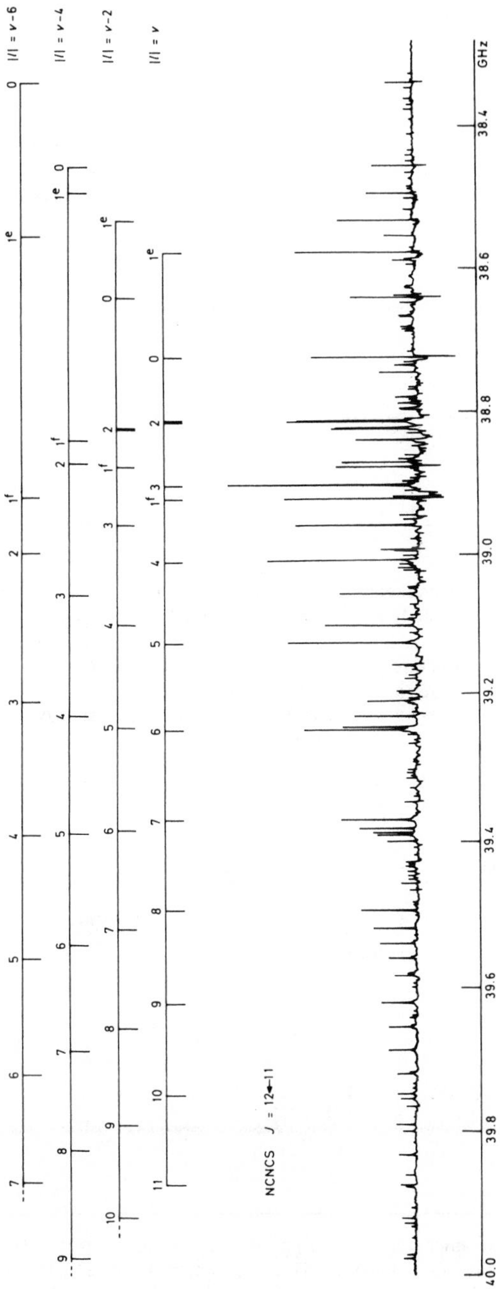

Figure P.22 The $J = 12 \leftarrow 11$ transition of NCNCS (see Figure 23)[22]

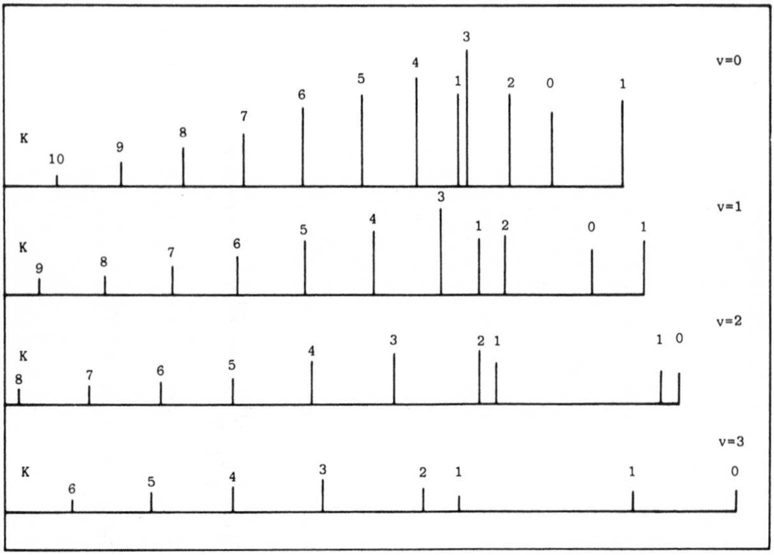

Figure P.23 Schematic diagram in which the vibrational satellite branches of Figure 22 have been collated separately to show how the characteristic spectroscopic patterns vary with increasing excitation of the bending vibration. What is elegantly depicted here is the way in which the K_A structure changes from a pattern characteristic of an asymmetric rotor at $\upsilon_7 = 0$ to a pattern more characteristic of a linear molecule when $\upsilon_7 = 4$. This spectrum is a textbook example of the way in which ambivalent character may be displayed by a quasi-linear molecule

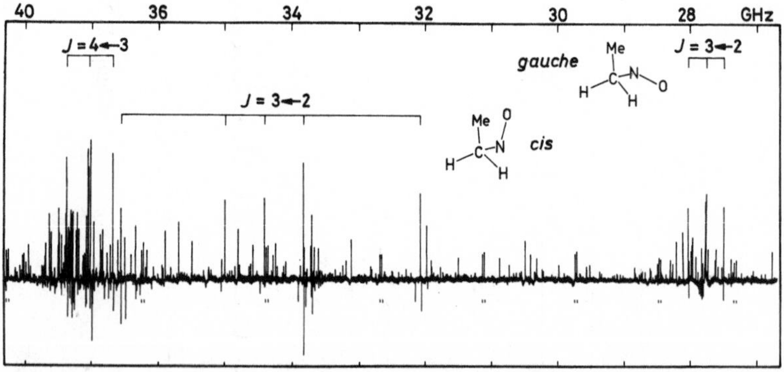

Figure P.24 The microwave spectrum of CH_3CH_2NO.[23] Two sets of spectra are identified: one for the gauche isomer and the other for the cis isomer. The lines of a Q branch of the gauche isomer (split by tunneling) are identified by markers below the spectrum

References

1. Kroto, H. W., *Chem. Soc. Revs.*, **11,** 435–491 (1982).
2. Balle, T. J., and Flygare, W. H., *Rev. Sci. Instrum.*, **52,** 33 (1981).
3. Klemperer, W., *Ber. Bunsen Phys. Chem.*, **78,** 128 (1974).
4. Dyke, T. R., *Topics in Curr. Chem.*, **120,** 85 (1984).
5. Legon, A. C., and Millen, D. J., *Chem. Soc. Revs.*, **21,** 71 (1992).
6. Winnewisser, B., in *Molecular Spectroscopy: Modern Research*, ed. K. N. Rao. Vol 3. Academic Press, N.Y. (1985).
7. Kroto, H. W., *Internat. Revs. Phys. Chem.*, **1,** 309–376 (1981).
8. Lequeux, J., and Roueff, E., *Physics Reports*, **200,** 241 (1991).
9. Kroto, H. W., Nixon, J. F., and Simmons, N. P. C., *J. Mol. Spectrosc.*, **82,** 185–192 (1980).
10. Alexander, A. J., Kroto, H. W., and Walton, D. R. M., *J. Mol. Spectrosc.*, **62,** 175–180 (1976).
11. Kroto, H. W., Heath, J. R., O'Brien, S. C., Curl, R. F., and Smalley, R. E., *Nature*, **318,** 162–163 (1985).
12. Kroto, H. W., *Angew. Chem.*(Int. Ed.), **31,** 111–129 (1992).
13. Hutchinson, M., Kroto, H. W., and Walton, D. R. M., *J. Mol. Spectrosc.*, **82,** 394–410 (1980).
14. Kirby, C., Kroto, H. W., and Walton, D. R. M., *J. Mol. Spectrosc.*, **83,** 261–265 (1980).
15. Kroto, H. W., Kirby, C., Walton, D. R. M., Avery, L.W., Broten, N. W., MacLeod, J. M., and Oka, T., *Astrophysics J.*, **219,** L133–L137 (1978).
16. Kirby, C., and Kroto, H. W. *J. Mol. Spectrosc.*, **83,** 130–147 (1980).
17. Kroto, H. W., Nixon, J. F., and Simmons, N. P. C., *J. Mol. Spectrosc.*, **77,** 270–285 (1979).
18. Kirby, C., and Kroto, H. W., *J. Mol. Spectrosc.*, **83,** 1–14 (1980).
19. Alexander, A. J., Kroto, H. W., Maier, M., and Walton, D. R. M., *J. Mol. Spectrosc.*, **70,** 84–90 (1978).
20. Alexander, A. J., Firth, S., Kroto, H. W., and Walton, D. R. M., *J. Chem. Soc. Farad. Trans.*, **88,** 531–533 (1992).
21. Kroto, H. W., Nixon, J. F., Ohashi, O., and Simmons, N. P. C., *J. Mol. Spectrosc.*, **103,** 113–124 (1984).
22. King, M. A., Kroto, H. W., and Landsberg, B. M., *J. Mol. Spectrosc.*, **113,** 1–20, (1985).
23. Hardy, J., Cox, A. P., Milverton, D., Maier, M., and Kroto, H. W., to be published.

Preface to the First Edition

I have written the sort of book that I wish had been available when I started to take an interest in spectroscopy and I have tried to answer some of the questions that I asked at that time. I have always found difficulty in formulating questions because a problem generally manifests itself as a kaleidoscope of confusing impressions. It now seems to me however, that once you have worked out how to ask the question, you also know how to answer it. As far as spectroscopy in particular is concerned it is the aim of this book to develop a blend of the physical understanding necessary to formulate problems and the rather abstract quantum mechanical expertise to solve them.

The Book is divided into two parts, In the first half, the basic theory of rotational spectroscopy is developed and applied to systems which are well described by a rigid-rotor treatment. The second half is devoted to the problems associated with non-rigidity, field effects and interactions involving electrons. It is the philosophy of the book that a true understanding requires a quantitative treatment. The spectra have thus been faithfully reproduced so that they can be used directly in analysis.

Acknowledgements

It is a pleasure to acknowledge the friendship and guidance of several people. Jim Watson has been a close friend and his advice and help have been invaluable. He has read the complete manuscript and has eliminated many obscurities, ambiguities and errors. I would like to thank Jon Hougen for his thorough Quantum Mechanical indoctrination. The absence of vector diagrams in this book is due to him. He has read the first half of the manuscript and made many helpful contributions. Takeshi Oka has somehow managed to find time in between, or more likely during, his experiments to plough through the manuscript and also made many valuable comments. Cec Costain first introduced me to microwave spectroscopy and I could not have had a more agreeable and thorough introduction. I would like to thank Richard Dixon for originally awakening my interest in spectroscopy. I am also grateful to Ian Mills for asking me to write this book in the first place.

I would also like to acknowledge the hard work and friendship of Allen Careless, Barry Landsberg and Terry Morgan, who have worked with me during the period that this book evolved. Much of the work presented in this book has been successful because of their perseverence and enthusiasm. They have all read parts of the book and helped to make it more easily understood.

I would like to thank my parents for the unselfish help they have always given. Finally I owe a very great debt to my wife Marg for the continual efforts she has made to ease the task of writing this book.

I am grateful to Professor Birnbaum and Dr Dowling for permitting me to reproduce Figures 5.3 and 11.3 respectively. The spectrum shown in Figure 7.6 was sent to me by Hewlett–Packard. All the microwave spectra were obtained using Hewlett–Packard spectrometers.

H. W. KROTO

Contents

Chapter 1

Quantum Mechanical Methods

1.1 The Description of Molecular Rotation

A detailed understanding of molecular rotation is only possible with the aid of quantum mechanics. As a consequence the quantized motion which gives rise to rotational spectra may not easily be related to the rotation of macroscopic objects with which we are familiar. In such cases the mathematical description may have to suffice. It is probably also worth pointing out that we rarely see an object rotating in free space.

In the next section we introduce the expression which describes the *overall* molecular energy. This expression can be separated to a good degree of approximation into several terms each of which can be treated individually using quantum mechanics. One term in particular, the vibrational energy, provides an ideal vehicle for a succinct summary of the various quantum mechanical steps we will need throughout this book. At the same time we develop results which are necessary for any detailed understanding of rotational spectra.

A complete derivation of the principles of quantum mechanics is given by Dirac,[67] Landau and Lifschitz[169] and Von Neumann.[312] Those aspiring to be professional theorists should note the books by Van der Waerden[308] and Ludwig[185] on the origins of quantum mechanics. There are numerous other texts; some offering concise introduction are Eyring, Walter and Kimball,[83] Dicke and Wittke,[65] and Atkins.[9]

1.2 Molecular Energy and the Quantum Mechanical Hamiltonian

A stable molecule is composed of a set of atoms and electrons arranged in some quite well-defined way. The general procedure for developing an understanding of molecular motion starts with the total energy expression, E, which is the sum of the total kinetic energy T and the total potential energy V. Thus

$$E = T + V \tag{1.1}$$

The kinetic energy can be written in terms of momenta of the nuclei $\boldsymbol{P}_n$ and electrons $\boldsymbol{P}_e$

$$T = \sum_n \frac{\boldsymbol{P}_n^2}{2m_n} + \sum_e \frac{\boldsymbol{P}_e^2}{2m_e} \tag{1.2}$$

The potential energy in this case can be written as the sum

$$V = V_{nn} + V_{ee} + V_{ne} \tag{1.3}$$

where V_{nn}, V_{ee} and V_{ne} are nucleus–nucleus, electron–electron and nucleus–electron interaction terms respectively; they are functions of the coordinates $\boldsymbol{q}_n$ and $\boldsymbol{q}_e$ and are close to Coulombic in form.

Schrödinger discovered that such expressions as (1.1) could be related to observable atomic and molecular energy-level patterns if the differential equation derived from (1.1) by replacing the vector components p_i and q_i by operators according to

$$p_i \rightarrow -i\frac{\partial}{\partial q_i} \qquad q_i \rightarrow q_i \tag{1.4}$$

could be solved.[270,185] Such a transformation applied to (1.1) yields the *Quantum Mechanical Hamiltonian.* This transformation is discussed briefly in Section A1.1. The momentum in (1.4) is written in units of $\hbar$ for the reasons discussed in Section A1.1.

$$H = -\sum_n \frac{\nabla_n^2}{2m_n} - \sum_e \frac{\nabla_e^2}{2m_e} + V_{nn} + V_{ne} + V_{ee} \tag{1.5}$$

We now discuss the very much simpler energy-level expression for a system which can be described by one momentum variable and its associated conjugate coordinate, the one and only *Simple Harmonic Oscillator.*

1.3 The Harmonic Oscillator

In this section we will not prove any of the theorems used; that is the job of the specialized texts on quantum mechanics. We will however attempt to show how they apply to this prototype example and invoke and describe the results we need.

A simple harmonic oscillator is defined as any system which can be described by a single coordinate, x, and whose motion is governed by Hooke's Law, that the force is $-kx$. Thus the associated equation of motion is

$$m\ddot{x} = -kx \tag{1.6}$$

where m is a parameter depending on the masses and structure of the system. From (1.6) we see that the classical Hamiltonian can be written as

$$H = \frac{p_x^2}{2m} + \frac{k}{2}x^2 \tag{1.7}$$

where $p_x = \partial T/\partial \dot{x} = m\dot{x}$ is the momentum conjugate to the coordinate x. This equation can be solved by Schrödinger Wave Mechanics[185] which requires that the vectors in (1.7) be transformed to operators according to (1.4) and the resulting differential equation solved. We will follow a somewhat different, though equivalent,[271,185] procedure and use the Matrix Mechanics techniques developed by Heisenberg, Born, Jordan and Dirac.[308]

The expression (1.7) is the Quantum Mechanical Hamiltonian for the system if the variables, p_x and x satisfy the *Quantum Condition* represented by the commutator relation

$$[q_i, p_j] = i\delta_{ij} \tag{1.8}$$

which they do if $p_x = -i\partial/\partial x$. We shall in this method apply the quantum condition (1.8) when necessary rather than make the substitution (1.4).

Two variants of our basic Hamiltonian (1.7) which we will need are

$$H = \tfrac{1}{2}(P^2 + \omega^2 Q^2) \tag{1.9a}$$

where $\sqrt{m}x = Q$ and $\omega = \sqrt{k/m}$, and

$$H = \tfrac{1}{2}\omega(p^2 + q^2) \tag{1.9b}$$

where $\sqrt{m\omega}\,x = q$ ($\sqrt{\omega}Q = q$). If we write a general simple harmonic oscillator expression as

$$H = \alpha p^2 + \beta q^2 \tag{1.9c}$$

Then it is worth noting that the associated frequency $\omega = 2\sqrt{\alpha\beta}$ or $\alpha\beta = \frac{1}{4}\omega^2$. All these forms satisfy the appropriate quantum condition related to (1.8). ω is essentially the classical frequency of the motion. We will in general use (1.9b), however we note that (1.9a) is closely related to the general expression for vibrational energy in terms of normal coordinates.[335]

It is our aim to determine a set of functions $|E_n\rangle$ (in Dirac notation) which satisfy an equation of the form

$$H|E_n\rangle = E_n|E_n\rangle \tag{1.10}$$

where E_n is a number or scalar, in this case the energy of the state represented by the function $|E_n\rangle$. The equation (1.10) is an *eigenvalue equation* where E_n is the eigenvalue and $|E_n\rangle$ is the eigenfunction.

The basic information we need is carried in the commutation properties which hold among the various operators representing the observables associated with the problem. In our case the main operators are H, p and q. We can add to these the family of powers p^m, q^n, $p^m q^n$, $p^m q^n p^o$, etc. as well as linear combinations of p and q and products. Of particular importance are two types of commutator

$$[\boldsymbol{A}, \boldsymbol{B}] = 0 \tag{1.11}$$

and

$$[A, B] = kB \tag{1.12}$$

where k is a scalar number. When (1.11) applies we say that the two operators A and B *commute* and it follows that functions exist which are simultaneously eigenfunctions of both operators. The associated quantities have *simultaneously* determinable eigenvalues. The second commutator (1.12) allows us to generate complete sets of functions which are eigenfunctions of A. Consider the eigenfunction $|A'\rangle$ defined by

$$A|A'\rangle = A'|A'\rangle \tag{1.13}$$

in general a primed symbol will specify the associated eigenvalue by (1.13). If we find an operator B such that (1.12) applies then we can apply both the left- and right-hand sides of (1.12) to $|A'\rangle$ and rearrange the resulting expression to give

$$A\{B|A'\rangle\} = (A' + k)\{B|A'\rangle\} \tag{1.14}$$

(1.14) implies that the effect of the operation of B on $|A'\rangle$ is to generate a *new eigenfunction* of A with eigenvalue $(A' + k)$. Repeated application of B will clearly generate the set of functions $B^n|A'\rangle$ with eigenvalues $(A' + nk)$. Operators of this type are called *Ladder* or *Shift* operators.

The basic commutation relations for the harmonic oscillator problem (1.9b) are

$$[q, p] = i, \qquad [H, q] = \omega(-ip), \qquad [H, p] = \omega iq \tag{1.15}$$

This set of relations indicates that the eigenvalues of q and p cannot be determined simultaneously with either, each other or the energy of the system, a result that is the basis of the Heisenberg Uncertainty Principle. Inspection of the relations (1.15) indicates the operators

$$F^{\pm} = q \pm ip \tag{1.16$^{a}_{b}$}$$

satisfy a commutator of the type (1.12) with H, and are thus shift operators. Thus

$$[H, F^{\pm}] = \mp\omega F^{\pm} \tag{1.17$^{a}_{b}$}$$

and H can be factorized using $F^{\pm}$ as

$$H = \tfrac{1}{2}\omega(F^{\pm}F^{\mp} \mp 1) \tag{1.18$^{a}_{b}$}$$

Shift operators usually have a form similar to (1.16) and their commutation relations are usually most conveniently derived from relations such as (1.18). For instance we obtain (1.17a) by post-multiplying (1.18a) by F^{+} and subtracting the relation obtained by pre-multiplying (1.18b) by F^+. Application of (1.17) to the eigenfunction defined by $H|E_n\rangle = E_n|E_n\rangle$ yields

$$H\{F^{\pm}|E_n\rangle\} = (E_n \mp \omega)\{F^{\pm}|E_n\rangle\} \qquad (1.19^{a}_{b})$$

which is of the form (1.14). If we concentrate on (1.19a) we see that F^+ ladders down through eigenfunctions which have eigenvalues successively reduced by *units* of ω. At some point we must come to a function $E_{\downarrow}$ where another application would yield a negative energy. The Hamiltonian is a sum of squared Hermitian operators which are defined as having only *real* eigenvalues and thus H may only have *real and positive* eigenvalues. In matrix language the Hermitian condition requires that a matrix operator be self-adjoint. The adjoint $\mathbf{A}^{\dagger}$ of a matrix $\mathbf{A}$ ($\equiv A_{mn}$) is the complex conjugate of the transpose i.e. $(\mathbf{A}^{\dagger})_{mn} = (A_{nm})^*$ and self-adjoint or Hermitian matrices are matrices for which $\mathbf{A}^{\dagger} = \mathbf{A}$.

The Hermitian requirement may only be reconciled with (1.19a) if

$$F^+|E_{\downarrow}\rangle = 0 \qquad (1.20)$$

i.e. F^+ annihilates $|E_{\downarrow}\rangle$. Application of both the left- and right-hand sides of (1.18b) on $|E_{\downarrow}\rangle$ indicates that

$$H|E_{\downarrow}\rangle = \tfrac{1}{2}\omega|E_{\downarrow}\rangle \qquad (1.21)$$

This procedure has enabled us to show that the set of functions $|E_n\rangle$ has a lower bound function with an energy $\frac{1}{2}\omega$, known as the zero point energy. Laddering up by successive application of F^- allows us to generate the infinite set of eigenfunctions each separated by energy quanta of ω. The general energy-level expression can be written as

$$E(v) = \omega(v + \tfrac{1}{2}) \qquad (1.22)$$

where we can now specify the eigenfunctions by a quantum number v. We have in this way determined the diagonal matrix of H where the elements $\langle v|H|v\rangle = E(v)$, which can be written

H	$\|0\rangle$	$\|1\rangle$	$\|2\rangle$	$\|3\rangle$	
$\langle 0\|$	$\frac{1}{2}\omega$	0	0	0	.
$\langle 1\|$	0	$\frac{3}{2}\omega$	0	0	.
$\langle 2\|$	0	0	$\frac{5}{2}\omega$	0	.
$\langle 3\|$	0	0	0	$\frac{7}{2}\omega$	.
	.	.	.	.	.

(1.23)

The matrices of p and q in the *basis* or *representation* $|v\rangle$ which diagonalizes H are not diagonal as is implied by the fact that neither p nor q commute with H (1.15). The matrix properties of p and q can be determined from those of $F^{\pm}$. As implied by (1.19)

$$F^{+}|v\rangle = N_v|v-1\rangle \tag{1.24}$$

where N_v is a number. A self-consistent set of numbers, N_v, can be determined for a normalized set of functions $|v\rangle$ (i.e. $\langle v|v\rangle = 1$).

$$\langle v|F^{-}F^{+}|v\rangle = N_v^{*}N_v \tag{1.25}$$

Note that $F^{\pm}$ are not Hermitian and that $(F^{\pm})^{\dagger} = F^{\mp}$. Using (1.18b) together with (1.25) we see that‡

$$|N_v|^2 = \left\langle v\left|\frac{2H}{\omega} - 1\right|v\right\rangle = 2v \tag{1.26}$$

A similar procedure can be applied to derive the appropriate factor for F^{-}. Thus we see that

$$\langle v-1|F^{+}|v\rangle = \sqrt{2v} \qquad \langle v+1|F^{-}|v\rangle = \sqrt{2(v+1)} \tag{1.27}$$

The matrix elements of p and q are now directly obtainable as $q = \frac{1}{2}(F^{-} + F^{+})$ and $p = \frac{1}{2}i(F^{-} - F^{+})$. The matrices thus can be written out as

$$q = \alpha\begin{bmatrix} 0 & \sqrt{1} & 0 & \cdot & \cdot \\ \sqrt{1} & 0 & \sqrt{2} & \cdot & \cdot \\ 0 & \sqrt{2} & 0 & \sqrt{3} & \cdot \\ 0 & 0 & \sqrt{3} & 0 & \cdot \\ \cdot & \cdot & \cdot & \cdot & \cdot \end{bmatrix} \qquad p = i\alpha\begin{bmatrix} 0 & -\sqrt{1} & 0 & 0 & \cdot \\ \sqrt{1} & 0 & -\sqrt{2} & 0 & \cdot \\ 0 & \sqrt{2} & 0 & -\sqrt{3} & \cdot \\ 0 & 0 & \sqrt{3} & 0 & \cdot \\ \cdot & \cdot & \cdot & \cdot & \cdot \end{bmatrix} \tag{1.28}$$

where $\alpha = 1/\sqrt{2}$.

The explicit expressions for the matrix elements are listed in Table 1.1. The matrices of p and q are not diagonal but they are Hermitian.

We should not pass on without a few words about phase conventions; this problem was glossed over at step (1.25). There is an arbitrariness in that (1.26) can be satisfied by any number of the form $N_v = A\,e^{i\delta}$, and in this case we have in fact chosen $\delta = 0$. A common alternative phase convention has the matrix of q imaginary, rather than that of p, the result we would obtain if we had chosen an imaginary phase convention.

‡ *Note* $\int \psi^{*}\psi\, d\tau \equiv (F^{+}|v\rangle)^{\dagger}F^{+}|v\rangle = \langle v|F^{-}F^{+}|v\rangle$.

Table 1.1 Matrix Elements of q, q^2 and q^3 [a]

	$\langle v\|q\|v'\rangle$	$\langle v\|q^2\|v'\rangle$	$\langle v\|q^3\|v'\rangle$
$\langle v\|q^n\|v+3\rangle$			$\alpha^3\sqrt{(v+1)(v+2)(v+3)}$
$\langle v\|q^n\|v+2\rangle$		$\alpha^2\sqrt{(v+1)(v+2)}$	
$\langle v\|q^n\|v+1\rangle$	$\alpha\sqrt{v+1}$		$\alpha^3 3(v+1)\sqrt{v+1}$
$\langle v\|q^n\|v\rangle$		$\alpha^2(2v+1)$	
$\langle v\|q^n\|v-1\rangle$	$\alpha\sqrt{v}$		$\alpha^3 3v\sqrt{v}$
$\langle v\|q^n\|v-2\rangle$		$\alpha^2\sqrt{v(v-1)}$	
$\langle v\|q^n\|v-3\rangle$			$\alpha^3\sqrt{v(v-1)(v-2)}$

[a] $\alpha = 1/\sqrt{2}$.

It is possible to develop an explicit form for the associated wavefunctions using (1.20), $F^+|0\rangle = 0$. The resulting differential equation is

$$\left(q + \frac{\partial}{\partial q}\right)\psi_0 = 0 \tag{1.29}$$

which has solutions $\psi_0 = A\,e^{-1/2q^2}$. Hence we can generate the complete set of wavefunctions according to $(q + \partial/\partial q)^v\psi_0 = N\psi_v$.

1.4 Non-diagonal Matrices

We are often confronted with matrices which are not diagonal and in these cases we in general need to diagonalize them. As an example we consider the perturbed harmonic oscillator.

1.4a The Perturbed Harmonic Oscillator Hamiltonian

We can expand a general potential function, V, such as that which governs the vibrational motion of a diatomic molecule in the form

$$V = \frac{1}{2}\left[\frac{\partial^2 V}{\partial q^2}\right]_0 q^2 + \frac{1}{6}\left[\frac{\partial^3 V}{\partial q^3}\right]_0 q^3 + \frac{1}{24}\left[\frac{\partial^4 V}{\partial q^4}\right]_0 q^4 + \cdots \tag{1.30}$$

The Hamiltonian for such a system can thus be written

$$H = H^0 + H' \tag{1.31a}$$

$$H^0 = \tfrac{1}{2}\omega(p^2 + q^2) \tag{1.31b}$$

$$H' = \tfrac{1}{6}\varphi_3 q^3 + \tfrac{1}{24}\varphi_4 q^4 + \cdots \tag{1.31c}$$

where $(\partial^3 V/\partial q^3)_0 = \varphi_3$ etc. It is usually convenient to set up the matrix of the Hamiltonian (1.31a) in a harmonic oscillator basis defined by H^0—because in many cases the expansion (1.30) converges quite rapidly and $H^0 \gg H'$. There are however cases when this is not so. For instance, in some small ring compounds such as trimethylene oxide a ring puckering motion occurs in which quartic potential terms can be dominant (see Section 9.4). Chan and Stelman have discussed this type of vibrational problem in terms of a quartic oscillator basis.[44] In general the solution of the problem is facilitated if a basis can be found for which on average the *mn*th off-diagonal elements are an order of magnitude smaller than the *differences* between the *mm*th and *nn*th diagonal elements. We now need to determine the matrix elements of H' in the $|v\rangle$ basis and add the resulting values to the matrix of H^0 given in (1.23). The matrix elements of q^2, q^3, q^4, etc. can be determined by straightforward matrix multiplication. Thus using

$$\langle v'|q^{m+n}|v''\rangle = \sum_{v'''} \langle v'|q^m|v'''\rangle\langle v'''|q^n|v''\rangle \tag{1.32}$$

we can develop the matrix elements listed in Table 1.1 from those just derived for q. The additional contributions, in general, involve off-diagonal entries and these extra terms when added to the matrix of H_0 yield a matrix of the form (quartic and higher potential terms have been dropped for simplicity)

$$\begin{bmatrix} \frac{1}{2}\omega & 3\beta & 0 & \sqrt{6}\beta & 0 & \cdot \\ 3\beta & \frac{3}{2}\omega & 6\sqrt{2}\beta & 0 & 2\sqrt{6}\beta & \cdot \\ 0 & 6\sqrt{2}\beta & \frac{5}{2}\omega & 9\sqrt{3}\beta & 0 & \cdot \\ \sqrt{6}\beta & 0 & 9\sqrt{3}\beta & \frac{7}{2}\omega & 48\beta & \cdot \\ \cdot & \cdot & \cdot & \cdot & \cdot & \cdot \end{bmatrix} \tag{1.33}$$

The row and column labels are the same as in (1.23) and have been dropped for simplicity. $\beta = \frac{1}{6}\varphi_3\alpha^3$ and $\alpha = 1/\sqrt{2}$.

1.4b Matrix Diagonalization

The Hamiltonian matrix can be diagonalized using a similarity transformation $\mathbf{U}^{-1}\mathbf{HU}$. $\mathbf{U}$ is a matrix the columns of which consist of the eigenvectors of $\mathbf{H}$ according to

$$\mathbf{HU} = \mathbf{UE} \tag{1.34}$$

where $\mathbf{E}$ is the *diagonal* eigenvalue matrix. We solve for the eigenvalues by determining the roots of the secular equation

$$|\mathbf{H} - \mathbf{EI}| = 0 \tag{1.35}$$

where $\mathbf{I}$ is the unit matrix. In matrix–vector notation the basis functions are defined by

$$H^0|v\rangle = E_v^0|v\rangle \tag{1.36}$$

(see equation (1.10)) and the transformation essentially corresponds to taking a new representation whose basis functions are linear combinations of those originally chosen, i.e.

$$|\tilde{v}\rangle = \sum_{v} c_{v}^{\tilde{v}}|v\rangle \tag{1.37}$$

As an example, on diagonalization of the matrix (1.33) we obtain modified eigenvalues and the columns of the transformation matrix are the set of coefficients $c_v^{\tilde{v}}$ in (1.37).

It is now commonplace to use computer library subroutines to diagonalize matrices numerically. In some cases the problem can to a good degree of approximation be reduced to the diagonalization of a 2×2 matrix. Consider the two-state system with basis states $|m\rangle$ and $|n\rangle$ defined by

$$H^0|m\rangle = E_m^0|m\rangle, \qquad H^0|n\rangle = E_n^0|n\rangle \tag{1.38}$$

If the total Hamiltonian is $H = H^0 + H'$ where H' has only off-diagonal matrix elements, $\alpha = \langle m|H|n\rangle$, then the secular determinant is

$$\begin{vmatrix} E_m^0 - E & \alpha \\ \alpha^* & E_n^0 - E \end{vmatrix} \tag{1.39}$$

which yields the two solutions

$$E = \tfrac{1}{2}(E_m^0 + E_n^0) \pm \tfrac{1}{2}\sqrt{4|\alpha|^2 + \delta^2} \tag{1.40}$$

where $\delta = E_m^0 - E_n^0$. The transformation $\mathbf{U}$ which diagonalizes (1.39) can be written in trigonometric form as the rotation matrix

$$\mathbf{U} = \begin{bmatrix} \cos\theta & \sin\theta \\ -\sin\theta & \cos\theta \end{bmatrix} \tag{1.41}$$

in terms of a phase angle θ. The requirement that the off-diagonal elements of the new matrix $\tilde{\mathbf{H}} = \mathbf{U}^{-1}\mathbf{H}\mathbf{U}$ vanish indicates that

$$\tan 2\theta = -2\alpha/\delta \tag{1.42}$$

The eigenvectors of H are therefore

$$|\tilde{m}\rangle = \cos\theta|m\rangle + \sin\theta|n\rangle \tag{1.43a}$$

$$|\tilde{n}\rangle = -\sin\theta|m\rangle + \cos\theta|n\rangle \tag{1.43b}$$

When the basis states are degenerate $E_m^0 = E_n^0$ and $\delta = 0$ then $\theta = 45°$ and the new wave functions are $(|m\rangle \pm |n\rangle)/\sqrt{2}$ and the energy is $E = E^0 \pm \alpha$. These results are illustrated in Figure 1.1.

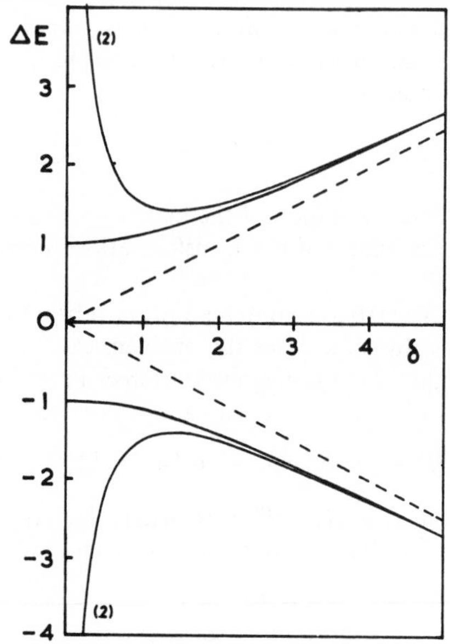

Figure 1.1 Perturbation of two levels as a function of their separation, δ. The line --- represents the positions of the unperturbed levels. The solid lines passing through $\Delta E = +1$ and -1 depict the results of evaluating the interaction energy shifts exactly using (1.40). The interaction term $\alpha = 1$, and δ is varied from 0 to 5. As a comparison, the curves labelled (2), which are the result of applying perturbation theory (1.48) to second order, are also shown. As $\delta \to 0$ the second-order perturbation treatment is no longer valid as the energy becomes infinite. When $\alpha/\delta \sim \frac{1}{2}$ the second-order perturbation result starts to become acceptable

1.4c Perturbation Theory

An approximate procedure which provides a powerful alternative to direct diagonalization is perturbation theory. The general theory is discussed by numerous authors.[83,169] In those cases where the first- and second-order theory is adequate, useful and succinct algebraic relations can often be developed which can present a useful aid for understanding problems. We can summarize the main results necessary for our purposes:

$$H = H^0 + H' \tag{1.44}$$

$$E = E^0 + E^1 + E^2 + \cdots \tag{1.45}$$

In this case H^0 is the zeroth-order Hamiltonian whose eigenfunctions $|m\rangle$ define the basis representation and whose eigenvalues are E^0 according to $H^0|m\rangle = E_m^0|m\rangle$. H' is a correction term to the basis Hamiltonian whose effect is to introduce correction terms $E^1, E^2, \ldots$ etc. to the zeroth-order eigenvalue E^0. These terms are

$$E_m^0 = \langle m|H^0|m\rangle \tag{1.46a}$$

$$E_m^1 = \langle m|H'|m\rangle \tag{1.46b}$$

$$E_m^2 = \sum_n{}' \frac{\langle m|H'|n\rangle\langle n|H'|m\rangle}{E_m^0 - E_n^0} \tag{1.46c}$$

There are two related provisos that must be mentioned. One is that the summation is limited to $n \neq m$, and the second is that when $E_m^0 \simeq E_n^0$ the denominator in (1.46c) $\rightarrow 0$ and the approximation will break down. When two levels become degenerate one can usually treat the two levels separately by the exact procedure given in Section 1.4b.

As a simple quantitative guide to the range of validity of second-order perturbation theory we can study the application to the two-state system considered previously. The exact result is given by the expression (1.40). If we rearrange this expression to

$$E = \tfrac{1}{2}(E_m^0 + E_n^0) \pm \tfrac{1}{2}(E_m^0 - E_n^0)\sqrt{1 + \frac{4\alpha^2}{\delta^2}} \tag{1.47}$$

and expand the square-root term we see that we obtain the result

$$E = \tfrac{1}{2}(E_m^0 + E_n^0) \pm \tfrac{1}{2}(E_m^0 - E_n^0) \pm \left[\frac{\alpha^2}{\delta} - \frac{\alpha^4}{\delta^3} + \cdots\right] \tag{1.48}$$

Dropping the terms higher than α^2/δ we see that two levels are 'pushed' apart by the term α^2/δ. This is exactly the result we would have obtained if we had applied perturbation theory to second order using the expression (1.46c). The result of this term is plotted in Figure 1.1 as a comparison with the exact result. If on the other hand $\alpha \gg \delta$ as is the case when $\delta \rightarrow 0$ we see that (1.47) becomes

$$E \simeq \tfrac{1}{2}(E_m^0 + E_n^0) \pm \alpha \tag{1.49}$$

Chapter 2

The Hamiltonian for a Rotating Molecule

2.1 The Born–Oppenheimer Separation of Electronic and Nuclear Motion

The complete molecular Hamiltonian given by relation (1.5) involves a summation of several terms over all the particles—nuclei as well as electrons—in the molecule. The solution of this Hamiltonian is in general simplified by the Born–Oppenheimer approximation which is valid, at least as a starting point, in the majority of molecular problems. This approximation[24] (an English version is given by Born and Huang[23]) takes into account the fact that the electrons are much lighter than the atomic nuclei and thus tend to move very much more quickly. For instance, if we consider a rigid nuclear framework tumbling in isotropic space, then the electronic distribution is assumed to follow the framework and not lag behind. When distortion of the nuclear framework takes place the electrons are assumed to adjust immediately to give the new resultant *equilibrium* distribution associated with the deformed nuclear environment. This situation is essentially fulfilled when small amplitude vibrational motion occurs.

The electrostatic forces which govern the motion of the electrons in a molecule are of the same order of magnitude as those which govern the motion of the nuclei. If we assume approximately simple harmonic oscillations of the various particles then $\omega = \sqrt{k/m}$, where ω is the frequency associated with the motion, k is the harmonic force constant which we assume to be approximately the same for electrons and atoms and m is the reduced mass (Chapter 1). This rough analysis implies that the ratio of the nuclear vibrational energy, E_v, to the electronic energy, E_e, is

$$\frac{E_v}{E_e} \sim \frac{\omega_n}{\omega_e} \sim \frac{\delta r_n^2}{\delta r_{el}^2} \sim \left[\frac{m_e}{m_n}\right]^{1/2} = \kappa^2 \qquad (2.1)$$

$\delta \boldsymbol{r}_{el}$ and $\delta \boldsymbol{r}_n$ are electronic and nuclear displacements and m_e and m_n are the electronic and nuclear masses respectively. Born and Oppenheimer systematically applied perturbation theory to this problem using $\kappa = (m_e/m_n)^{1/4}$ as defined in (2.1) as the expansion parameter. They deduced that the vibrational energy was separable from the electronic energy because $m_e \ll m_n (\kappa \sim 0{\cdot}1)$ and that $E_v \sim \kappa^2 E_e$. An extension of the treatment to fourth order yields the result that $E_r \sim \kappa^2 E_v \sim \kappa^4 E_e$ where E_r is the energy associated with rotational motion. The approximation is valid because in general the non-separable terms are of order $\kappa^2 E_r$ or less. This further order-of-magnitude separation of rotation and vibration is discussed in more detail in Chapter 6.

The result of the separation of electronic from nuclear motion can be summarized as follows. Equation (1.5) can to a good approximation be written as

$$H = H_e(\boldsymbol{q}_e\boldsymbol{q}_n) + H_n(\boldsymbol{q}_n) \tag{2.2}$$

where

$$H_e(\boldsymbol{q}_e\boldsymbol{q}_n) = -\frac{1}{2m_e}\sum_e \nabla_e^2 + V_{ee} + V_{ne} + V_{nn} \tag{2.3}$$

$$H_n(\boldsymbol{q}_n) = \sum_n -\frac{1}{2m_n}\nabla_n^2 \tag{2.4}$$

If the eigenfunction of H_e is $\psi_e(\boldsymbol{q}_e\boldsymbol{q}_n)$ then the approximate eigenfunction of H can be written as the product function

$$\Psi(\boldsymbol{q}_e\boldsymbol{q}_n) = \psi_e(\boldsymbol{q}_e\boldsymbol{q}_n)\psi_n(\boldsymbol{q}_n) \tag{2.5}$$

This will be a good approximation as long as terms such as $\nabla_n^2\psi_e(\boldsymbol{q}_e\boldsymbol{q}_n)$ can be neglected. This is in general the case in rotational spectroscopy where we are usually dealing with the ground electronic state and electronic degeneracy is rare.

For a particular internuclear configuration specified by $\boldsymbol{q}_n$ the electronic energy $E_e(\boldsymbol{q}_n)$ would be given by

$$H_e(\boldsymbol{q}_e\boldsymbol{q}_n)\psi_e(\boldsymbol{q}_e\boldsymbol{q}_n) = E_e(\boldsymbol{q}_n)\psi_e(\boldsymbol{q}_e\boldsymbol{q}_n) \tag{2.6}$$

and thus

$$[H_n(\boldsymbol{q}_n) + E_e(\boldsymbol{q}_n)]\psi_n(\boldsymbol{q}_n) = E_{ne}\psi_n(\boldsymbol{q}_n) \tag{2.7}$$

Note that if we consider a motion during which $\boldsymbol{q}_n$ varies, equation (2.7) is only valid if the electronic distribution equilibrates instantly (called by Van Vleck the 'clamped nuclei' approximation.[310]) Then $E_e(\boldsymbol{q}_n)$ depends on $\boldsymbol{q}_n$ and can be considered as the potential which governs nuclear motion. In the context of this approximation the nuclear wavefunctions are independent of the electronic coordinates for a particular electronic state. When electronic states are close in energy the electronic wave function may change rapidly with nuclear configuration, this implies that the term $\nabla_n^2\psi_e(\boldsymbol{q}_e\boldsymbol{q}_n)$ is large and may not be neglected. The Born–Oppenheimer approximation then breaks down. This usually only happens when the electronic state is degenerate or nearly so.

We shall in this book assume that the Hamiltonian of equation (2.7) is the correct basis Hamiltonian for nuclear motion. A simple justification of this assumption is given by Eyring, Walter and Kimball.[83] We now go on to consider H_n in detail and how this may itself be separated.

2.2 The Partition of the Nuclear Kinetic Energy

The classical nuclear kinetic energy is a sum over the kinetic energies of the individual atoms which make up the system.

$$T = \tfrac{1}{2} \sum_n m_n \dot{\boldsymbol{\rho}}_n \cdot \dot{\boldsymbol{\rho}}_n \tag{2.8}$$

where $\boldsymbol{\rho}_n$ is the position vector of the nth particle relative to some arbitrary set of axes and m_n the mass. In the case of an isolated molecule the nuclei are in general constrained relative to one another such that they deviate only by small displacements from an equilibrium structure. Some molecules possess internal degrees of freedom such as internal rotation; these flexible molecules are considered separately in Chapter 9. This equilibrium structure is the one we *should* mean when we discuss the structure of a molecule and in Chapter 6 the problem of deducing its parameters from experiment is considered. The coordinates, $\boldsymbol{\rho}_n$, and thus molecular motion itself may usefully be described in terms of the displacements from the equilibrium structure taken as a reference frame ($\boldsymbol{r}_n^e$) and the behaviour of this frame, relative to a space-fixed coordinate system.

The relation

$$\boldsymbol{\rho}_n = \boldsymbol{R} + \mathbf{S}^{-1}(\boldsymbol{r}_n^e + \boldsymbol{d}_n) \tag{2.9}$$

takes the above structural considerations neatly into account,[131,133,28] and with the aid of Figure 2.1, can be interpreted as follows. We wish to relate the overall motion of a quivering molecule, which is also tumbling, to arbitrary *space-fixed* cartesian axes x, y and z. A new set of axes, a, b and c is introduced which moves with reference frame of the molecule and is called the *molecule-fixed axis system.* We shall assume in this book that the equilibrium and reference structures coincide, though this is not necessary.[120] The coordinates $\boldsymbol{r}_n^e$ describe an equilibrium nuclear reference geometry and $\boldsymbol{d}_n$ the displacements from this reference geometry in terms of molecule-fixed coordinates. The $\boldsymbol{d}_n$ of course are not fixed and vary with time.

The origin of the molecule-fixed axis system is chosen to lie at the centre-of-mass of the framework, which is defined by the condition

$$\sum_n m_n \boldsymbol{r}_n^e = \boldsymbol{0} \tag{2.10}$$

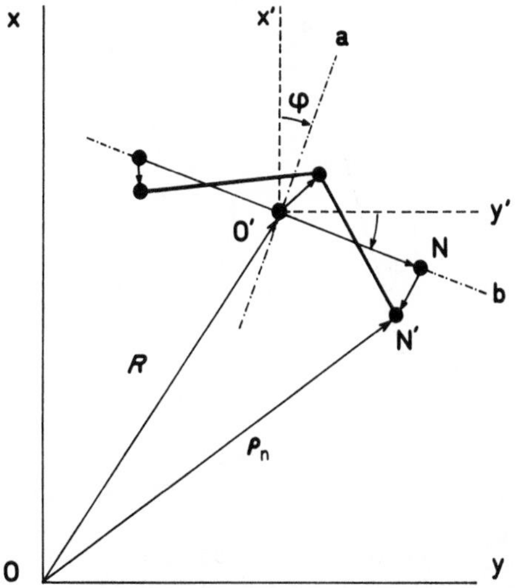

Figure 2.1 Born–Oppenheimer coordinates. A simplified diagram of the relation defined in (2.9). For simplicity the case shown here is a symmetric linear triatomic molecule moving in two dimensions only. OO′ ≡ $\boldsymbol{R}$, O′N ≡ $\boldsymbol{r}_n^e$, NN′ ≡ $\boldsymbol{d}_n$, ON′ ≡ $\boldsymbol{\rho}_n$ and O′N′ ≡ $\boldsymbol{r}_n$

$\mathbf{S}^{-1}$ in (2.9) is the inverse of the 3 × 3 rotation matrix, **S**, which relates the coordinates of the atoms in the molecule-fixed axis system to their coordinates in an axis system parallel to the space-fixed axes, x, y and z. $\boldsymbol{R}$ is a vector in three dimensions whose cartesian coordinates represent the origin of the molecule-fixed axes. The definition of **S** must now be considered in more detail. In two dimensions, the relation between the coordinates of a point in a rotated-axis system a, b to the coordinates in an original coordinate system, x, y as depicted in Figure 2.2 is given by

$$\begin{bmatrix} a \\ b \end{bmatrix} = \begin{bmatrix} \cos\varphi & \sin\varphi \\ -\sin\varphi & \cos\varphi \end{bmatrix} \begin{bmatrix} x \\ y \end{bmatrix} \tag{2.11}$$

The convention of defining the positive angle by the rotation which produces the advance of a right-hand screw into the paper is adhered to. (See also Figure 2.3.) Here a one-dimensional rotation has been performed about an axis perpendicular to both two-dimensional axis systems. In the three-dimensional general case **S** can be derived by making three successive similar rotations. There are several ways of doing this; we shall however use the convention given by Wilson, Decius and Cross.[335] In Figure 2.3 the

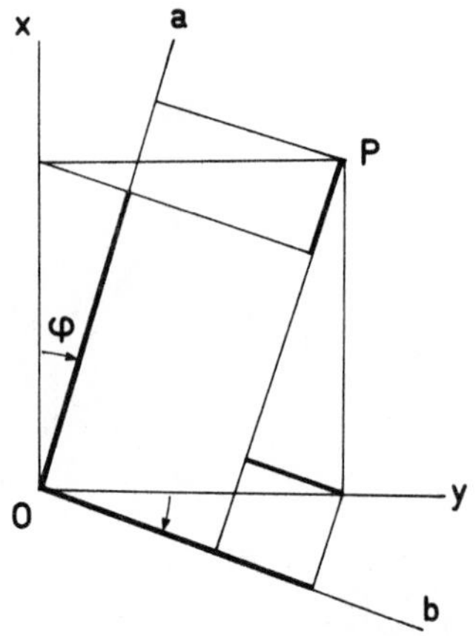

Figure 2.2 Right-handed rotation of the x- and y-axes about the z-axis. This diagram relates to the rotational transformation (2.11)

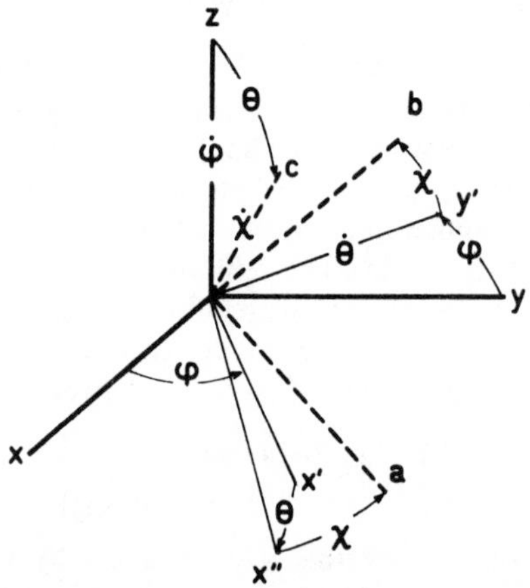

Figure 2.3 The Eulerian angles χ, θ and φ are here defined diagrammatically according to the convention of Wilson, Decius and Cross.[335] They relate the space-fixed x-, y- and z-axes to the molecule-fixed a-, b- and c-axes (2.12)

way in which the molecule-fixed axes a, b and c are related to the space-fixed axes x, y and z is depicted in terms of the Eulerian angles χ, θ and φ:

(i) The plane in which z and c lie intercepts the xy plane in x'.
(ii) Rotate x and y by an angle φ about z into x' and y' respectively.
(iii) Rotate x' and z by an angle θ about y' into x'' and c respectively.
(iv) Rotate x'' and y' by an angle χ about c into a and b respectively.

$$\mathbf{S} = \begin{bmatrix} \cos\chi & \sin\chi & 0 \\ -\sin\chi & \cos\chi & 0 \\ 0 & 0 & 1 \end{bmatrix} \begin{bmatrix} \cos\theta & 0 & -\sin\theta \\ 0 & 1 & 0 \\ \sin\theta & 0 & \cos\theta \end{bmatrix} \begin{bmatrix} \cos\varphi & \sin\varphi & 0 \\ -\sin\varphi & \cos\varphi & 0 \\ 0 & 0 & 1 \end{bmatrix} \tag{2.12}$$

Thus by multiplying the three matrices which represent the above three consecutive rotations, we obtain **S**. Note that $\mathbf{S}^\dagger$, the transpose of **S**, is equal to the inverse; $\mathbf{S}^\dagger = \mathbf{S}^{-1}$. Thus

$$\begin{bmatrix} a \\ b \\ c \end{bmatrix} = \begin{bmatrix} \cos\theta\cos\varphi\cos\chi - \sin\varphi\sin\chi & \cos\theta\sin\varphi\cos\chi + \cos\varphi\sin\chi & -\sin\theta\cos\chi \\ -\cos\theta\cos\varphi\sin\chi - \sin\varphi\cos\chi & -\cos\theta\sin\varphi\sin\chi + \cos\varphi\cos\chi & \sin\theta\sin\chi \\ \sin\theta\cos\varphi & \sin\theta\sin\varphi & \cos\theta \end{bmatrix} \begin{bmatrix} x \\ y \\ z \end{bmatrix} \tag{2.13}$$

This matrix is in fact the matrix of direction cosines which relate molecule-fixed to space-fixed axes in terms of the Eulerian angles.

It is important to note that when defined in this way, (2.12), the first rotation was performed about the space-fixed z-axis and the final one about the molecule-fixed c-axis. We shall return to this point later when the wave-functions are considered.

The final point that we must consider is that there is a superficial indeterminacy in relation (2.9) which should be noted. On the left-hand side there are $3N$ degrees of freedom and on the right-hand side there are three contained in $\boldsymbol{R}$ and three implicit in **S**; this implies that six constraints operate on the $\boldsymbol{d}_n$ so that the two sides balance. There are thus only $3N - 6$ degrees of freedom allowed for the internal motion represented by $\boldsymbol{d}_n$. The necessary constraints are imposed by the Eckart conditions

$$\sum_n m_n \boldsymbol{d}_n = \boldsymbol{0} \tag{2.14}$$

$$\sum_n m_n \boldsymbol{r}_n^e \times \boldsymbol{d}_n = \boldsymbol{0} \tag{2.15}$$

The first condition ensures that no particular distortion of the molecule represented by $\boldsymbol{d}_n$ constitutes a translation. The second similarly excludes the possibility of a contribution by the $\boldsymbol{d}_n$ to rotation. The first, (2.14), essentially requires that the centre of mass does not change during the displacements.

We can now proceed to evaluate T by taking the time derivative of $\boldsymbol{\rho}_n$ (note that the derivative of a matrix is obtained by differentiating the matrix elements),

$$\dot{\boldsymbol{\rho}}_n = \dot{\boldsymbol{R}} + \dot{\mathbf{S}}^{-1}(\boldsymbol{r}_n^e + \boldsymbol{d}_n) + \mathbf{S}^{-1}\dot{\boldsymbol{d}}_n \tag{2.16}$$

and substituting this expression into (2.8). Of the nine terms so obtained, five involve translation and may, in homogeneous space, be neglected. Those which remain are:

$$\tfrac{1}{2}\sum_n m_n[\mathbf{S}^{-1}\dot{\boldsymbol{d}}_n]\cdot[\mathbf{S}^{-1}\dot{\boldsymbol{d}}_n] \tag{2.17a}$$

$$\sum_n m_n[\dot{\mathbf{S}}^{-1}(\boldsymbol{r}_n^e + \boldsymbol{d}_n)]\cdot[\mathbf{S}^{-1}\dot{\boldsymbol{d}}_n] \tag{2.17b}$$

$$\tfrac{1}{2}\sum_n m_n[\dot{\mathbf{S}}^{-1}(\boldsymbol{r}_n^e + \boldsymbol{d}_n)]\cdot[\dot{\mathbf{S}}^{-1}(\boldsymbol{r}_n^e + \boldsymbol{d}_n)] \tag{2.17c}$$

We can simplify the first term (2.17a) by noting that the scalar product of two vectors is independent of a rotation of axes (Section A1.5). Thus we see that (2.17a) is just the kinetic energy associated with molecular distortion, i.e. the vibrational kinetic energy, T_v.

$$T_v = \frac{1}{2}\sum_n m_n\dot{\boldsymbol{d}}_n\cdot\dot{\boldsymbol{d}}_n \tag{2.18}$$

The second term when simplified can also be written using summation convention (Section A1.5) over cartesian axes

$$\sum_n m_n(\mathbf{S}\dot{\mathbf{S}}^{-1})_{\alpha\beta}(r_n^e + d_n)_\beta \dot{d}_{n\alpha} \tag{2.19}$$

The matrix $\mathbf{S}\dot{\mathbf{S}}^{-1}$ has quite a simple form and we can get a good feel for its properties by restricting ourselves to a rotation about a single axis as in Figure 2.2. In this case if we rotate about the c-axis,

$$\mathbf{S}\dot{\mathbf{S}}^{-1} = \begin{pmatrix} \cos\chi & \sin\chi \\ -\sin\chi & \cos\chi \end{pmatrix}\begin{pmatrix} -\dot{\chi}\sin\chi & -\dot{\chi}\cos\chi \\ \dot{\chi}\cos\chi & -\dot{\chi}\sin\chi \end{pmatrix} = \begin{pmatrix} 0 & -\omega_c \\ \omega_c & 0 \end{pmatrix} \tag{2.20}$$

where $\dot{\chi} = \omega_c$ (two dimensions only). In the rather more complicated three-dimensional case if ω_a, ω_b and ω_c are defined as the angular velocities about the molecule-fixed axes a, b and c (Section A2) then the correct form for the full three-dimensional matrix is (A2.7)

$$\mathbf{S}\dot{\mathbf{S}}^{-1} = \begin{pmatrix} 0 & -\omega_c & \omega_b \\ \omega_c & 0 & -\omega_a \\ -\omega_b & \omega_a & 0 \end{pmatrix} \tag{2.21a}$$

$$(\mathbf{S}\dot{\mathbf{S}}^{-1})_{\beta\gamma} = -e_{\alpha\beta\gamma}\omega_\alpha \tag{2.21b}$$

(The permutation symbol, $e_{\alpha\beta\gamma}$, has been introduced to simplify this equation and it will be used from now on wherever it can simplify the treatment (Section A1.6). Using the above result we see that we obtain from (2.17b) a term which involves both the time variation of the molecular displacements, $\dot{\boldsymbol{d}}$, and also the angular velocity of the frame. This is the coriolis or vibration–rotation interaction term. If we substitute the result (2.21) into (2.19) and apply the Eckart conditions (2.14) and (2.15) then we can write

$$T_{vr} = \sum_n m_n \omega_\alpha e_{\alpha\beta\gamma} d_{n\beta} \dot{d}_{n\gamma} \tag{2.22}$$

Note that terms in $\boldsymbol{r}^e \times \boldsymbol{d}$ vanish because of the Eckart condition (2.15). The effect of this term will be discussed further in Chapter 6.

To evaluate the last term (2.17c) we shall first of all define the instantaneous coordinate of the nth atom with respect to a, b and c as $\boldsymbol{r}_n$, i.e.

$$\boldsymbol{r}_n = \boldsymbol{r}_n^e + \boldsymbol{d}_n \tag{2.23}$$

We can make use of the important relation (2.21) again as shown in (A1.6) to obtain a term from (2.17c) that involves only $\dot{\chi}$, $\dot{\theta}$ and $\dot{\varphi}$ as velocities. This is thus a pure rotational kinetic energy, T_r, which can be written as

$$T_r = \tfrac{1}{2}\sum_n m_n e_{\varepsilon\alpha\gamma} e_{\varepsilon\beta\delta} \omega_\alpha \omega_\beta r_{n\gamma} r_{n\delta} \tag{2.24}$$

This expression can be more neatly written if we separate out the $\boldsymbol{\omega}$ independent term which we can write as the tensor

$$\begin{aligned} I_{\alpha\beta} &= \sum_n m_n e_{\varepsilon\alpha\gamma} e_{\varepsilon\beta\delta} r_{n\gamma} r_{n\delta} \\ &= \sum_n m_n (\delta_{\alpha\beta} r_{n\gamma} r_{n\gamma} - r_{n\alpha} r_{n\beta}) \end{aligned} \tag{2.25}$$

If we write out the elements of the $I_{\alpha\beta}$ tensor we see that we have derived the form of the *Moment of Inertia Tensor* **I**. Thus

$$\begin{aligned} I_{aa} &= \sum_n m_n [r_{nb}^2 + r_{nc}^2], & I_{ab} &= -\sum_n m_n r_{na} r_{nb} \\ I_{bb} &= \sum_n m_n [r_{na}^2 + r_{nc}^2], & I_{bc} &= -\sum_n m_n r_{nb} r_{nc} \\ I_{cc} &= \sum_n m_n [r_{na}^2 + r_{nb}^2], & I_{ca} &= -\sum_n m_n r_{nc} r_{na} \end{aligned} \tag{2.26}$$

We can thus write

$$T_r = \tfrac{1}{2} I_{\alpha\beta} \omega_\alpha \omega_\beta \tag{2.27}$$

or the expanded form:

$$T_r = \tfrac{1}{2}I_{aa}\omega_a^2 + \tfrac{1}{2}I_{bb}\omega_b^2 + \tfrac{1}{2}I_{cc}\omega_c^2 + I_{ab}\omega_a\omega_b + I_{bc}\omega_b\omega_c + I_{ca}\omega_c\omega_a \quad (2.28)$$

The quantities $I_{\alpha\beta}$ are components of the *instantaneous* moment of inertia tensor; they vary in time as they involve the $\boldsymbol{d}_n$ and are thus functions of the vibration coordinates. This aspect is discussed in Chapter 6 where the methods of determining the parameters of the reference configuration r_e are considered. This problem is a generalization of the familiar problem of determining r_e, the equilibrium bond length, for a diatomic molecule.

If we consider the quantities ω_α as a three-dimensional column vector and write $I_{\alpha\beta}$ as a tensor we can collect (2.18), (2.22) and (2.27) and write the total kinetic energy as:

$$T = \tfrac{1}{2}\boldsymbol{\omega}^\dagger \mathbf{I}\boldsymbol{\omega} + \sum_n m_n\boldsymbol{\omega}\cdot(\boldsymbol{d}_n \times \dot{\boldsymbol{d}}_n) + \tfrac{1}{2}\sum_n m_n\dot{\boldsymbol{d}}_n\cdot\dot{\boldsymbol{d}}_n \quad (2.29)$$

2.3 Angular Momenta

As discussed in Chapter 1 the next step is to write (2.29) in terms of momenta. The rigorous procedure for deriving the quantum mechanical Hamiltonian from (2.29) is however complicated because momentum type terms occur which are not conjugate to any coordinates. The standard transformation procedure, discussed in Chapter 1, involving the use of the quantum conditions (1.8) is thus not valid. The problem is considered in Chapter 6. We are however particularly interested in rotational motion and as an introduction to the problem we need, for the present, only consider the rather less general but much simpler procedure which treats T_r separately. The quantum mechanical Hamiltonian for 'pure' rotation can be derived by writing T_r (2.27) in terms of momenta conjugate to the three coordinates associated with the components of $\boldsymbol{\omega}$. These are the angles of rotation about the molecule-fixed axes *a*, *b* and *c*.

Thus from (2.27) we obtain the classical definition of the molecule-fixed angular momentum, J_α as

$$J_\alpha = \partial T/\partial\omega_\alpha = I_{\alpha\beta}\omega_\beta \quad (2.30)$$

Using tensor notation $\boldsymbol{J} = \mathbf{I}\boldsymbol{\omega}$ and thus $T_r = \tfrac{1}{2}(\mathbf{I}\boldsymbol{\omega})^\dagger\mathbf{I}^{-1}(\mathbf{I}\boldsymbol{\omega}) = \tfrac{1}{2}\boldsymbol{J}^\dagger\mathbf{I}^{-1}\boldsymbol{J}$ which allows us to write the Hamiltonian (in free space the potential energy is zero)

$$H_r = \tfrac{1}{2}\mu_{\alpha\beta}J_\alpha J_\beta \quad (2.31)$$

where $\mu_{\alpha\beta}$ is an element of inverse of the inertia tensor ($\boldsymbol{\mu} = \mathbf{I}^{-1}$). The tensor $\mathbf{I}$ is symmetric and can be diagonalized by the 3×3 unitary transformation matrix analogous to the matrix given in (1.41). There is thus an orientation of the axes *a*, *b* and *c* for which the cross-product terms in (2.27) and (2.31) vanish. In this special orientation, the axes are known as the *Principal Axes* and the three diagonal elements of **I** are called the *Principal Moments of Inertia*. This unique set of parameters will carry the subscripts A, B and C (capital letters). Note that the order of A, B and C is determined by the convention that $I_A \leqslant I_B \leqslant I_C$ (Section 3.6) and that *in this book* no order will be assumed for a, b and c (Section A1.4). Thus T_r (2.27) can be reduced to

$$T_r = \tfrac{1}{2}(I_A\omega_A^2 + I_B\omega_B^2 + I_C\omega_C^2) \qquad (2.32)$$

From (2.31) we see that $J_A = \partial T/\partial\omega_A$ etc. and yields a reduced 'instantaneous' rotational Hamiltonian

$$H_r = AJ_A^2 + BJ_B^2 + CJ_C^2 \qquad (2.33)$$

where A, B and C are the *Principal Rotational Constants* $A = \mu_A/2 = 1/2I_A$, $B = \mu_B/2 = 1/2I_B$ and $C = \mu_C/2 = 1/2I_C$.

In this chapter we have been careful to specify that the inertial constants and thus also the rotational constants depend on vibrational motion. In the next chapter we develop general expressions for the energy levels and assume that this variation can be neglected. This assumption is the essence of the *Rigid-Rotor Approximation* that A, B and C in (2.33) are constant parameters. This is a good initial assumption especially in those cases where large amplitude motions (such as free rotation) are not involved. In these systems which are defined in Chapter 9 as semi-rigid molecules the corrections due to non-rigidity are small and can be adequately taken into account by perturbation theory as they are smaller than the rigid-rotor energy by at least an order of magnitude. The corrections do however present the most important obstacle to the evaluation of accurate structural information from the *observed* rotational parameters and this problem is discussed in Chapter 6.

Chapter 3

Angular Momentum and the Energy Levels of a Rigid Rotor

3.1 Angular Momentum and Axes of Reference

The rigid-rotor Hamiltonian developed in the previous chapter has been written in terms of molecule-fixed coordinates, and as $d_{n\alpha} = 0$ the inertial constants do not vary in time. It could just as easily have been written in terms of space-fixed coordinates, but of course the inertial constants would then be time dependent even though $d_{n\alpha} = 0$. It is clear that we need the properties of the angular momentum $\boldsymbol{J}$, defined in (2.3), with respect to the molecule-fixed axes. For a complete understanding of rotational spectroscopy however, we also need to understand the properties of $\boldsymbol{J}$ with respect to the space-fixed axes, essentially because we make our 'observations' from this frame of reference.

In this chapter the matrix elements of an arbitrary angular momentum operator, $\boldsymbol{J}$, are first determined with respect to the space-fixed axes. The procedure is the same as that used in the study of atomic problems. In fact one can relate them to well known angular momentum coupling cases in atomic spectroscopy as was shown by Van Vleck.[310] When the matrix elements of $\boldsymbol{J}$ in the space-fixed representation (xyz) have been derived one can proceed to determine the analogous expressions in the molecule-fixed representation in a similar way. There is, however, a subtle but important difference between the two procedures and this difference must be carefully considered whenever molecular rotational motion is involved.

3.2 Space-fixed Angular Momentum

Classically the angular momentum of a particle about an arbitrary origin can be written as

$$J_i = e_{ijk} q_j p_k \tag{3.1}$$

where $\boldsymbol{q}$ is the position vector and $\boldsymbol{p}$ the conjugate momentum.

We wish to study the stationary-state problem and thus we must determine the eigenvalues of the quantum mechanical operator $\boldsymbol{J}$. As indicated in

Chapter 1 we can obtain the quantum mechanical form of $\boldsymbol{J}$ by replacing $\boldsymbol{q}$ and $\boldsymbol{p}$ in (3.1) by their quantum operator equivalents.

The rigid-rotor Hamiltonian (2.33) involves the various components of an angular momentum and to obtain the eigenvalues of H_r we need to determine the commutation relations which govern the quantum properties of this angular momentum.

For instance, the commutator $[J_x, J_y]$, which can with the aid of (3.1) be written as

$$[J_x, J_y] = [q_y p_z - q_z p_y, q_z p_x - q_x p_z] \tag{3.2}$$

reduces, if we take account of (1.8), to

$$q_y p_x[p_z, q_z] - q_y q_x[p_z, p_z] - p_y p_x[q_z, q_z] + p_y q_x[q_z, p_z] = iJ_z \tag{3.3}$$

Similar expressions obviously can be deduced for $[J_y, J_z]$ and $[J_z, J_x]$ and the three relations can be written as

$$[J_i, J_j] = i\, e_{ijk} J_k \tag{3.4}$$

Note that summation over the space-fixed axes, xyz is represented by the indices ijk and summation over molecular-fixed axes, abc, by the indices $\alpha\beta\gamma$. The commutation relations (3.4) can be considered as the quantum conditions which define the space-fixed components of a generalized angular-momentum operator $\boldsymbol{J}$. Using (3.4) one can show that any component of $\boldsymbol{J}$ commutes with $\boldsymbol{J}^2$ the square of the angular momentum. In index notation, $\boldsymbol{J}^2 \equiv J_i J_i$ and thus:

$$\begin{aligned}
[J_i J_i, J_j] &= J_i J_i J_j - J_j J_i J_i \\
&= J_i[J_i, J_j] + J_i J_j J_i + [J_i, J_j]J_i - J_i J_j J_i \\
&= ie_{ijk}(J_i J_k + J_k J_i) = 0
\end{aligned} \tag{3.5}$$

This result is perhaps fairly obvious from the fact that $\boldsymbol{J}^2$ is a scalar quantity and thus independent of orientation. We shall use this result, (3.5), and the relation (3.4) as the coathanger from which to suspend the theory of angular momentum.

3.3 The Eigenvalues of Angular-momentum Operators

As shown in Section 3.2, $\boldsymbol{J}^2$ and J_i commute, and thus there exist functions which are simultaneously eigenfunctions of both $\boldsymbol{J}^2$ and one component, by convention J_z. These eigenfunctions can be represented by $|\boldsymbol{J}^{2\prime}\, J_z'\rangle$, where the eigenfunction is specified by the eigenvalues, and the eigenvalue of an operator is represented by the primed operator, as defined in relation (1.13). Because of (3.4) these functions are not eigenfunctions of J_x or J_y. It is however necessary to obtain the properties (matrix elements) of J_x and J_y

with respect to these functions and to do this we introduce the shift operators (Section 1.3),

$$J_+ = J_x + iJ_y \qquad J_- = J_x - iJ_y \tag{3.6}$$

and generate from (3.4) the commutation relation:

$$[J_z, J_\pm] = \pm J_\pm \tag{3.7}$$

Note that $(J_+)^\dagger = J_-$.

If we operate on $|\boldsymbol{J}^{2\prime}\, J_z'\rangle$ with both the left- and right-hand sides of (3.7) we obtain the relation:

$$J_z J_\pm |\boldsymbol{J}^{2\prime}\, J_z'\rangle = (J_z' \pm 1) J_\pm |\boldsymbol{J}^{2\prime}\, J_z'\rangle \tag{3.8}$$

This expression indicates that when J_+ and J_- operate on an eigenfunction of J_z (with eigenvalue J_z') two new eigenfunctions are generated with eigenvalues one unit of angular momentum more or less respectively, i.e. $J_z' \pm 1$. Thus, starting with a particular eigenfunction and using these two operators we can obviously generate a whole set of eigenfunctions of $\boldsymbol{J}^2$ and J_z with the *same* value of $\boldsymbol{J}^{2\prime}$ but *different* values of J_z'. In essence one can consider J_+ and J_- as operators which can change the spatial orientation (specified by J_z') of an angular momentum vector by one unit, i.e. to the next adjacent allowed orientation. This set must have both an upper and lower bound, as a component of $\boldsymbol{J}$ clearly cannot be greater in magnitude than $\boldsymbol{J}$ itself, i.e.

$$|\boldsymbol{J}_z'| \leqslant \sqrt{\boldsymbol{J}^{2\prime}} \tag{3.9}$$

Thus upper and lower boundary functions exist above and below which the set must terminate. We can specify the value of J_z' for the upper and lower bound functions by $J_z^\uparrow$ (upper) and $J_z^\downarrow$ (lower) respectively. The generating relation (3.8) and the boundary condition (3.9), both of which must be valid under all conditions, can only be reconciled with regard to the boundary functions if J_- annihilates the lower, and J_+ annihilates the upper one according to:

$$J_- |\boldsymbol{J}^{2\prime}\, J_z^\downarrow\rangle = 0 \tag{3.10a}$$

$$J_+ |\boldsymbol{J}^{2\prime}\, J_z^\uparrow\rangle = 0 \tag{3.10b}$$

If we pre-multiply (3.10a) by J_+ and (3.10b) by J_- and take into account the relation

$$\boldsymbol{J}^2 - J_z^2 = J_\pm J_\mp \mp J_z \tag{3.11}$$

we find that

$$(\boldsymbol{J}^{2\prime} - (J_z^\downarrow)^2 + J_z^\downarrow)|\boldsymbol{J}^{2\prime}\, J_z^\downarrow\rangle = 0 \tag{3.12a}$$

and

$$(\boldsymbol{J}^{2\prime} - (J_z^\uparrow)^2 - J_z^\uparrow)|\boldsymbol{J}^{2\prime}\, J_z^\uparrow\rangle = 0 \tag{3.12b}$$

By definition the two eigenfunctions are non-zero and thus in equations (3.12) the multipliers must be identically zero. The difference of these two multipliers can be factorized and equated to zero:

$$(J_z^{\uparrow} + J_z^{\downarrow})(J_z^{\uparrow} - J_z^{\downarrow} + 1) = 0 \tag{3.13}$$

The left-hand bracket in (3.13) must be zero as the definition of the boundary functions yields the condition $J_z^{\downarrow} \leqslant J_z^{\uparrow}$ which ensures that the right-hand bracket is >0. Thus we see that

$$J_z^{\uparrow} = -J_z^{\downarrow} \quad \text{and therefore} \quad J_z^{\uparrow} - J_z^{\downarrow} = 2j \tag{3.14}$$

where $2j$ must be an integral positive number. If we now *sum* the eigenvalues of equations (3.12a and b) for a particular $\boldsymbol{J}^{2\prime}$ we obtain the result

$$\boldsymbol{J}^{2\prime} = j(j + 1) \tag{3.15}$$

The results in the relations (3.14) indicate that J_z' is quantized and can take on only integral or half-integral values; i.e. $J_z' = m$ where

$$m = j, j - 1, \ldots, 1, 0, -1, \ldots, -j$$

or

$$m = j, j - 1, \ldots, \tfrac{1}{2}, -\tfrac{1}{2}, \ldots, -j.$$

We can now use the quantum numbers j and m to specify the eigenfunctions in place of $\boldsymbol{J}^{2\prime}$ and J_z', i.e. we shall represent $|\boldsymbol{J}^{2\prime}\, J_z'\rangle$ by $|j\, m\rangle$.

The matrix elements of J_x and J_y can be obtained in the following way using J_+ and J_-. From (3.8) we can see that from a normalized function $|j\, m\rangle$ we can generate the functions $|j\, m \pm 1\rangle$ according to the relation

$$J_{\pm}|j\, m\rangle = n_{\pm}|j\, m \pm 1\rangle \tag{3.16}$$

Normalization can be carried out as in (1.25) which indicates that

$$\langle j\, m|J_{\pm}^{\dagger} J_{\pm}|j\, m\rangle = n_{\pm}^{*} n_{\pm} \tag{3.17}$$

and thus from (3.11) we see that

$$n_{\pm}^{*} n_{\pm} = [j(j + 1) - m^2 \mp m] \tag{3.18}$$

As we wish to determine the numbers $n_{\pm}$ we see that an arbitrary phase factor of the type $e^{i\delta}$ must be considered at this point for completeness.‡ Condon and Shortley[49] have used the convention that $\delta = 0$ and thus $n_{\pm}^{*} = n_{\pm}$ is real. We now obtain the following matrix elements.

$$J_{\pm}|j\, m\rangle = \sqrt{j(j + 1) - m(m \pm 1)}\,|j\, m \pm 1\rangle \tag{3.19}$$

‡ Phase factors crop up wherever normalization occurs and must be carefully taken into account. Here we see that $(e^{i\delta})^{*} e^{i\delta} = 1$.

and consequently

$$\langle j\, m \pm 1 | J_x | j\, m \rangle = \tfrac{1}{2}\sqrt{(j \mp m)(j \pm m + 1)} \tag{3.20a}$$

$$\langle j\, m \pm 1 | J_y | j\, m \rangle = \mp\tfrac{1}{2}i\sqrt{(j \mp m)(j \pm m + 1)} \tag{3.20b}$$

3.4 Molecule-fixed Angular Momentum

We have developed a theory which yields the matrix elements of a generalized angular momentum $\boldsymbol{J}$ in terms of space-fixed components. It now remains to develop the relations which apply when this same angular momentum is related to the molecule-fixed axes. This problem is discussed in the review article by Van Vleck[310] where a number of problems in molecular dynamics are considered.

We must now clarify exactly what we mean by space-fixed and molecule-fixed angular momentum. If we observe a rotating body in a laboratory (i.e. determine $\boldsymbol{J}$ in a space-fixed frame of reference) then we can describe its motion in terms of the three components of $\boldsymbol{J}$ relative to axes fixed in the laboratory. These are the space-fixed components. At *any moment in time* we could relate these components to any other set of axes we choose. If we relate them to the molecular reference axes of Chapter 2, we would obtain molecular-fixed components of $\boldsymbol{J}$. It is clear that one requires such a two-way process, via an external space-fixed reference, to obtain these components.

We can carry out the 'back reference' process using the direction cosines which relate the two axis systems. For instance, the component J_a is given by

$$J_a = \cos(ax)J_x + \cos(ay)J_y + \cos(az)J_z \tag{3.21}$$

where $\cos(\alpha i)$ represents the angle between the molecule-fixed α-axis and the space-fixed i-axis. These direction cosines are in fact just the elements of the **S** matrix discussed in Chapter 2, i.e. $S_{ax} \equiv \cos(ax)$ etc. The rows and columns of the **S** matrix possess properties which indicate that they belong to an important family of vector operators known as T class operators which bear a special relationship with $\boldsymbol{J}$.

The general properties of T class operators are discussed in Chapter 4, where their properties are developed from the commutation relations which define them:

$$[J_i, T_j] = i\, e_{ijk} T_k \tag{3.22}$$

Clearly, $\boldsymbol{J}$ is itself such an operator (3.4) as are also $\boldsymbol{r}$ and $\boldsymbol{p}$. In fact any polar vector belongs to this class. The elements of **S** also obey this relation‡

$$[J_x, S_{ax}] = 0, \qquad [J_x, S_{ay}] = iS_{az}, \qquad [J_x, S_{az}] = -iS_{ay} \tag{3.23}$$

‡ The rows of **S** are T class operators with respect to J_i the space-fixed angular momentum, whereas the columns are T class operators with respect to J_α, the molecule-fixed angular momentum.

or more succinctly

$$[J_i, S_{\alpha j}] = ie_{ijk}S_{\alpha k} \tag{3.24}$$

We can use (3.24) to show the important result that the space- and molecule-fixed components of $\boldsymbol{J}$ commute. If we write (3.21) in the general form

$$J_\alpha = S_{\alpha i}J_i \tag{3.25}$$

we see that

$$\begin{aligned}[J_i, J_\alpha] &= J_iS_{\alpha j}J_j - S_{\alpha j}J_jJ_i \\ &= [J_i, S_{\alpha j}]J_j + S_{\alpha j}(J_iJ_j - J_jJ_i) \\ &= ie_{ijk}(S_{\alpha k}J_j + S_{\alpha j}J_k) = 0\end{aligned} \tag{3.26}$$

This result is perhaps almost obvious as the two axis systems are independent.

The next thing is to determine the commutation relations which hold between the various molecule-fixed components; J_a, J_b and J_c. Using (3.25) we see that

$$[J_\alpha, J_\beta] = [S_{\alpha i}J_i, S_{\beta j}J_j] = S_{\alpha i}J_iS_{\beta j}J_j - S_{\beta j}J_jS_{\alpha i}J_i \tag{3.27}$$

The commutation relation (3.24) can now be used to yield

$$[J_\alpha, J_\beta] = ie_{ijk}(S_{\alpha i}S_{\beta k}J_j + S_{\alpha i}S_{\beta j}J_k + S_{\alpha k}S_{\beta j}J_i) \tag{3.28}$$

In the determinant of direction cosines each element is equal to its co-factor (Section A1.6) and thus‡

$$e_{ijk}S_{\alpha i}S_{\beta j} = e_{\alpha\beta\gamma}S_{\gamma k} \tag{3.29}$$

By substituting this into (3.28) we obtain

$$[J_\alpha, J_\beta] = -ie_{\alpha\beta\gamma}J_\gamma \tag{3.30}$$

We thus find that the molecule-fixed components of $\boldsymbol{J}$ obey a *similar* commutation relation to the space-fixed components, *the sign of i has now however been reversed.* Only the *overall* angular momentum behaves in this peculiar way. As discussed in Chapter 10 other molecule-fixed momenta either obey the standard relation of the form (3.4) or one which is somewhat more complicated.[318]

Because $[J_i, J_\alpha] = 0$ and $\boldsymbol{J}^2$ commutes with not only its space-fixed components but also its molecule-fixed components there exist functions, $|J^{2\prime}\, J_c'\, J_z'\rangle$, which are simultaneously eigenfunctions of $\boldsymbol{J}^2$, J_c and J_z. We can follow the same procedure as laid down in Section 3.3 and generate the relevant matrix elements. If we carry out this procedure a new quantum

‡ One can look upon the e_{ijk} and $e_{\alpha\beta\gamma}$ symbols as ensuring that the signs of the elements of the determinant are correct.

number corresponding to J'_c is generated and usually given the symbol k. We can thus specify the wavefunction by $|j\ k\ m\rangle$ where

$$J_c|j\ k\ m\rangle = k|j\ k\ m\rangle \tag{3.31}$$

The main point to note is that now $J_a \pm iJ_b \equiv J^\pm$ obey the relations

$$J^\pm|j\ k\ m\rangle = \sqrt{j(j+1) - k(k \mp 1)}|j\ k \mp 1\ m\rangle \tag{3.32}$$

whereas before

$$J_\pm|j\ k\ m\rangle = \sqrt{j(j+1) - m(m \pm 1)}|j\ k\ m \pm 1\rangle \tag{3.33}$$

The complete set of angular momentum matrix elements is given in Table 3.1.

Table 3.1 Matrix Elements of Angular Momentum Operators[a]

$$J^2|j\,k\,m\rangle = j(j+1)|j\,k\,m\rangle$$
$$J_c|j\,k\,m\rangle = k|j\,k\,m\rangle$$
$$J_z|j\,k\,m\rangle = m|j\,k\,m\rangle$$
$$(J_a \mp iJ_b)|j\,k\,m\rangle = \sqrt{(j \mp k)(j \pm k + 1)}|j\,k \pm 1\,m\rangle$$
$$(J_x \pm iJ_y)|j\,k\,m\rangle = \sqrt{(j \mp m)(j \pm m + 1)}|j\,k\,m \pm 1\rangle$$

[a] Note that $J^\pm = J_a \pm iJ_b$ is a shift (down/up) operator in k, whereas $J_\pm = J_x \pm iJ_y$ is a shift (up/down) operator in m. The phase convention chosen here is that which requires the normalization coefficients in (3.42) to be real and positive.[129] It is not the same as that of King, Hainer and Cross[159] who choose $J_b \mp iJ_a$ to have real matrix elements. We will use j, k and m (lower case) when dealing with generalized angular momenta and J, K and M (upper case) when dealing with molecular rotational functions and eigenvalues. Note that Herzberg[123] uses the convention $K = |k|$ and $M = |m|$.

3.5 The Eigenfunctions of the Angular Momentum Operators

We have so far represented the eigenfunctions of an operator symbolically by Dirac kets. It is sometimes useful to be aware of certain properties of the explicit wavefunctions associated with rotational motion. For this purpose we shall briefly discuss the Schrödinger representation of angular momentum.

If we transfer from cartesian to the usual polar coordinates, r, θ and φ[169] then the angular momentum operators for *a single mass point* become

$$J_\pm = e^{\pm i\varphi}\left(i\cot\theta\frac{\partial}{\partial\varphi} \pm \frac{\partial}{\partial\theta}\right) \tag{3.34a}$$

$$J_z = -i\frac{\partial}{\partial\varphi} \tag{3.34b}$$

$$J^2 = -\left[\frac{1}{\sin^2\theta}\frac{\partial^2}{\partial\varphi^2} + \frac{1}{\sin\theta}\frac{\partial}{\partial\theta}\left(\sin\theta\frac{\partial}{\partial\theta}\right)\right] \tag{3.34c}$$

We have just solved the eigenvalue problem associated with J^2 and J_z and we can thus write down the associated Schrödinger equations. If we conform to tradition and write $|j\ m\rangle$ as

$$\Psi = \Theta_{jm}(\theta)\Phi_m(\varphi) \equiv |j\ m\rangle \tag{3.35}$$

we obtain for J_z the relation

$$-i\frac{\partial}{\partial\varphi}\Phi_m(\varphi) = m\Phi_m(\varphi) \tag{3.36}$$

Using this result we see that (3.34c) yields the relation

$$-\left[\frac{1}{\sin\theta}\frac{\partial}{\partial\theta}\left(\sin\theta\frac{\partial}{\partial\theta}\right) - \frac{m^2}{\sin^2\theta}\right]\Theta_{jm}(\theta) = j(j+1)\Theta_{jm}(\theta) \tag{3.37}$$

when it operates on Ψ. The eigenfunctions of (3.36) are immediately obtained as

$$\Phi_m(\varphi) = \frac{1}{\sqrt{2\pi}}e^{im\varphi} \tag{3.38}$$

where the functions have been normalized according to

$$\int_0^{2\pi} \Phi^*(\varphi)\Phi(\varphi)\, d\varphi = 1 \tag{3.38a}$$

The eigenfunctions $\Theta_{jm}(\theta)$ are somewhat more complicated to obtain. When $m = 0$ they are the *Legendre Polynomials* and when $m \neq 0$ they are the *Associated Legendre Polynomials.*[169,8] We shall need only to consider the properties of the Legendre polynomials explicitly.

Substituting $\cos\theta = x$ and setting $m = 0$ in (3.37) we obtain the Legendre equation

$$\left[(1 - x^2)\frac{d^2}{dx^2} - 2x\frac{d}{dx} + j(j+1)\right]y = 0 \tag{3.39}$$

The Legendre Polynomial solutions $P_n(x)$ or $P_n(\cos\theta)$ can be represented by the Rodriguez formula[169,8]

$$P_n(x) = \frac{1}{2^n n!}\frac{d^n}{dx^n}(x^2 - 1)^n \tag{3.40}$$

which generates the familiar expressions

$$P_0(\cos\theta) = 1$$

$$P_1(\cos\theta) = \cos\theta$$

$$P_2(\cos\theta) = \tfrac{1}{2}(3\cos^2\theta - 1)$$

$$P_3(\cos\theta) = \tfrac{1}{2}(5\cos^3\theta - 3\cos\theta)$$

We can thus write

$$\Theta_{j0}(\theta) = \sqrt{j + 1/2}\,P_j(\cos\theta) \tag{3.41}$$

using the normalization choice of Condon and Shortley. These functions are clearly equivalent to the ket $|j\ 0\ 0\rangle$. We could obtain the explicit expressions for the associated Legendre functions by direct solution of (3.37) with $m \neq 0$, or by application of $J_{\pm}$ as many times as necessary to (3.41) according to (3.19).[169] We shall find the following expressions for the functions useful[129] though we shall not need to evaluate them.

$$|j\ \ +k\ \ +m\rangle = N'_{+}(J_a - iJ_b)^k(J_x + iJ_y)^m|j\ 0\ 0\rangle \tag{3.42a}$$

$$|j\ \ -k\ \ -m\rangle = N'_{-}(J_a + iJ_b)^k(J_x - iJ_y)^m|j\ 0\ 0\rangle \tag{3.42b}$$

$$|j\ \ +k\ \ -m\rangle = N''_{+}(J_a - iJ_b)^k(J_x - iJ_y)^m|j\ 0\ 0\rangle \tag{3.42c}$$

$$|j\ \ -k\ \ +m\rangle = N''_{-}(J_a + iJ_b)^k(J_x + iJ_y)^m|j\ 0\ 0\rangle \tag{3.42d}$$

where $N'_{\pm}, N''_{\pm} > 0$, and the quantum numbers k and m are defined.[129]

The functions can thus be written as

$$\Psi = \Theta_{jkm}(\theta)\,e^{im\varphi}\,e^{ik\chi} \tag{3.43}$$

and are closely related to the familiar spherical harmonics, Y_{lm}. If however we treat a system such as a molecule with several mass points we should also add the χ dependent terms into the relations (3.34) such as $J_c = -i\partial/\partial\chi$. These yield an equation similar to (3.37) but with an added k dependence. This equation can be related to the *hypergeometric equation* and thus solved.[303]

3.6 General Approach to Solving the Rigid-rotor Hamiltonian

Our aim is the solution of the eigenvalue equation

$$H_r|E_r\rangle = E_r|E_r\rangle \tag{3.44}$$

where H_r is given by (2.33). H_r contains only a sum of powers of the components of $\boldsymbol{J}$ and thus commutes with $\boldsymbol{J}^2$. The matrix of H_r can thus be set up in a representation involving the basis states $|J\ K\rangle$.‡ The matrix will be diagonal

‡ As discussed in Section A1.2 we use lower-case j k and m for generalized angular momenta in line with standard quantum mechanical texts and upper-case J, K and M for molecular problems in line with standard spectroscopic convention.

in J but not necessarily so in K. Note that in homogeneous space the energy is independent of M and for convenience we have dropped this quantum number. Once we have set up the matrix of H_r (which involves generating the complete set of elements $\langle J\ K'|H_r|J\ K''\rangle$ for a given J) if it is not already diagonal we must seek a new set of functions in which it is diagonal; $|J\ \Gamma\rangle$. These will be linear combinations of the original basis functions, i.e.

$$|J\ \Gamma\rangle = \sum_K a^J_K |J\ K\rangle \tag{3.45}$$

such that

$$H|J\ \Gamma\rangle = E(J\ \Gamma)|J\ \Gamma\rangle \tag{3.46}$$

where $E(J\ \Gamma)$ is the energy of the rotational state $|J\ \Gamma\rangle$.

It is convenient to classify the general problem into several distinct cases by means of the relations which hold among the inertial and rotational constants by virtue of symmetry. By convention $I_A \leqslant I_B \leqslant I_C$ and thus $A \geqslant B \geqslant C$. Two of the four main groups are further classified according to mass distribution to facilitate the solution of (3.44). Thus we have

(1) Spherical tops	$I_A = I_B = I_C$	CH_4, SF_6
(2) Linear molecules	$I_A = 0, I_B = I_C$	AlF, OCS, C_3O_2
(3) Symmetric tops (a) prolate	$I_A < I_B = I_C$	$CH_3Br, CH_3{-}C{\equiv}C{-}H$
(b) oblate	$I_A = I_B < I_C$	$Mn(CO)_5H, NH_3, C_6H_6$
(4) Asymmetric rotor‡	$I_A < I_B < I_C$	H_2O, NH_2D, C_6H_5F

3.7 The Energy Levels of Spherical Top Molecules

In this particular case, H_r can be written as

$$H_r = BJ^2 \quad \text{and} \quad E(J\ K) = BJ(J+1) \tag{3.47}$$

The Hamiltonian is thus already diagonal and the energy is independent of K. The *rigid* spherical top cannot however possess a dipole moment and thus electric dipole radiation (as we shall see in the next chapter) cannot induce transitions among the rotational energy levels. It should however be noted that Watson has shown that centrifugal distortion effects can for a semi-rigid system allow transitions to occur[320,60a] (see Chapter 6). The energy levels follow the same pattern as the diatomic molecule which we shall discuss next. Note however that each level is $2J + 1$ fold degenerate in K as well as $2J + 1$ fold degenerate in M.

3.8 The Energy Levels of Linear Molecules

There are subtle problems associated with the Hamiltonian for a linear molecule which have been discussed by Hougen[128] and Watson.[319] We will

‡ The subclassification of the asymmetric rotor is discussed in Section 3.10.

sidestep the issue and treat linear molecules in the normal way which the more careful approach has shown to be correct.

We can to some extent treat linear molecules as a special case of a prolate symmetric top. For a real molecule such as a diatomic molecule the angular momentum J_A is associated with electronic motion and can be treated as part of the electronic energy which has been separated off.

For a linear polyatomic molecule there may be vibrational contributions which can be treated along with the vibrational problem and again in the rigid-rotor approximation can be neglected. We discuss these problems further in Chapter 6.

With these provisos we can write

$$H_r = B(J_B^2 + J_C^2) = BJ^2 \tag{3.48}$$

where the problem lies in the fact that $\boldsymbol{J}$ does not appear to obey the usual commutation relation (3.30). This arises because we need only two degrees of freedom to describe the rotation of a linear molecule. It is however valid to assume that the normal commutation relations apply.[319] This allows us to use the matrix elements we have developed (Table 3.1), and obtain the solution

$$E(J) = BJ(J + 1) \tag{3.49}$$

with the apparent proviso that $K = 0$. We thus obtain the energy level pattern shown in Figure 3.1. The levels are non-degenerate as $K = 0$. The

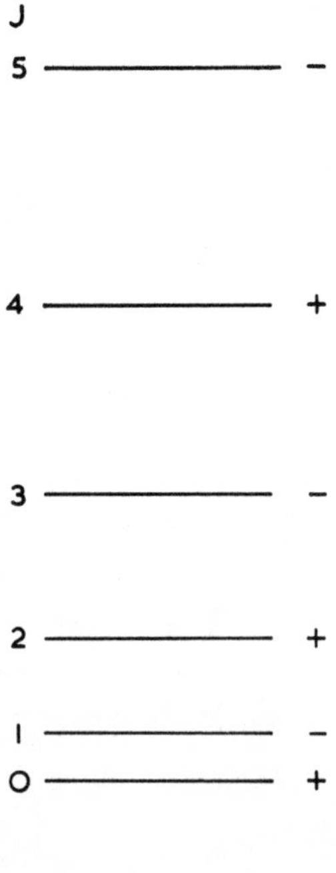

Figure 3.1 The energy levels of a rigid linear molecule. The + and − symmetry labels are discussed in Section 3.11c

levels are however $2J + 1$ fold degenerate in M which is important to remember when we discuss intensities and field effects. Note the characteristic spacing, $2B$, $4B$, $6B$ etc.

3.9 The Energy Levels of Symmetric Top Molecules

Symmetric tops are defined as having two moments of inertia equal and different from the third. This condition is fulfilled by symmetry for all molecules with a C_n rotation axis with $n \geqslant 3$. The 'possible but non-existent accidental symmetric top' has now essentially been discovered and can be treated as a special case. For the molecule DSSD, $A - B = 69\,916{\cdot}28$ MHz, $A - \frac{1}{2}(B + C) = 69\,916{\cdot}42$ MHz, $C - B = -0{\cdot}293$ MHz and $b_A = -0{\cdot}210 \times 10^{-5}$ and $\kappa = -0{\cdot}999\,999\,34$.[338] It is useful to use the prolate–oblate subclassification. In the prolate case $A > B = C$ and such molecules, when composed of nuclei of similar mass, are usually long and described as 'cigar shaped'. We should note that we are really discussing the properties of the moment of inertia tensor.

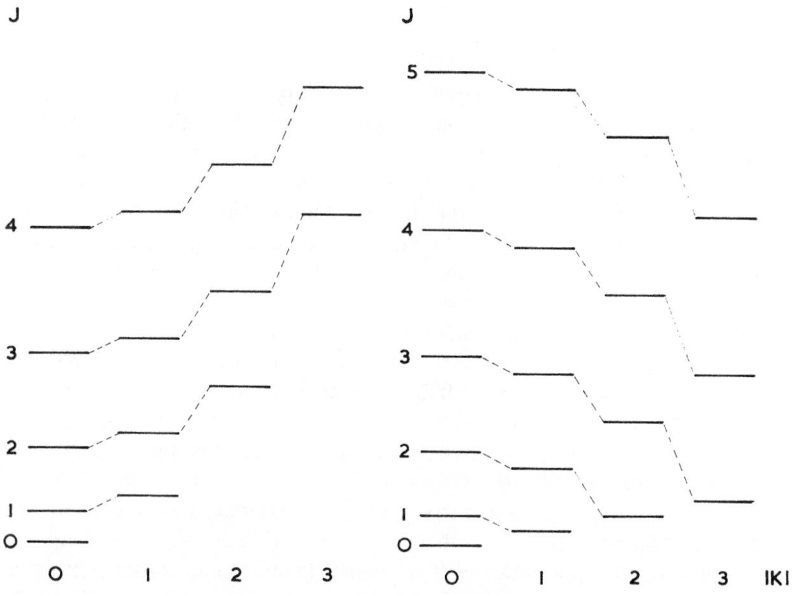

Figure 3.2 The lower energy levels of a prolate symmetric top (left) and an oblate symmetric top (right)

In the prolate case we can write the Hamiltonian as

$$H_r = AJ_A^2 + B(J_B^2 + J_C^2) = BJ^2 + (A - B)J_A^2 \tag{3.50}$$

where only $\boldsymbol{J}^2$ and one component, here J_A, are involved. H_r thus commutes with $\boldsymbol{J}^2$ and J_A and it is already diagonal in the $|J\ K\rangle$ basis. The solutions are

$$E(J, K) = BJ(J + 1) + (A - B)K^2 \tag{3.51}$$

A similar treatment for the oblate limit yields the result

$$E(J, K) = BJ(J + 1) + (C - B)K^2 \tag{3.52}$$

Note that now the coefficient of K^2 is negative. The associated energy levels are depicted in Figure 3.2. Note that the $K = 0$ stack are non-degenerate in K whereas the $K \neq 0$ levels are doubly degenerate. Each level as usual is $2J + 1$ fold degenerate in M.

If we consider for a moment the physical significance of these symmetric top formulae we see that overall rotational motion of the top can be described by the rotation of a molecule about an axis perpendicular to the symmetry axis (two degrees of freedom) plus a rotation about this symmetry axis. That the energy is independent of the direction of rotation about this axis is implied by the fact that the energy is independent of the sign of K.

3.10 The General Asymmetric Rotor Problem

In this case it is no longer possible to rearrange the Hamiltonian so that it is comprised solely of $\boldsymbol{J}^2$ and one component of $\boldsymbol{J}$. H_r is thus not diagonal in K for the asymmetric rotor. We must thus find the transformation which will diagonalize H_r in the basis $|J\ \Gamma\rangle$ according to (3.46). This essentially means that we cannot describe the rotational motion in terms of a conserved motion about a particular axis of the molecule. This point is physically apparent on a macroscopic scale where asymmetric objects do not spin in the same way as tops do.

In practice the mode of attacking the problem depends to some extent on 'how asymmetric the molecule actually is'. For instance one would not expect the overall pattern of a 'slightly asymmetric' molecule such as H_2CO ($A = 282.1$, $B = 38.8$ and $C = 34.0$, GHz) to depart too far from that of a symmetric top at least semi-quantitatively. There is, however, an important qualitative difference.

It turns out to be convenient to rearrange the Hamiltonian to take advantage of any near symmetric top character present and set the problem up in a symmetric rotor representation. The general Hamiltonian is thus written as

$$H_r = \tfrac{1}{2}(a + b)(J_a^2 + J_b^2) + cJ_c^2 + \tfrac{1}{2}(a - b)(J_a^2 - J_b^2) \tag{3.53}$$

which can usefully be rewritten as

$$H_r = \alpha \boldsymbol{J}^2 + \beta J_c^2 + \gamma(J^{+2} + J^{-2}) \tag{3.54}$$

and

$$\alpha = \tfrac{1}{2}(a + b), \qquad \beta = c - \tfrac{1}{2}(a + b), \qquad \gamma = \tfrac{1}{4}(a - b) \tag{3.54a}$$

The third term in γ tends to spoil the symmetric top form of this expression and, as we might expect, the ratio of γ to the other constants (usually β) tends to serve as a gauge of how bad the situation really is. Just as it was wise in the symmetric top case to be aware of whether the molecule was prolate or oblate because the ensuing energy-level patterns are rather different (Figure 3.2) so it is now wise to know whether the molecule is near-oblate, near-prolate or very asymmetric. Thus for a particular molecule one will usually map the ordered set of constants A, B and C onto the unordered set a, b and c of our general equation (2.53) such that γ/β is as small as possible. There are six possible mappings, i.e.

	I^R	I^L	II^R	II^L	III^R	III^L
a	B	C	C	A	A	B
b	C	B	A	C	B	A
c	A	A	B	B	C	C

(3.55)

Clearly there are only three essentially different sets because the handedness of the coordinate system does not affect the eigenvalues. The labelling is that devised by King, Hainer and Cross.[159]

We shall now discuss the various methods of solving the problem. The matrix elements we shall need according to (3.53) are obtained from Table 3.1. as

$$\langle J\, K|J^2|J\, K\rangle = J(J + 1) \tag{3.56a}$$

$$\langle J\, K|J_c^2|J\, K\rangle = K^2 \tag{3.56b}$$

$$\begin{aligned}\langle J\, K + 2|J^-J^-|J\, K\rangle &= \langle J\, K|J^+J^+|J\, K + 2\rangle \\ &= \sqrt{(J - K)(J + K + 1)(J - K - 1)(J + K + 2)}\end{aligned}$$

$$\begin{aligned}\langle J\, K - 2|J^+J^+|J\, K\rangle &= \langle J\, K|J^-J^-|J\, K - 2\rangle \qquad (3.56c) \\ &= \sqrt{(J + K)(J - K + 1)(J + K - 1)(J - K + 2)}\end{aligned} \tag{3.56d}$$

3.10a Algebraic Solution for Low J Values

For very low J values it is possible to solve H_r algebraically. As examples let us calculate the solutions for the cases $J = 1$ and $J = 3$. The Hamiltonian matrix is diagonal in J and it thus reduces to a set of blocks of differing J, the dimension of each block being $(2J + 1) \times (2J + 1)$, the range of K for any particular J value. We obtain the following 3×3 matrix for $J = 1$.

$$\begin{array}{c|ccc|}
K & |-1\rangle & |0\rangle & |+1\rangle \\
\hline
\langle -1| & 2\alpha+\beta & 0 & 2\gamma \\
\langle 0| & 0 & 2\alpha & 0 \\
\langle +1| & 2\gamma & 0 & 2\alpha+\beta \\
\hline
\end{array} \tag{3.57}$$

This particular matrix is so simple that diagonalization presents no difficulty, however, we shall note a few properties of the matrix which will prove useful in general. First, the off-diagonal matrix elements in (3.57) connect basis states which differ by 2 in K. The matrix could thus be separated into two blocks, one consisting only of even K values, the other only of odd K values. A second important property of this matrix is that it is symmetrical about *both* diagonals. This is related to the fact that the basis vectors are degenerate in $\pm K$. This suggests the use of a transformation to a new representation involving basis functions of the type

$$\frac{1}{\sqrt{2}}\{|J \; +K\rangle \pm |J \; -K\rangle\} \tag{3.58}$$

This transformation can be carried out by the Wang matrix $\mathbf{U}_J$ which has the form

$$\mathbf{U}_J = \frac{1}{\sqrt{2}}\begin{bmatrix} \ddots & & & & & & \iddots \\ & -1 & 0 & 0 & 0 & 1 & \\ & 0 & -1 & 0 & 1 & 0 & \\ & 0 & 0 & \sqrt{2} & 0 & 0 & \\ & 0 & 1 & 0 & 1 & 0 & \\ & 1 & 0 & 0 & 0 & 1 & \\ \iddots & & & & & & \ddots \end{bmatrix} \tag{3.59}$$

$\mathbf{U}_J$ is of order $(2J + 1) \times (2J + 1)$ and note also that $\mathbf{U}_J = \mathbf{U}_J^{-1}$. The transformation using this matrix is

$$\mathbf{U}_J\mathbf{H}\mathbf{U}_J = \tilde{\mathbf{H}} \tag{3.60}$$

which thus yields a further factorization. In general one obtains four submatrices each of order $\sim\frac{1}{2}J$.

In the case of the small matrix (3.57) above for $J = 1$, the Wang transformation $\mathbf{U}_3\mathbf{H}\mathbf{U}_3$ immediately diagonalizes H_r yielding

$$\begin{bmatrix} 2\alpha + \beta - 2\gamma & 0 & 0 \\ 0 & 2\alpha & 0 \\ 0 & 0 & 2\alpha + \beta + 2\gamma \end{bmatrix} \tag{3.61}$$

We thus obtain the three solutions for $J = 1$ as

$$a + b, \qquad a + c, \qquad b + c$$

It is also worth noting that the axes are essentially all equivalent and we could have obtained these three results by cyclic permutation of a, b and c in the linear solution which can be obtained directly from (3.57), i.e. 2α.

If we now specify the axes A, B and C according to the Mulliken convention that $A \geqslant B \geqslant C$ we can specify the eigenstates to which the eigenvalues belong as follows

$$E(1_{10}) = A + B$$
$$E(1_{11}) = A + C$$
$$E(1_{01}) = B + C$$

Where the states are designated by $J_{K_A K_C}$.‡ K_A is the value of $|K|$ for the limiting polate symmetric rotor with which the particular level correlates and K_C the value of $|K|$ in the oblate rotor limit. The correlation is carried out in Figure 3.3 for a few of the lower states. A second useful way of specifying the states uses the symbol J_τ where $\tau = K_A - K_C$. Thus the above states would then be designated as 1_1, 1_0 and 1_{-1}. The designation has the useful property that if states of a given J are labelled in decreasing τ value (τ runs from J to $-J$) then this is also the order of decreasing energy. Using the alternative $J_{K_A K_C}$ labelling K_A runs from J to 0 and K_C from 0 to J with decreasing energy, as shown in Table 3.9 for the particular case of $J = 3$.

It is instructive before considering the general case to carry out the calculation for $J = 3$. In Figure 3.4(a) the tri-diagonal matrix has been written. In Figure 3.4(b) the result of the Wang transformation (3.60) is given and in Figure 3.4(c) the final projection out of odd and even K states is shown. We thus obtain a typical factorization into four sub-blocks.

We have taken into account the symmetry properties of the matrix and the following abbreviations have been used for the diagonal elements.

$$d_K = d_{-K} = \alpha J(J + 1) + \beta K^2 \tag{3.62}$$

‡ In the literature K_A and K_C are often written as K_{-1} and K_1 respectively or less often K_p and K_o respectively.

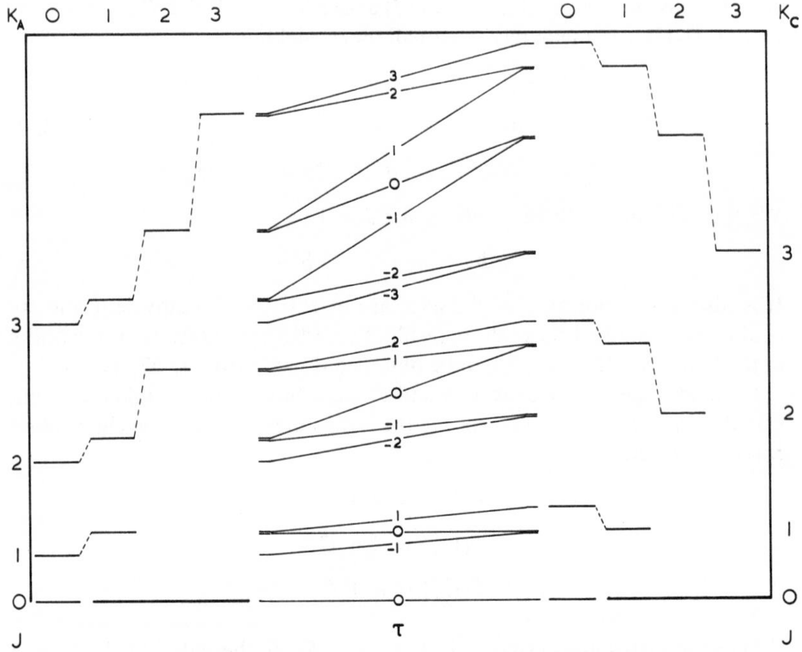

Figure 3.3 Schematic correlation diagram of the asymmetric-rotor energy levels ($J = 0, 1, 2$ and 3) between the prolate (on the left) and the oblate (on the right) limits. Note that the levels of a given J do not cross

For the off-diagonal elements, $K' \neq K''$ and these are labelled by $\lambda_{|K'||K''|}$.

$$\lambda_{31} = \sqrt{60}\gamma, \qquad \lambda_{20} = \sqrt{120}\gamma, \qquad \lambda_{11} = \sqrt{144}\gamma$$

We see that the problem is easily soluble as there are three quadratic solutions and one linear one. The linear root is trivial and can be written down immediately as

$$4a + 4b + 4c$$

We choose to solve the simplest of the three quadratic equations which is

$$\begin{bmatrix} d_0 & \sqrt{2}\lambda_{20} \\ \sqrt{2}\lambda_{20} & d_2 \end{bmatrix} = \begin{bmatrix} 12\alpha & 0 \\ 0 & 12\alpha \end{bmatrix} + \begin{bmatrix} 0 & \sqrt{240}\gamma \\ \sqrt{240}\gamma & 4\beta \end{bmatrix} \tag{3.63}$$

If we now use (1.40) we see that the solution is

$$12\alpha + 2\beta \pm \sqrt{240\gamma^2 + 4\beta^2} \tag{3.64}$$

d_3		λ_{31}				
	d_2		λ_{20}			
λ_{13}		d_1		λ_{11}		
	λ_{02}		d_0		λ_{02}	
		λ_{11}		d_1		λ_{13}
			λ_{20}		d_2	
				λ_{31}		d_3

(a)

d_3		λ_{31}				
	d_2					
λ_{31}		$d_1-\lambda_{11}$				
			d_0		$\sqrt{2}\lambda_{20}$	
				$d_1+\lambda_{11}$		λ_{31}
			$\sqrt{2}\lambda_{20}$		d_2	
				λ_{31}		d_3

(b)

d_2						
	d_3	λ_{31}				
	λ_{31}	$d_1-\lambda_{11}$				
			d_0	$\sqrt{2}\lambda_{20}$		
			$\sqrt{2}\lambda_{20}$	d_2		
					$d_1+\lambda_{11}$	λ_{13}
					λ_{13}	d_3

(c)

Figure 3.4 Matrix representation of the asymmetric rotor Hamiltonian for $J = 3$. (a) The $J = 3$ matrix in the symmetric rotor representation. The dimension of the matrix is 7×7 in this case and the rows and columns are labelled by the value of K which runs from -3 (top *left* hand corner) to $+3$. (b) The matrix which results after applying the Wang transformation (3.60). (c) The matrix which results after the interleaved even and odd K elements have been separated. The overall factorization is represented in general by (3.73). The blocks (starting from the top left-hand corner) can thus be related as E^-, O^-, E^+, and O^+

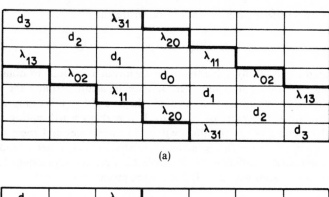

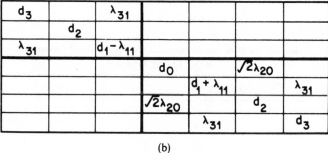

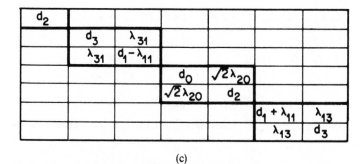

which with a little rearrangement can be written as

$$5a + 5b + 2c \pm 2\sqrt{4(a-b)^2 + (a-c)(b-c)} \qquad (3.65)$$

One could algebraically solve the other quadratic equations to obtain the remaining solutions; it is however, simpler to again permute the set of rotational constants a, b and c as we did in the case of $J = 1$. We immediately obtain the other two pairs of solutions. It is perhaps worth noting at this point that certain of the solutions will go in threesomes and some will be symmetric in a, b and c. This behaviour is related to the symmetry properties of the associated eigenfunctions. In Table 3.2 algebraic expressions for the energies of the levels with $J = 1$, 2 and 3 are given.

Table 3.2 Table of Algebraic Relations for the Rigid Asymmetric Rotor Energy Levels

$J_{K_A K_C}$	$E_r(JK_AK_C)$	$E_{J\tau}(\kappa)$[a]
0_{00}	0	0
1_{10}	$A + B$	$\kappa + 1$
1_{11}	$A + C$	0
1_{01}	$B + C$	$\kappa - 1$
2_{20}	$2A + 2B + 2C + 2\sqrt{(B-C)^2 + (A-C)(A-B)}$	$2(\kappa + \sqrt{\kappa^2 + 3})$
2_{21}	$4A + B + C$	$\kappa + 3$
2_{11}	$A + 4B + C$	4κ
2_{12}	$A + B + 4C$	$\kappa - 3$
2_{02}	$2A + 2B + 2C - 2\sqrt{(B-C)^2 + (A-C)(A-B)}$	$2(\kappa - \sqrt{\kappa^2 + 3})$
3_{30}	$5A + 5B + 2C + 2\sqrt{4(A-B)^2 + (A-C)(B-C)}$	$5\kappa + 3 + 2\sqrt{4\kappa^2 - 6\kappa + 6}$
3_{31}	$5A + 2B + 5C + 2\sqrt{4(A-C)^2 - (A-B)(B-C)}$	$2(\kappa + \sqrt{\kappa^2 + 15})$
3_{21}	$2A + 5B + 5C + 2\sqrt{4(B-C)^2 + (A-B)(A-C)}$	$5\kappa - 3 + 2\sqrt{4\kappa^2 + 6\kappa + 6}$
3_{22}	$4A + 4B + 4C$	4κ
3_{12}	$5A + 5B + 2C - 2\sqrt{4(A-B)^2 + (A-C)(B-C)}$	$5\kappa + 3 - 2\sqrt{4\kappa^2 - 6\kappa + 6}$
3_{13}	$5A + 2B + 5C - 2\sqrt{4(A-C)^2 - (A-B)(B-C)}$	$2(\kappa - \sqrt{\kappa^2 + 15})$
3_{03}	$2A + 5B + 5C - 2\sqrt{4(B-C)^2 + (A-B)(A-C)}$	$5\kappa - 3 - 2\sqrt{4\kappa^2 + 6\kappa + 6}$

[a] $E_{J\tau}(\kappa)$ is defined in equation (3.69).

3.10b Numerical Procedures

The procedure described in the previous section lends itself to numerical computation also. However such calculations commonly employ the following form for the rigid-rotor Hamiltonian

$$H = \tfrac{1}{2}(A + C)\boldsymbol{J}^2 + \tfrac{1}{2}(A - C)H(\kappa) \tag{3.66}$$

$$H(\kappa) = J_A^2 + \kappa J_B^2 - J_C^2 \tag{3.67}$$

κ is Ray's asymmetry parameter[258,343]

$$\kappa = (2B - A - C)/(A - C) \tag{3.68}$$

which is essentially $\beta/2\gamma$ (see relations (3.54)). The value of κ ranges from -1 in the prolate symmetric rotor limit to $+1$ in the oblate symmetric rotor limit. The eigenvalues of (3.66) are usually written as

$$E(J\,\tau) = \tfrac{1}{2}(A + C)J(J + 1) + \tfrac{1}{2}(A - C)E_{J\tau}(\kappa) \tag{3.69}$$

where $E_{J\tau}(\kappa) = \langle J\,\tau|H(\kappa)|J\,\tau\rangle$. With a little algebraic manipulation and the use of (3.56) the matrix elements of $H(\kappa)$ in (3.66) can be written as

$$\langle J\,K|H(\kappa)|J\,K\rangle = FJ(J + 1) + (G - F)K^2 \tag{3.70}$$

$$\langle J\,K|H(\kappa)|J\,K \pm 2\rangle = H[f(J\,K \pm 1)]^{1/2} \tag{3.71a}$$

$$f(JK \pm 1) = \tfrac{1}{4}[J(J + 1) - K(K \pm 1)][J(J + 1) - (K \pm 1)(K \pm 2)] \tag{3.71b}$$

where the way in which F, G and H are identified with the rotational constants depends on the mapping chosen. Taking the right-handed mappings from (3.55) we see that

	I^r	II^r	III^r
F	$\tfrac{1}{2}(\kappa - 1)$	0	$\tfrac{1}{2}(\kappa + 1)$
G	1	κ	-1
H	$-\tfrac{1}{2}(\kappa + 1)$	1	$\tfrac{1}{2}(\kappa - 1)$

(3.72)

It is usual to use the following mappings

near prolate	$\kappa \sim -1$	use I
very asymmetric	$\kappa \sim 0$	use II
near oblate	$\kappa \sim 1$	use III

Note that for the limiting symmetric rotor cases these choices ensure that H, the off-diagonal scale factor, is small which facilitates diagonalization. For the left-handed mappings the sign of H is reversed.

The matrix of $H(\kappa)$ can be factorized into four submatrices in essentially the same way as was arrived at in Figure 3.4, where we have however retained an extra diagonal term. The factorization can be written

$$\mathbf{H}(\kappa) = \mathbf{E}^+ + \mathbf{E}^- + \mathbf{O}^+ + \mathbf{O}^- \tag{3.73}$$

as specified by King, Hainer and Cross.[159] In their notation

$$\mathbf{E}^+ = \left[\begin{array}{c|cccc} E_{00} & \sqrt{2}E_{02} & 0 & \cdot & \cdot \\ \hline \sqrt{2}E_{02} & E_{22} & E_{24} & 0 & \cdot \\ 0 & E_{24} & E_{44} & \cdot & \cdot \\ \cdot & \cdot & \cdot & \cdot & \cdot \end{array}\right] \tag{3.74}$$

$$\mathbf{E}^- = \begin{bmatrix} E_{22} & E_{24} & \cdot & \cdot & \cdot \\ E_{24} & E_{44} & E_{46} & \cdot & \cdot \\ \cdot & E_{46} & E_{66} & \cdot & \cdot \\ \cdot & \cdot & \cdot & \cdot & \cdot \end{bmatrix} \tag{3.75}$$

$$\mathbf{O}^\pm = \begin{bmatrix} E_{11} \pm E_{-11} & E_{13} & 0 & \cdot & \cdot \\ E_{13} & E_{33} & E_{35} & \cdot & \cdot \\ 0 & E_{35} & E_{55} & \cdot & \cdot \\ \cdot & \cdot & \cdot & \cdot & \cdot \end{bmatrix} \tag{3.76}$$

Note that $\mathbf{E}^+$ differs from $\mathbf{E}^-$ only by the additional first row and column whereas $\mathbf{O}^+$ and $\mathbf{O}^-$ differ only in the first element. Compare this with the matrices given for $J = 3$.

Tabulations of $E_{J\tau}(\kappa)$ compiled initially by King, Hainer and Cross[159] have been extended.[346] $E_{J\tau}(\kappa)$ values for only one sign of κ need be calculated because

$$E_{J\tau}(\kappa) = -E_{J\,-\tau}(-\kappa) \tag{3.77}$$

3.10c Perturbation Solutions

If the system under consideration is a near symmetric top then the Hamiltonian (3.53) can be written so that the coefficients of the off-diagonal terms are small and the Hamiltonian may be solved by perturbation theory. The levels which in the symmetric rotor limit have $K = \pm 1$ are degenerate and linked by $\Delta K = \pm 2$ matrix elements as indicated by the γ term in (3.57) for the $J = 1$ case and also in Figure 3.4 for $J = 3$. Relation (1.40) can be applied with $\delta = 0$ to yield a first-order correction using (3.56c, d) with $K = \pm 1$

$$\Delta E = \pm \gamma J(J + 1) \tag{3.78}$$

The splitting of these two levels is, thus to first order of approximation, $\frac{1}{2}(B - C)J(J + 1)$ in the prolate limit and $\frac{1}{2}(A - B)J(J + 1)$ in the oblate limit. All other levels are subjected to second- and higher-order corrections only and the splittings are much smaller in magnitude as indicated in Figure 3.5. To some extent one can see that if the angular momentum component along the axis is small the molecule will tend to wobble more than if this component is large. In the high $|K|$ case the molecule rotates with more symmetric top character and this manifests itself in the energy-level pattern (Figure 3.5). A convenient tabulation of the perturbation solutions is given by Townes and Schawlow.[303] These tables give solutions in terms of the asymmetry parameters

$$b_A = b_p = \frac{C - B}{2A - B - C}, \qquad b_C = b_o = \frac{A - B}{2C - B - A} \tag{3.79}$$

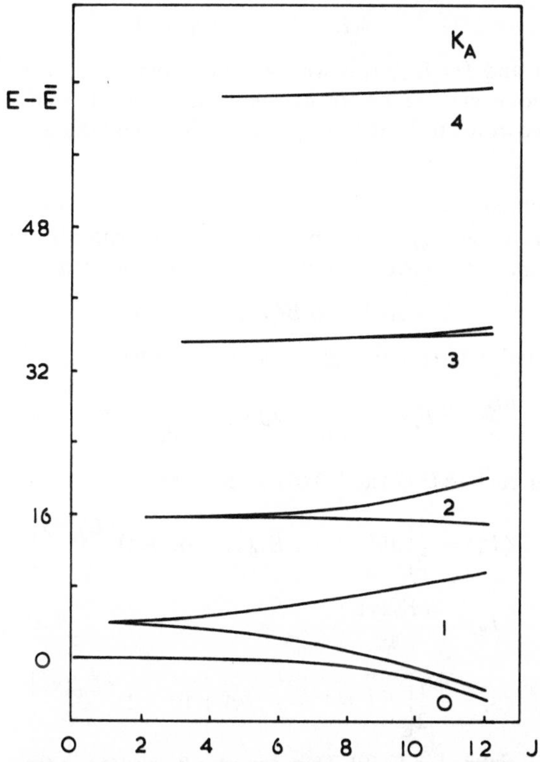

Figure 3.5 The rotational energy levels of a slightly asymmetric rotor with $b = 0{\cdot}025$. The term $\bar{E} = \frac{1}{2}(B + C)J(J + 1)$ has been subtracted, so that deviations from the horizontal represent the effects of asymmetry

for the near polate and near oblate limits respectively, and the associated energy-level expressions in terms of powers of the b coefficients up to 5 are

$$E = \tfrac{1}{2}[B + C]J(J + 1) + [A - \tfrac{1}{2}(B + C)][K_A^2 + C_1 b_A + C_2 b_A^2 + C_3 b_A^3 + C_4 b_A^4 + C_5 b_A^5] \quad (3.80a)$$

$$E = \tfrac{1}{2}[A + B]J(J + 1) + [C - \tfrac{1}{2}(A + B)][K_C^2 + C_1 b_C + C_2 b_C^2 + C_3 b_C^3 + C_4 b_C^4 + C_5 b_C^5] \quad (3.80b)$$

The general expressions for the C coefficients are given by Gordy and Cook.[100,2] We have just shown that the coefficient C_1 is $\pm\tfrac{1}{2}J(J + 1)$ for K_A or $K_C = 1$ and zero for all other cases. C_2 is obtained by second-order perturbation theory as

$$C_2 = \frac{f(J\,K - 1)}{4(K - 1)} - \frac{f(J\,K + 1)}{4(K + 1)} \quad (3.81)$$

when $|K| \neq 1$ and $f(J\,K \pm 1)$ is with various provisos[100] given by (3.71b). Transition frequencies are directly calculated from the tables by subtracting coefficients according to $\sum_n (C'_n - C''_n)b^n$ and if b is small the series converges rapidly.

3.10d Expectation Values $\langle J_\alpha^2 \rangle$

The rigid-rotor energy-level expression can be written in terms of the expectation values $\langle J_\alpha^2 \rangle$ averaged over the asymmetric rotor eigenfunctions

$$E = A\langle J_A^2 \rangle + B\langle J_B^2 \rangle + C\langle J_C^2 \rangle \quad (3.82)$$

which on partial differentiation gives us to first order

$$\frac{\partial E}{\partial A} = \langle J_A^2 \rangle, \qquad \frac{\partial E}{\partial B} = \langle J_B^2 \rangle, \qquad \frac{\partial E}{\partial C} = \langle J_C^2 \rangle \quad (3.83)$$

and if we refer back to (3.66) and (3.67) we see that

$$\langle J_A^2 \rangle = \frac{1}{2}\left[J(J + 1) + E_{J\tau}(\kappa) - (\kappa + 1)\frac{\partial E_{J\tau}(\kappa)}{\partial \kappa}\right] \quad (3.84a)$$

$$\langle J_B^2 \rangle = \frac{\partial E_{J\tau}(\kappa)}{\partial \kappa} \quad (3.84b)$$

$$\langle J_C^2 \rangle = \frac{1}{2}\left[J(J + 1) - E_{J\tau}(\kappa) + (\kappa - 1)\frac{\partial E_{J\tau}(\kappa)}{\partial \kappa}\right] \quad (3.84c)$$

where $E_{J\tau}(\kappa)$ is defined in (3.69). They can also be written in terms of b_A and b_C.[100]

These expressions are useful whenever molecular properties are averaged over the asymmetric rotor functions. Average value relations are often used to determine centrifugal distortion, quadrupole, internal rotation parameters and intensities.

3.11 The Rotational Eigenfunctions

It is useful, especially when we come to discuss spectral intensities, to be aware of certain properties of the rotational eigenfunctions—in particular their symmetry properties. As we have seen, the Hamiltonian for the symmetric top (3.50) commutes with J^2, J_c and J_z and thus the eigenfunctions discussed in Section 3.5 are simultaneously the eigenfunctions for the symmetric top. We can thus write

$$\Psi_{JKM} = \Theta_{JKM}(\theta)\, e^{iM\varphi}\, e^{iK\chi} \tag{3.85}$$

The only change from (3.43) is to substitute J, K and M, the quantum numbers of molecular rotation convention for j, k and m, the quantum numbers of general angular momentum theory convention. The eigenfunctions of the linear molecule are obtained by setting $K = 0$ and thus they are the Legendre Polynomials (3.40). We can neglect the M dependence when the molecule is in free space at least as far as the energy is concerned—though not as we shall see the degeneracy. As the asymmetric top Hamiltonian was set up in a symmetric rotor basis the eigenfunctions which diagonalize H can be written as linear combinations of symmetric rotor eigenfunctions. They can also be written as linear combinations of the Wang functions which are themselves linear combinations of symmetric rotor functions of the same $|K|$ (3.58).

$$|J\,\tau\,\pm\rangle = \sum_K a^{JK}(|J\;K\rangle \pm |J\;-K\rangle) \tag{3.86}$$

Note that the summation is carried out with two restrictions: K either all even or all odd, and Wang functions either all symmetric (+) or all antisymmetric (−).

3.11a Symmetry and the Complete Hamiltonian

In Chapter 1 we pointed out that a consequence of the commutation of two operators A and B according to (1.11) is that there exist functions simultaneously eigenfunctions of both operators. We now wish to discuss some general implications of this theorem with regard to the complete Hamiltonian and its eigenfunctions.

If an operator A exists such that

$$[H, A] = 0 \tag{3.87}$$

then the eigenfunctions of H are automatically eigenfunctions of A. If we read the fine print associated with the derivation of this Quantum Mechanical theorem we see that the reverse does not necessarily hold because the eigenfunctions of H (*complete*) are not degenerate whereas those of A often are.

Theoretically at least there will be some term in H which will split any apparent degeneracy. As a consequence we can extend the classification of ψ_r to include signatures which represent the properties of other operators. There may be many types of operators which fall into this category and a set which are particularly important are what we can call *fundamental transformations* which leave the Hamiltonian invariant according to (3.87). In the case of some simple molecules these fundamental transformations are closely related to the geometric symmetry operations which characterize the point groups of the molecules. In what follows we shall use the conventions and tables of Wilson, Decius and Cross,[335] Hougen[129] and Longuet–Higgins.[184]

Depending on the problem, some transformations lead to more useful results than others.[184] We will restrict ourselves to those which are particularly suitable for an understanding of the rotational spectra of rigid molecules. Those sets suitable for characterizing rovibronic wavefunctions of semi-rigid molecules have been discussed by Hougen.[129] This scheme is based on the symmetry operations which characterize the full molecular point group and is more versatile than the traditional scheme based on rotational subgroups of the full group. Longuet–Higgins has developed a procedure based on permutation symmetry which is particularly useful for classifying the eigenfunctions of flexible molecules.[184]

There are three fundamental transformations[184] which we need to consider. They are:

(1) R, the operator which rotates the positions and spins of all molecular particles (electrons and nuclei) about any axis through the centre of mass.

(2) I, the space-fixed inversion operator which simultaneously changes all coordinates into their negatives.

(3) P_{nm}, the set of all permutations of positions *and* spins of all identical particles n and m.

We shall discuss these now in a little more detail.

If space is isotropic we mean that the Hamiltonian is invariant to rotation about the centre of mass. If the operator which carries out this transformation is R, then quantum mechanically the isotropy of space is specified by the commutator

$$[H, R] = 0 \tag{3.88}$$

For an infinitesimal rotation $R_{\delta\varphi}$ operating on $f(q_{ni})$

$$R_{\delta\varphi} f(q_{ni}) = f(q_{ni}) + \delta q_{ni} \frac{\partial f(q_{ni})}{\partial q_{ni}} \tag{3.89}$$

$\boldsymbol{\delta\varphi}$ represents an infinitesimally small rotation vector along the axis about which the rotation occurs and $\boldsymbol{q}_n$ is the coordinate of the nth particle. Now,

$\delta q_{ni} = e_{ijk}\delta\varphi_j q_{nk} (\equiv \delta \boldsymbol{q}_n = \delta\boldsymbol{\varphi} \times \boldsymbol{q}_n)$ and thus

$$R_{\delta\boldsymbol{\varphi}} = 1 + \delta\varphi_j\, e_{jki} q_{nk} \frac{\partial}{\partial q_{ni}} \equiv 1 + \delta\boldsymbol{\varphi} \cdot (\boldsymbol{q}_n \times \boldsymbol{\nabla}_n) \tag{3.90}$$

substitution of (3.90) into (3.88), multiplication by $-i$ and summation over the n particles yields

$$[H, e_{ijk} q_j p_k] = [H, \boldsymbol{J}] = 0 \tag{3.91}$$

note that $\delta\varphi_j$ is a constant vector and commutes with H. Thus what was true for the particular case of H_r is in fact true for H (complete). The main point that follows is that the conservation of angular momentum is a consequence of the isotropy of space. One can note that in an analogous way we can show that the conservation of linear momentum is a consequence of the homogeneity (or translational isotropy) of space. Finally we note that a finite rotation $\boldsymbol{R} = (R_{\delta\boldsymbol{\varphi}})^n$ and that $\boldsymbol{J}^n$ must therefore commute with H.

Inversion of all *space-fixed* coordinates through the origin is defined by

$$I f(\boldsymbol{q}_1 \boldsymbol{q}_2 \cdots \boldsymbol{q}_n) \rightarrow f(-\boldsymbol{q}_1 - \boldsymbol{q}_2 \cdots -\boldsymbol{q}_n) \tag{3.92}$$

where the coordinates $\boldsymbol{q}_n$ have been transformed into their negatives. The total kinetic energy (1.2) consists of a sum of squared terms involving the $\boldsymbol{q}_n$ and the potential energy depends only on the moduli of the internuclear separations. As a consequence H is invariant to, and commutes with, I. The eigenfunctions of H are also eigenfunctions of I. The allowed eigenvalues of I are easily derived because two successive applications of I upon Ψ must return Ψ.

$$I^2\Psi = I(I'\Psi) = I'^2\Psi = \Psi \tag{3.93}$$

I' may therefore only be ± 1. The eigenfunctions may therefore be labelled by the value of I' or more commonly just the sign $+$ or $-$. A level is said to possess even parity $(+)$ if $I' = +1$ and odd parity $(-)$ if $I' = -1$. I is also called the parity operator.

The permutation operator P_{nm} which exchanges the set of position variables $\boldsymbol{q}$ *and* spin coordinate $\boldsymbol{I}$ of particle n for those of particle m and vice versa is defined by

$$P_{nm} f(\boldsymbol{q}_m \boldsymbol{I}_m \boldsymbol{q}_n \boldsymbol{I}_n) \rightarrow f(\boldsymbol{q}_n \boldsymbol{I}_n \boldsymbol{q}_m \boldsymbol{I}_m) \tag{3.94}$$

In a similar way to that which applied in the case of the inversion operator we can show that the eigenvalue $P'_{nm} = \pm 1$ for all pairwise exchanges of *identical particles*. If the function is the *total* wavefunction it has been found that when the particles exchanged possess half-integral spin, $P'_{nm} = -1$ and when they possess integral spin, $P'_{nm} = +1$. This is perhaps the most basic general form of the Exclusion Principle formulated in a somewhat different, but equivalent, way by Pauli for electrons.

In more general cases we may need to consider some further transformations.[184]

3.11b Geometric Symmetry Operations

We now need to discuss the properties of the rotational eigenfunctions with respect to the symmetry operations which characterize the more familiar point groups. As has been pointed out, in some cases they can be related to the fundamental transformations.[184] For our purposes we will consider the operations which characterize the groups; $D_{\infty h}$ $C_{\infty v}$ C_{nv} and D_2.

For an asymmetric rotor the Hamiltonian is rather special because it is a sum of square powers of the three components of $\boldsymbol{J}$ and invariant to the simultaneous change in sign of any two components (equivalent to C_2^α). The set of three two-fold rotations C_2^a, C_2^b and C_2^c together with the identity, E, constituting the set of operations for the group D_2 may thus be used. This is the case even if the point group of the molecule has lower symmetry.

The main symmetry operations which characterize the above point groups are: the identity, E; reflection in the $\alpha\beta$ plane, $\sigma^{\alpha\beta}$; C_n^γ which represents an n-fold rotation about the axis α and the *molecule-fixed* inversion, i. Not all may always occur, for instance i only exists if the molecule has a center of inversion. On the other hand, I, the space-fixed inversion operator always exists and is not the same as i.

The rotational eigenfunctions are functions of the Eulerian angles χ, θ and φ and thus the next step is to examine the behaviour of these variables when subjected to the various types of symmetry operation. The behaviour under rotation is fairly straightforward and can be determined by studying Figure 2.3 (see below). The so-called sense-reversing operations such as i and σ are rather more obscure in their effect on $f(\chi\,\theta\,\varphi)$, because they change the handedness of the axis system. This type of transformation cannot be described by some arbitrary set of values for χ, θ and φ, for these can only represent relative axis orientations. Hougen[134] has shown that a valid prescription is:

The transformation properties of $f(\chi\,\theta\,\varphi)$ under a sense reversing operation is the same as that of the pure rotation obtained by multiplying the operation by i. As a consequence, as far as $f(\chi\,\theta\,\varphi)$ is concerned; i behaves like E, $\sigma^{\alpha\beta}$ behaves like C_2^γ and $S_\infty^\gamma(\varepsilon)$ behaves like $C_\infty^\gamma(\pi+\varepsilon)$ (because $ii \rightarrow E$, etc). The rotation operation can be defined as an anticlockwise rotation of coordinates[134]

$$C_\infty^c(\varepsilon)\begin{bmatrix} a \\ b \\ c \end{bmatrix} \equiv \begin{bmatrix} \cos\varepsilon & \sin\varepsilon & 0 \\ -\sin\varepsilon & \cos\varepsilon & 0 \\ 0 & 0 & 1 \end{bmatrix}\begin{bmatrix} a \\ b \\ c \end{bmatrix} = \begin{bmatrix} a' \\ b' \\ c \end{bmatrix} \tag{3.95}$$

and this is equivalent to the χ dependent part of $\mathbf{S}$, (2.12), which is a clockwise rotation of axes. For the asymmetric rotor we need to know the behaviour of

χ, θ and φ with respect to $C_2^\alpha(\alpha = a, b \text{ and } c)$. Thus we find from Figure 2.3 that the resulting behaviour is as summarized in Table 3.3.

Table 3.3 The Properties of χ, θ and φ When Subjected to C_2^γ

	χ	θ	φ
C_2^a	$-\chi$	$\pi - \theta$	$\pi + \varphi$
C_2^b	$\pi - \chi$	$\pi - \theta$	$\pi + \varphi$
C_2^c	$\pi + \chi$	θ	φ

According to the above discussion (3.95) and Table 3.3 we see that

$$C_n^c f(\chi\,\theta\,\varphi) \to f\left(\frac{2\pi}{n} + \chi\,\theta\,\varphi\right) \tag{3.96a}$$

$$\sigma^{ac} f(\chi\,\theta\,\varphi) \to f(\pi - \chi\;\pi - \theta\;\pi + \varphi) \tag{3.96b}$$

$$\sigma^{ab} f(\chi\,\theta\,\varphi) \to f(\pi + \chi\;\theta\;\varphi) \tag{3.96c}$$

This set is particularly useful for a discussion of the properties of symmetric rotor functions.[129]

The ways in which these geometric symmetry operations relate to the fundamental transformations depends on the system in question. In the case of the rotation operator, R and C_n are directly related. I is particularly useful for classifying the *total* wavefunctions of linear molecules including electrons. One can show[169,158,134] that when σ^{ac} operates on the variables of a diatomic molecule the *spatial* coordinates x, y and z change into their negatives and thus σ^{ac} is equivalent in this case to I. $C_2(b)$ turns out to be equivalent to P_{nm} if the system is homonuclear. Thus we come to the conclusion that $i \equiv \sigma^{ac} C_2^b \equiv P_{nm} I$.

We are now in a position to consider the eigenfunctions explicitly. Using (3.96) in conjunction with the Legendre functions (3.40) we see that (with the convention $\sigma_h = \sigma^{ab}$ and $\sigma_v = \sigma^{ac}$)

$$C_n^c|J\,0\,0\rangle = |J\,0\,0\rangle \tag{3.97a}$$

$$\sigma_v|J\,0\,0\rangle = (-1)^J|J\,0\,0\rangle \tag{3.97b}$$

$$\sigma_h|J\,0\,0\rangle = |J\,0\,0\rangle \tag{3.97c}$$

This set of relations yields the properties of the eigenfunctions of the linear molecule and the $K = 0$ stack for the symmetric top. As the $K \neq 0$ symmetric top functions can be generated from $|J\,0\,0\rangle$ by using $J^\pm$ according to

(3.42) we thus need the relations[129]

$$C_n^c J^{\pm} = \varepsilon^{\mp 1} J^{\pm} \tag{3.98a}$$

$$\sigma_v J^{\pm} = -J^{\mp} \tag{3.98b}$$

$$\sigma_h J^{\pm} = -J^{\pm} \tag{3.98c}$$

where $\varepsilon = \exp(2\pi i/n)$. From these we obtain the following set of transformation properties of the symmetric rotor functions.[129]

$$C_n^c |J\ K\ M\rangle = \varepsilon^K |J\ K\ M\rangle \tag{3.99a}$$

$$\sigma_v |J\ K\ M\rangle = (-1)^{J-K} |J\ -K\ M\rangle \tag{3.99b}$$

$$\sigma_h |J\ K\ M\rangle = (-1)^K |J\ K\ M\rangle \tag{3.99c}$$

3.11c The Symmetry of the Rotational Eigenfunctions

The rotational energy levels of molecules are usually specified by the symmetry properties of the *total* eigenfunction (apart from nuclear spin).

$$\Psi = \psi_e \psi_v \psi_r \tag{3.100}$$

and *not* just the rotational part. We shall assume that ψ_e is totally symmetric which is almost always the case for stable molecules in their electronic ground state. We shall also restrict ourselves to the totally symmetric vibrational ground state. In this case $\Gamma(\Psi) \equiv \Gamma(\psi_r)$; otherwise we must take into account the symmetry properties of ψ_e and ψ_v by the usual procedures of group theory, i.e. $\Gamma(\Psi) = \Gamma(\psi_e) \times \Gamma(\psi_v) \times \Gamma(\psi_r)$.

In the case of the diatomic molecule, the rotational wavefunctions are specified as positive or negative depending on whether ψ_r remains unchanged or changes sign under $I = \sigma_v$.[122] Note that $J_c = 0$ in this case. We thus see from (3.97b) that if J is even, the levels are +, and if J is odd, the levels are − as shown in Figure 3.1.

As far as the symmetric tops are concerned, we shall concentrate on C_{3v} molecules which are fairly representative and by far the most common. The C_{3v} character table is given in Table 3.4 and the rotational subgroup given in Table 3.5.

Table 3.4

C_{3v}	E	$2C_3$	$3\sigma_v$	
A_1	1	1	1	c
A_2	1	1	−1	R_c
E	2	−1	0	a, b, R_a, R_b

Table 3.5[a]

C_3	E	C_3	C_3^2
A	1	1	1
E	1	ε	ε^2
	1	ε^2	ε

[a] $\varepsilon = \exp(2\pi i/3)$.

We can determine the species of the C_{3v} symmetric top functions by examining their behaviour under E, C_3 and σ_v. For instance we see that

$$C_3 \begin{bmatrix} |J \quad K \quad M\rangle \\ |J \quad -K \quad M\rangle \end{bmatrix} = \begin{bmatrix} \varepsilon^K & 0 \\ 0 & \varepsilon^{-K} \end{bmatrix} \begin{bmatrix} |J \quad K \quad M\rangle \\ |J \quad -K \quad M\rangle \end{bmatrix} \tag{3.101}$$

where $\varepsilon = \exp(2\pi i/3)$. The trace of the transformation matrix is thus $\varepsilon^K + \varepsilon^{-K} = 2\cos(2\pi K/3)$. The behaviour under σ_v can be determined using (3.99b). As a result we can generate the species of the various types of C_{3v} symmetric top functions as given in Table 3.6.

Table 3.6 Symmetry Properties of the Rotational Functions of a Molecule with C_{3v} Symmetry[a]

	E	$2C_3$	$3\sigma_v$	
$K = 0$	1	1	$(-1)^J$	J even A_1 J odd A_2
$\lvert K\rvert \neq 3n$	2	-1	0	E
$\lvert K\rvert = 3n$	2	2	0	$A_1 + A_2$

[a] n is an integer $\geqslant 1$.

We can classify the *rotational* functions for an asymmetric top by their behaviour with respect to the operations which characterize the group D_2 or V, as given in Table 3.7, even when the molecular point group is of lower symmetry.[169] An asymmetric rotor wavefunction is specified by the values of $|K|$ of the limiting prolate and oblate symmetric top functions with which the particular function correlates Figure 3.3. As the species of the function is independent of the asymmetry we can determine the species by studying the behaviour of the two limiting cases. We can use (3.99) with the

Table 3.7 Character Table for the Point Group D_2

	E	C_2^a	C_2^b	C_2^c
A	1	1	1	1
B_a	1	1	-1	-1
B_b	1	-1	1	-1
B_c	1	-1	-1	1

assignments $a \sim A$, $b \sim B$ and $c \sim C$ and if we note that $C_2^c \equiv \sigma^{ab} \equiv \sigma_h$, we obtain

$$C_2^A|J\ K_A\ M\rangle = (-1)^{K_A}|J\ K_A\ M\rangle \tag{3.102a}$$

$$C_2^C|J\ K_C\ M\rangle = (-1)^{K_C}|J\ K_C\ M\rangle \tag{3.102b}$$

We can complete the classification relations by noting that two successive two-fold rotations about two axes are equivalent to one two-fold rotation about the third, i.e. $C_2^\alpha C_2^\beta \equiv C_2^\gamma$. Thus we obtain the general symmetry properties given in Tables 3.8, and those specific to the $J = 3$ set of eigenfunctions in Table 3.9.

Table 3.8 Symmetry Properties of the Asymmetric Rotor Levels as a Function of K_A and K_C

K_A	K_C	
even	even	A
even	odd	B_A
odd	odd	B_B
odd	even	B_C

Table 3.9 Symmetries of the $J = 3$ set of Asymmetric Rotor Levels

τ	K_A	K_C	$J_{K_A K_C}$	$\Gamma(\psi_r)$
3	3	0	3_{30}	B_C
2	3	1	3_{31}	B_B
1	2	1	3_{21}	B_A
0	2	2	3_{22}	A
-1	1	2	3_{12}	B_C
-2	1	3	3_{13}	B_B
-3	0	3	3_{03}	B_A

It is worth noting how the symmetries of the various levels relate the symmetries of the four block matrices into which the asymmetric rotor was separated in Section 3.10.

3.11d Nuclear Spin and Statistical Weights

We must now discuss the symmetry properties of the nuclear spin functions because they form an integral part of Ψ, *the total eigenfunction*, and exert a curious influence on Ψ through the medium of the generalized Pauli Exclusion Principle. The permutation operator, P_{nm}, was defined in (3.94), and when it operates on the total eigenfunction of a molecule with two identical particles (n and m) its eigenvalue is $+1$ if the particles have integral spin, and -1 if the particles have half-integral spin. Half-integral spin particles such as electrons (1/2), H(1/2), Cl(3/2) and Bi(9/2) follow Fermi–Dirac statistics and are called Fermions.[83] Integral spin particles such as photons D(1), O(0) and V(6) follow Bose–Einstein statistics and are called Bosons.[83] The anomalous specific heat of molecular hydrogen at low temperature can be explained by understanding how the Exclusion Principle applies to the total molecular wavefunction.

In dealing with nuclear permutation symmetry it is convenient to introduce the procedure set up by Longuet–Higgins for handling the eigenfunctions of flexible molecules. The procedure is not limited, as is the geometric point group method, to semi-rigid molecules whose internal motions consist of only small displacements from a single equilibrium configuration. It can handle flexible systems which tunnel between configurations of varying geometric symmetry. In ethane the molecular point group is D_{3d} when staggered and D_{3h} when eclipsed and the molecule can tunnel between the equilibrium configurations changing its symmetry in the process.

The prescription for setting up the permutation group is as follows:[184]

(1) Let P be any permutation of positions and spins of identical nuclei or any product of such permutations.

(2) Let E be the identity.

(3) Let E^* be the inversion of all particle positions.

(4) Let P^* be the product $PE^* \equiv E^*P$.

The permutation group of a molecule (other than a linear one) consists of

(a) all feasible P including E,

(b) all feasible P^* not necessarily including E^*.

A feasible operation is one which the molecule may undergo within the time scale of the experiment.

Thus a single nuclear permutation is, as we shall see, feasible in ammonia for which umbrella inversion occurs rapidly and *not* feasible in PH_3 for which inversion appears to be very slow, if it occurs at all.

...ill not in general be the same as a geometric ... may often be isomorphic with one. We finally ... Principle governs only transformations described ...g to Longuet–Higgins, 'non-commital' about permuta- ...ch as P^* and E^*.

... the procedure into sharper focus by applying it to NH_3 and ... ectra of NH_3 can be reconciled with a model in which the mole- cu... ...els between two C_{3v} (mirror-image) configurations, where the poten...al energy is a minimum, through an intermediate planar (D_{3h}) configuration where the potential is a maximum.[303] The tunneling frequency is of the order of $2{\cdot}4 \times 10^{10}$ Hz and the barrier height is $\sim 2020\ \text{cm}^{-1}$ [293] The inversion frequency of PH_3 is very much smaller[63] and has not so far been detected. Such a motion may occur but because the experimental time scale is much too short the effects are not observed and we can effectively neglect this motion.

In Figure 3.6 the various transformations which, according to the above recipe, comprise the permutation group for NH_3 are defined.

We can clarify the meaning of the term feasible with reference to NH_3 and PH_3 and the set of permutations given in Figure 3.6. E will always be a feasible operation, however E^* will be *in*feasible for PH_3 and feasible for NH_3. Note that the so-called inversion of ammonia is not an operation, but a term describing the umberella motion and should not be confused with the rigidly defined inversion operations. E^* is feasible in the case of NH_3 because from an initial configuration it can tunnel through the barrier and rotate according to

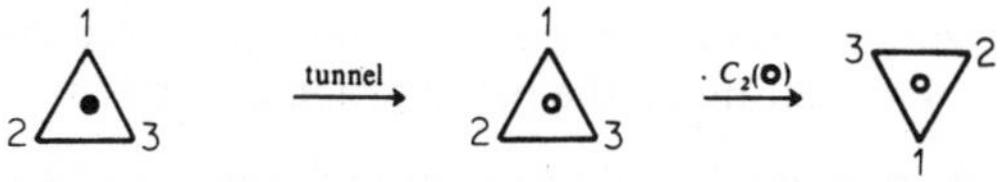

(3.103)

and thus physically accomplish E^* within the experimental time scale. The permutations (123) and (321) are essentially equivalent to the geometric operation $\pm C_3$ and thus, as there is no potential barrier to circumnavigate, these are feasible operations for both molecules. PH_3 can transform according to (12)*, etc., by simple rotations whereas (12) is not feasible. For NH_3 both are feasible operations. As a consequence the permutation group for PH_3 consists of the set of operations E, 2(123) and 3(12)* and is isomorphous with, though not equivalent to, C_{3v}. For NH_3 on the other hand, the group operations are: E, E^*, 2(123), 2(123)*, 3(12) and 3(12)* and the group is isomorphous with D_{3h}.

Let us now examine the properties of the symmetric top functions $|J\ K\ M\rangle$ with respect to the operations of the new group. We already know their properties with respect to the geometric operations which constitute the

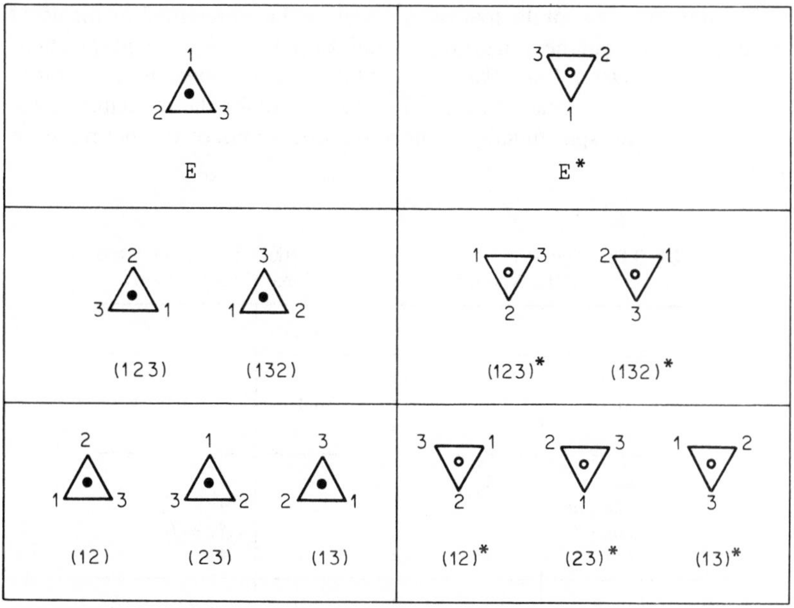

Figure 3.6 The set of operations which comprise the permutation group of NH_3 together with their resulting transformations. The equivalent H atoms are labelled 1, 2 and 3. When the N atom is on one side of the plane containing the H atoms it is represented by a point and when it is on the other side it is represented by a circle. The permutation symbol (12) represents the single permutation in which particles 1 and 2 are exchanged. The double permutation (12) (23) operates on a function at its right and thus can be represented by the cyclic permutation symbol (123) in which particle 1 is replaced by 2, 2 by 3 and 3 by 1. In this particular case it is equivalent to C_3—the rotation by 120°

group C_{3v} as these are given by the relations (3.99). It is fairly clear that E and C_3, in C_{3v} are equivalent to E and (123) respectively in the new group. We can also show that $C_2^b \equiv (23)^*$ quite simply:

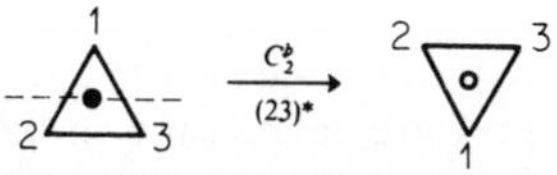

(3.104)

Now if we recall the transformation properties of $f(\chi\theta\varphi)$ under sense-reversing operations, as discussed in Section 3.11b, we see that *as far as* $f(\chi\ \theta\ \varphi)$ *is concerned* $(23)^* \equiv C_2^b \equiv \sigma_v$. We thus come to the conclusion that the characters of the symmetric top functions are the same in the new group as in C_{3v}, Table 3.4. These results are summarized in Table 3.10, using the same nomenclature as is used in C_{3v}. Note however that $(23)^*$ is *not* equivalent to σ_v in general.

The spin functions of the hydrogen nuclei can be represented by the usual product functions of the spin symbols α and β. *If we note that the spin functions are invariant under E^* and thus $P^* \equiv P$ (as far as the spins are concerned)*, we can generate the characters given in Table 3.10 for the reducible representations of the spin functions under the operations of the permutation group.

Table 3.10 The Permutation Group Equivalent to C_{3v} and the Properties of the Nuclear Spin and Rotational Eigenfunctions[a]

	E	2(123)	3(12)*	
A_1	1	1	1	
A_2	1	1	-1	
E	2	-1	0	
$\alpha\alpha\alpha$	1	1	1	A_1
$\alpha\alpha\beta$, $\alpha\beta\alpha$, $\beta\alpha\alpha$	3	0	1	$A_1 + E$
$\alpha\beta\beta$, $\beta\alpha\beta$, $\beta\beta\alpha$	3	0	1	$A_1 + E$
$\beta\beta\beta$	1	1	1	A_1
$E \times E$	4	1	0	$A_1 + A_2 + E$
$K = 0$	1	1	$(-1)^J$	A_1 or A_2
$\|K\| \neq 3n$	2	-1	0	E
$\|K\| = 3n$	2	2	0	$A_1 + A_2$

[a] n is an integer $\geqslant 1$.

We are now in a position to decide, with the aid of the Exclusion Principle, which of the product functions $\Psi = \psi_r\psi_{ns}$ are allowed. The Exclusion Principle allows only those Ψ to occur which have either A_1 or A_2 symmetry, because (123) is a double permutation.[331] As a result we obtain the statistical weights and symmetries as summarized in Table 3.11.

Table 3.11 Statistical Weights of Rotational Levels of a Rigid C_{3v} Molecule Containing Three Equivalent Spin-1/2 Nuclei

	$\Gamma(\psi_r)$	$\Gamma(\psi_r) \times \Gamma(\psi_{ns})$	$\Gamma(\Psi)$	Statistical weight
$K = 0$, J even	A_1	$4A_1 + 2E$	$4A_1$	4
$K = 0$, J odd	A_2	$4A_2 + 2E$	$4A_2$	4
$\|K\| \neq 3n$	E	$2A_1 + 2A_2 + 6E$	$2A_1 + 2A_2$	4
$\|K\| = 3n$	$A_1 + A_2$	$4A_1 + 4A_2 + 4E$	$4A_1 + 4A_2$	8

The general expressions for the statistical weights of the levels of C_{3v} molecules when inversion effects are *not* resolved are

$$g_I(K = 3n) = \tfrac{1}{3}(2I + 1)(4I^2 + 4I + 3) \tag{3.105a}$$

$$g_I(K \neq 3n) = \tfrac{1}{3}(2I + 1)(4I^2 + 4I) \tag{3.105b}$$

for each individual K level. For degenerate vibrational levels K must be replaced by $K - l$. Further expression have been given by Townes and Schawlow[303] which apply when inversion effects are evident, and for C_{4v} molecules.

In the case of phosphine we were able to arbitrarily halve the size of the group because the molecule does not show any evidence that it can penetrate the barrier separating the two configurations depicted under E and E^* in Figure 3.6. There are two sets of energy levels associated with these two distinct non-superposable configurations. They are degenerate and the resulting spectra are superimposed when tunneling is not possible. No useful information is to be obtained in this case by using the full group rather than the group reduced by using the feasibility clause. On the other hand, when the effects of inversion are spectroscopically detectable the full group yields pertinent extra information. The permutation group for ammonia which is generated when all the operations catalogued in Figure 3.6 are feasible is given in Table 3.12. That of phosphine, Table 3.8, is a *subgroup* of this more complete group. The new group is isomorphous with D_{3h} and the symmetry species have, for convenience, been labelled in accordance with the symbols traditionally used to classify D_{3h} functions.

Table 3.12 The Permutation Group Equivalent to D_{3h}

	E	2(123)	3(12)*	E^*	2(123)*	3(12)
A_1'	1	1	1	1	1	1
A_2'	1	1	−1	1	1	−1
A_1''	1	1	−1	−1	−1	1
A_2''	1	1	1	−1	−1	−1
E'	2	−1	0	2	−1	0
E''	2	−1	0	−2	1	0

In ammonia the inversion splitting is much smaller ($\sim$0·79 cm^{-1} in the 'ground' state) than the rotational spacings ($A \sim$ 9·94 cm^{-1} and $B \sim$ 6·20 cm^{-1}) and it is most convenient to consider how the inversion split levels correlate with those which have been determined in the non-tunneling limit. The symmetry species of these levels can be neatly determined using Watson's correlation rule.[313] As is the case with many group-theoretical procedures their simplicity is often obscured by the terminology. The ammonia problem presents a simple example of how to determine the symmetry properties and statistical weights in flexible molecules (Chapter 9).

Consider a state which has a particular symmetry in the C_{3v} subgroup. Then determine the *set of species* which in the full D_{3h} group exhibit the same behaviour under the operations which characterize C_{3v}. The original state splits into levels with this set of symmetry labels when tunneling occurs. Comparison of Table 3.8 with Table 3.12 indicates that the correlation is as given in Table 3.13.

Table 3.13 Correlation Table for a C_{3v} Molecule in which Tunneling can Occur

C_{3v}	D_{3h}
A_1	$A_1' + A_2''$
A_2	$A_2' + A_1''$
E	$E' + E''$

As a result we see that the inversion wavefunctions can be labelled by the species A_1' and A_2''. As far as the statistical weights are concerned we can now use the operation (12) to determine which functions may or may not occur. Thus using the Exclusion Principle we see that the *total* eigenfunctions of NH_3 must possess A_2' or A_2'' symmetry. As a consequence we see that rotational A_1 states do not occur for 0^+ states and rotational A_2 states do not occur for the 0^- states. This is further discussed in Section 9.3a.

Using the same procedure as in the case of PH_3 we can derive the statistical weights of the various levels and these are given in Table 3.14.

Table 3.14 Symmetries and Statistical Weights of the Eigenfunctions of NH_3

$\psi_r(C_{3v})$	$\psi_{inv}\psi_r$	$\psi_{inv}\psi_r\psi_{ns}$	Statistical weight
A_1	A_1'	$4A_1' + 2E'$	0
	A_2''	$4A_2'' + 2E''$	4
A_2	A_2'	$4A_2' + 2E'$	4
	A_1''	$4A_1'' + 2E''$	0
E	E'	$6E' + 2A_1' + 2A_2'$	2
	E''	$6E'' + 2A_1'' + 2A_2''$	2

We shall now discuss the way in which permutation symmetry governs the statistical weights for molecules such as H_2CO, F_2CO and H_2CCS; molecules which contain two equivalent nuclei. Consider the molecule

H_2CO; it is a prolate asymmetric rotor with C_{2v} symmetry. The operation C_2^A is essentially equivalent to the H nucleus permutation operation, as far as the rotational and nuclear spin functions are concerned. The rotational subgroup of C_{2v}, C_2, is given in Table 3.15, and the C_2 rotation (in this case C_2^A) operating on Ψ essentially effects the required permutation. Thus, as we are dealing with spin-$\frac{1}{2}$ particles, the Exclusion Principle requires that Ψ transform as B under the operations which characterize the group C_2.

Table 3.15 The Rotational Subgroup C_2

C_2	E	C_2
A	1	1
B	1	−1

There are $(2I + 1)^2 = 4$ basic spin functions which we can write in the form of a vector $(\alpha\alpha\ \alpha\beta\ \beta\alpha\ \beta\beta)$. Operating with E and C_2^A we see that the characters are $\chi^{(E)} = 4$ and $\chi^{(C_2)} = 2$ respectively, which indicates that the basis vector can be transformed into a vector consisting of three A and one B functions. There are, of course, the familiar sets

$$A \text{ species} \qquad \alpha\alpha \qquad \frac{1}{\sqrt{2}}(\alpha\beta + \beta\alpha) \qquad \beta\beta \tag{3.106a}$$

$$B \text{ species} \qquad \frac{1}{\sqrt{2}}(\alpha\beta - \beta\alpha) \tag{3.106b}$$

If we now inspect the behaviour of the asymmetric top functions under C_2^A as given in the D_2 group, Table 3.7, we can determine the conditions under which

$$\Gamma(\Psi) = \Gamma(\psi_r) \times \Gamma(\psi_{ns}) = B \tag{3.107}$$

Thus when ψ_r has species A or B_C then the associated spin function may only have species B. On the other hand, when ψ_r has species B_A or B_B then the spin function must have A species.

As the ratio of the statistical weights of the A nuclear spin states to the B states is 3 to 1, this factor must be included when we discuss the degeneracy of each individual level.

It is fairly simple to determine the relative statistical weights of the symmetric and antisymmetric spin functions in a molecule with two identical particles. Consider the case where the spin quantum number is I; there will be $(2I + 1)^2$ simple product functions of which $(2I + 1)$ are already symmetric under P_{12} as the values of M_I match. There remain $[(2I + 1)^2 - (2I + 1)]$

Table 3.16 Symmetry and Statistical Weight Table for H_2CO and D_2CO[a]

	K_A	$\chi(\psi_v)$	$\chi(\psi_r)$	$\chi(\psi_{ns})$	$\chi(\Psi)$	Statistical weight
H_2CO	even	+1	+1	−1	−1	1
	odd	+1	−1	+1	−1	3
	even	−1	+1	+1	−1	3
	odd	−1	−1	−1	−1	1
D_2CO	even	+1	+1	+1	+1	2
	odd	+1	−1	−1	+1	1
	even	−1	+1	−1	+1	1
	odd	−1	−1	+1	+1	2

[a] $\chi(\psi_r)$ is determined by whether K_A is even or odd and $\chi(\Psi)$ by the Exclusion Principle. In the vibrational ground state ψ_v is totally symmetric and $\chi(\psi_v) = +1$.

products which can be arranged into equal numbers of symmetric and antisymmetric linear combinations. As a result we see that the ratio

$$\frac{\text{number of symmetric spin functions}}{\text{number of antisymmetric spin functions}} = \frac{I+1}{I} \tag{3.108}$$

In the simple case of H where $I = \frac{1}{2}$ we obtain the familiar result 3 to 1 and in the case of D where $I = 1$ the ratio is 2 to 1.

If we now use these results in conjunction with (3.102a), we can determine the dependence of the statistical weights of the rotational levels of these types of molecules on K_A. We see that when K_A is even, ψ_r is unchanged under C_2^A and when K_A is odd, it changes sign. A similar procedure clearly applies to K_C using (3.102b). If we consider the total eigenfunction $\Psi = \psi_e\psi_v\psi_r\psi_{ns}$ and assume that ψ_e is symmetric under permutation, we can generate the symmetry and statistical weight Table 3.16 which applies to formaldehyde type molecules. Note that H nuclei are Fermions and D nuclei are Bosons.

Chapter 4

Interaction of Radiation with a Rotating Molecule

4.1 The Energy of a Molecule in an External Field

Now that the basic theory governing the rotational energy-level pattern of a molecule has been developed we can go on to consider the various mechanisms by which electromagnetic radiation can interact with the levels. The first step is to consider how a system in a particular eigenstate will respond to an external field. The treatment given here highlights the essential features of the *semiclassical* approach which describes the field classically and the molecular system quantum mechanically. This approach yields the theory of such field-induced processes as photon-absorption and the Raman effect in which photons are scattered with a shifted frequency, both of which are important in the study of rotational spectra. A more complete treatment based on quantum electrodynamical considerations[113] yields not only terms which correspond to these *stimulated* processes but also extra terms which correspond to *spontaneous* processes.

The essential result of the theory is that under certain conditions an oscillating field, such as that associated with photons, can induce a transition from one stationary state to another. This interaction causes a corresponding modulation of the power in the radiation field as the total energy must be conserved. The pattern of modulation (the spectrum) is characteristic of the molecule, the states involved and the interaction mechanism. It is by analysis of this pattern that one can reconstruct the array of energy levels and thus determine the molecular constants, such as the moments of inertia, that parametrize the energy-level expressions.

Though this chapter is mainly concerned with the response of the system to oscillating fields we start with general expressions which encompass also static field terms (corresponding to the Stark effect etc.). This seems reasonable as such effects can be considered as due to fields at zero frequency and give the true picture that all these effects, though they may at first sight seem quite distinct, are in fact related in a general overall scheme. It is advantageous to treat the problem generally not only to obtain a unified theory but also because one can deduce or *predict* almost automatically that hitherto unobserved higher-order processes might be detected given the correct conditions.

We can use the following expansion to describe the energy of the system in an electric field E_i (summation convention assumed over x, y, z).

$$U = U^0 + \left[\frac{\partial U}{\partial E_i}\right]^0 E_i + \frac{1}{2}\left[\frac{\partial^2 U}{\partial E_i \partial E_j}\right]^0 E_i E_j + \frac{1}{6}\left[\frac{\partial^3 U}{\partial E_i \partial E_j \partial E_k}\right]^0 E_i E_j E_k + \cdots \tag{4.1}$$

Here U^0 may be considered as the energy of the molecule in free space and is obtained by solving the field-free Hamiltonian. The second term corresponds to an additional term, linear in the electric field, the subsequent terms correspond to higher-order, so called, non-linear terms. If the change in the energy dU due to a change in the field is written as‡

$$dU = -m_i\, dE_i \tag{4.2}$$

then the expression we obtain for m_i, by comparing (4.2) with (4.1) is

$$m_i = \mu_i^0 + \alpha_{ij}^0 E_j + \tfrac{1}{2}\beta_{ijk}^0 E_j E_k + \cdots \tag{4.3}$$

m_i is called the electric dipole moment and the coefficient $\mu_i^0 = -(\partial U/\partial E_i)^0$ is the permanent dipole moment, $\alpha_{ij}^0 = -(\partial^2 U/\partial E_i \partial E_j)^0$ is the linear polarizability (or electric susceptibility) and the higher terms correspond to non-linear higher-order polarizabilities. The above equations relate to electric field interactions, on which we are concentrating, however parallel expressions to (4.1–4.3) hold for magnetic field interactions and some of the following results are transferable. One proviso which causes the correspondence not to be straightforward should however be noted; whereas the electric dipole operator is a *polar* vector the magnetic dipole operator is an *axial* vector and invariant to inversion.

4.2 Field Time Dependence

The final result of an interaction represented by (4.2) depends qualitatively on the dynamic properties of the field. We can use the expression

$$E_i = E_i^0 \cos \bar{\omega} t = \tfrac{1}{2}E_i^0 (e^{i\bar{\omega}t} + e^{-i\bar{\omega}t}) \tag{4.4}$$

to represent an oscillating electric vector such as that associated with a beam of monochromatic photons polarized in the ith plane. The energy of the photons is $E = \hbar\bar{\omega}$.§ Note that (4.4) implies that a plane polarized beam can be described in terms of left and right circularly polarized light waves.

‡ Note that the associated energy is given by:

$$U = \int_0^U dU = -\int_0^{E_i} m_i\, dE_i$$

§ We aim in this chapter to obtain absolute intensity expressions and it is convenient to write the energy, as here, using h or $\hbar$.

Manipulation of the above expressions yields numerous terms, the most important of which we will now briefly review (also included below are spontaneous effects not directly obtainable by this procedure). Those effects which are particularly important in rotational spectroscopy are considered in more detail in the next section.

The terms may be classified into essentially three groups; static field terms with $\bar{\omega} = 0$, resonant terms where $\bar{\omega} = \bar{\omega}_{nm}$ and non-resonant terms where $\bar{\omega} \neq \bar{\omega}_{nm}$. $\bar{\omega}_{nm}$ corresponds to a characteristic resonant frequency of the molecule, $\equiv (E_n - E_m)/\hbar$.

4.2a *Static Field Interactions,* $\bar{\omega} = 0$

The static field interactions represented by $\mu_i^0 E_i$ and $\frac{1}{2}\alpha_{ij}^0 E_i E_j$ in (4.1) (with $\bar{\omega} = 0$ in (4.4) so that $E_i = E_i^0$) correspond to the first- and second-order Stark energies. These static terms tend to shift the relative positions of the energy levels and are discussed in more detail in Chapter 7.

4.2b *Resonant Interactions with Oscillating Fields,* $\bar{\omega} = \bar{\omega}_{nm}$

Quantum mechanically resonant terms are only important when the energy of a photon happens to coincide exactly with the difference between two eigenstates $|m\rangle$ and $|n\rangle$ and the Bohr condition is fulfilled

$$\Delta E = E_n - E_m = \hbar\bar{\omega}_{nm} \tag{4.5}$$

The photon frequency must coincide with $\bar{\omega}_{nm} (= \Delta E/\hbar)$ which can to some extent be considered as the classical characteristic frequency of the molecular motion involved. The resonant terms represent interactions which, as we shall see, induce transitions between the eigenstates. The leading term is obtained by neglecting all but the first term in (4.3). On substitution of this term in (4.2) we obtain the field-dependent energy (i.e. $U - U^0 = V$)

$$V = -E_i^0 \mu_i^0 \cos \bar{\omega}t \tag{4.6}$$

By convention this term is written with a minus sign. This interaction term gives rise to induced absorption and induced emission.

(i) *Induced Absorption.* The interaction of the oscillating electric vector of the radiation field with a molecular dipole moment may under certain conditions induce a transition between two levels. There is a consequent gain or attenuation of the power in the radiation field. In Section 4.3 the most important of these conditions are determined. This process is the main one by which rotational information on polar molecules is obtained and gives rise to rotational absorption spectra in the far infra-red and microwave regions.

(ii) *Induced Emission.* In this process, complementary to induced absorption, a photon induces the emission of a second photon from a previously excited system causing light amplification. Masers and lasers are systems in which this process is dominant.

(iii) *Spontaneous Emission.* In this process a free excited system decays to a lower state with the emission of a photon. The process does not require perturbation by an applied field. Spontaneous emission may be understood by application of quantum electrodynamical arguments. At relatively high field energy densities the semi-classical approach given in the next section is adequate. At low or 'zero' field values, however, quantum electrodynamical techniques which quantize the radiation field indicate that there is a zero point energy density which is responsible for 'spontaneous' emission. The situation is similar to that which exists in mechanics. At high energies, classical mechanics is adequate, at low energies we need quantum mechanics. We thus see that spontaneous and stimulated processes are really related.

There are other terms which can contribute to the induced and spontaneous transition rates which originate in magnetic dipole, electric quadrupole and other interactions. They can usually be neglected in the microwave region except in the case of magnetic dipole interactions when the molecule is paramagnetic, as is the case with O_2.

4.2c Non-resonant Interactions with Oscillating Fields, $\bar{\omega} \neq \bar{\omega}_{nm}$

These terms, which are handled by dispersion theory, correspond to interactions which operate even when the energy of the photon does not coincide with an energy difference between two eigenstates of the system.

We can generate these terms by substituting (4.4) into (4.3) yielding

$$m_i = \mu_i^0 + \alpha_{ii}^0 E_i^0 \cos \bar{\omega} t + \tfrac{1}{2}\beta_{iii}^0 E_i^{0\,2} \cos^2 \bar{\omega} t + \cdots \tag{4.7}$$

If we now inspect these terms we see that the dipole moment of the system contains field induced components. The terms involving $\boldsymbol{\alpha}$, the linear polarizability tensor, oscillate at the *same* frequency as the exciting radiation. This term is associated with refraction in a continuous medium, as the linear polarization radiates into the medium modifying the way in which a wave propogates. If we consider an inhomogeneous system such as a free molecule this term can be thought of as representing a secondary oscillator which can radiate like a dipole aerial in an arc and thus tend to change the direction of polarization of the incident ray. This is essentially the classical description of the Rayleigh scattering process.

Interesting effects are associated with the higher terms and many have been observed with the advent of the high power obtainable from Q-switched pulse lasers. For instance, if we substitute $\frac{1}{2}(1 + \cos 2\bar{\omega} t)$ for $\cos^2 \bar{\omega} t$ in (4.7), we see that two terms are obtained from the quadratic polarizability. One corresponds to a d.c. polarization, the other an induced secondary oscillation at twice the frequency of the incident radiation. Both effects have been observed; the a.c. term is responsible for second harmonic generation.

Similarly third and higher harmonic generation can occur and these processes are associated with the higher terms in (4.7).

If we now take into account the fact that α_{ij} may be a periodic function of χ, θ and φ (the period classically corresponding to the rotational frequency) we find that photons may also be scattered by a rotating system with a shifted frequency, the frequency shift being related to this rotation frequency. This process is called Raman scattering.

4.3 Absorption and Emission Induced by Electric Dipole Interactions

Some of the consequences of the interaction between a quantized system and the oscillating electric field associated with electromagnetic radiation can be determined by time-dependent perturbation theory. The time-dependent Schrödinger equation for the system is

$$i\hbar\frac{\partial\Psi}{\partial t} = (H_0 + V)\Psi \tag{4.8}$$

H_0 is the time independent Hamiltonian for the free-molecule, solution of which yields the set of stationary states $|m\rangle$, $|n\rangle$, ... etc. V represents a time-dependent interaction potential. The system may be described at a particular time t by

$$\Psi = \sum_m a_m(t)\psi_m \tag{4.9}$$

here

$$i\hbar\frac{\partial}{\partial t}\psi_m = H_0\psi_m \quad \text{and} \quad \psi_m \equiv |m\rangle\, e^{-(i/\hbar)E_m t} \tag{4.10}$$

The coefficients, $a_m(t)$, can be considered as time-dependent population factors. If (4.9) is substituted into (4.8) then

$$i\hbar\sum_m \dot{a}_m(t)\psi_m = V\sum_m a_m(t)\psi_m \tag{4.11}$$

If we now multiply through by ψ_n^* and integrate over the wave-function space, we obtain

$$i\hbar\dot{a}_n(t) = \sum_m a_m(t)\langle n|V|m\rangle\, e^{+i\omega_{nm}t} \tag{4.12}$$

where

$$(E_n - E_m)/\hbar = \bar{\omega}_{nm} \tag{4.12a}$$

We now need to consider the form of the terms in V and if we take the leading term we see that we can write it according to (4.4) and (4.6) as

$$V = V^0(e^{i\bar{\omega}t} + e^{-i\bar{\omega}t}) \tag{4.13}$$

and

$$V^0 = -\tfrac{1}{2}E_i^0\mu_i^0 \tag{4.13a}$$

We shall now concentrate on a two-state system to determine the conditions under which (4.13) can induce a transition between a state $|m\rangle$ and another state $|n\rangle$. Substitution of (4.13) into (4.12) yields the following result for a two-state system

$$i\hbar\dot{a}_m(t) = a_n(t)[e^{i(-\bar{\omega}_{nm}+\bar{\omega})t} + e^{i(-\bar{\omega}_{nm}-\bar{\omega})t}]V_{nm}^0 \tag{4.14a}$$

$$i\hbar\dot{a}_n(t) = a_m(t)[e^{i(\bar{\omega}_{nm}+\bar{\omega})t} + e^{i(\bar{\omega}_{nm}-\bar{\omega})t}]V_{mn}^0 \tag{4.14b}$$

where $V_{nm}^0 = \langle n|V^0|m\rangle$. Now we consider that at time $t = 0$, $a_m(0) = 1$ and $a_n(0) = 0$ and we can obtain $a_n(t)$ at time t by integration of (4.14b) from $0 \rightarrow t$, assuming that a_n varies little, as

$$a_n(t) = -\frac{1}{\hbar}V_{mn}^0\left[\frac{e^{i(\bar{\omega}_{nm}+\bar{\omega})t} - 1}{\bar{\omega}_{nm} + \bar{\omega}} + \frac{e^{i(\bar{\omega}_{nm}-\bar{\omega})t} - 1}{\bar{\omega}_{nm} - \bar{\omega}}\right] \tag{4.15}$$

In general only the second term within the brackets in (4.15) will possess a non-negligible value. This occurs as the resonance condition $\bar{\omega} = \bar{\omega}_{nm}$ is approached and the denominator $\rightarrow 0$. Under these conditions $a_n^*(t)a_n(t)$ represents the probability that the system which was in the state $|m\rangle$ at time $t = 0$ will have made a transition to the state $|n\rangle$ by the time t. At high frequencies we can drop the non-resonant term in (4.15), as $\bar{\omega}_{nm} + \bar{\omega} \gg \bar{\omega}_{nm} - \bar{\omega}$, and thus

$$a_n^*(t)a_n(t) = \hbar^{-2}V_{nm}^{0*}V_{nm}^0 f(\bar{\omega}\,\bar{\omega}_{nm}\,t)\cdot \tag{4.16}$$

where

$$f(\bar{\omega}\,\bar{\omega}_{nm}\,t) = \frac{\sin^2[\frac{1}{2}(\bar{\omega}_{nm} - \bar{\omega})t]}{[\frac{1}{2}(\bar{\omega}_{nm} - \bar{\omega})]^2} \tag{4.16a}$$

If the frequencies contained in the incident radiation span the range for which (4.16) has a significant value then we can integrate over the frequencies. In the microwave region we integrate over the frequency in Hz, in the infrared it is usually in cm^{-1}.

We need to evaluate

$$\int_{-\infty}^{\infty} f(\bar{\omega}\,\bar{\omega}_{nm}\,t)\,d(\omega_{nm} - \omega)$$

where $d\omega = d\bar{\omega}/2\pi(\text{Hz})$. This integral has the same form as the standard integral

$$\int_{-\infty}^{\infty}\frac{\sin^2 ax}{x^2}\,dx = a\pi \tag{4.17}$$

whose form has been discussed by Schiff.[267] The function is very sharply peaked about ω_{nm} and has the properties of a delta function. We thus find that

$$f(\bar{\omega}\,\bar{\omega}_{nm}\,t) \rightarrow t\delta(\omega_{nm} - \omega)\cdot \tag{4.18}$$

where $\delta(\omega_{nm} - \omega)$ is a normalized delta function. Note that the bar has been dropped as ω is now in Hz. We can thus write (4.16) as

$$|a_n(t)|^2 = \hbar^{-2}|V^0_{nm}|^2 t\delta(\omega_{nm} - \omega) \tag{4.19}$$

where the delta function essentially represents the resonance condition. We can now relate (4.18) to the Einstein transition coefficient $B_{n\leftarrow m}$ for stimulated absorption ($B_{n\rightarrow m}$ for emission). $B_{n\leftarrow m}$ is defined by the relation

$$R_{n\leftarrow m} \equiv \frac{d}{dt}|a_n(t)|^2 = B_{n\leftarrow m}\rho(\omega_{nm}) \tag{4.20}$$

$R_{n\leftarrow m}$ is the transition rate for the process $n \leftarrow m$ amd $\rho(\omega_{nm})$ is the radiation energy density. Note that as $V^{0*}_{nm}V^0_{nm} \equiv V^{0*}_{mn}V^0_{mn}$ and thus $B_{n\leftarrow m} = B_{n\rightarrow m}$— the stimulated absorption and emission coefficients are equal for non-degenerate states $|m\rangle$ and $|n\rangle$.‡

We can relate the amplitude of the electromagnetic radiation E^0_i (4.6) to the energy density $\rho(\omega_{nm})$ by using the basic electromagnetic theory relation

$$\rho = \frac{1}{8\pi}(\boldsymbol{E}^2 + \boldsymbol{H}^2) = \frac{1}{4\pi}\boldsymbol{E}^2 \tag{4.21}$$

The second step of (4.21) follows because $\boldsymbol{E}^2 = \boldsymbol{H}^2$. We need the time average of $\boldsymbol{E}^2$ which for a homogeneous radiation field allows us to write:

$$\overline{\boldsymbol{E}^2} = \overline{E_x^2} + \overline{E_y^2} + \overline{E_z^2} \equiv 3\overline{E_x^2} \tag{4.22}$$

From (4.4) we see that $\overline{E_x^2} = \frac{1}{2}E_x^{0\,2}$ and thus

$$\rho(\omega) = \frac{3}{8\pi}[E_x^0(\omega)]^2 \tag{4.23}$$

If we substitute (4.23) in (4.20) and (4.13a) into the time derivative of (4.19) we see that a comparison of the resulting relations for $R_{n\leftarrow m}$ yields the form of $B_{n\leftarrow m}$ for unpolarized radiation

$$B_{n\leftarrow m} = \frac{2\pi}{3\hbar^2}|\langle n|\boldsymbol{\mu}|m\rangle|^2 \tag{4.24}$$

Note that $\delta(\omega_{nm} - \omega)\,\rho(\omega) = \rho(\omega_{nm})$.

‡ If the combining states are degenerate then this relation is $g_m B_{n\leftarrow m} = g_n B_{n\rightarrow m}$ where g_m and g_n are degeneracies.

The relations (4.19) together with (4.24) are important results. They imply that the following conditions must be fulfilled for a transition between two states to occur.

(1) The delta function of (4.19) indicates that the transition may occur only when resonance conditions apply, i.e. $E_n - E_m = \hbar\bar{\omega}$ (the energy of the photon must exactly match the energy separation).

(2) Relation (4.24) indicates that the transition occurs only if the matrix elements of at least one component of the space-fixed electric dipole moment is non-zero. Thus if we can find the conditions under which $\langle n|\boldsymbol{\mu}|m\rangle$ is non-zero we have determined the *selection rules*.

4.3a The Absorption Coefficient

In the previous section we have shown that a molecule can, if certain conditions are fulfilled, make a transition from one state to another. Given that these conditions are satisfied the next step is to use (4.24) to obtain a general expression which will quantitatively describe the overall spectral intensity pattern. Experimentally we need to know the fractional loss of intensity as radiation passes through a molecular medium. The most useful quantity for this purpose is the *absorption coefficient* $\gamma(\omega)$ which is defined by

$$dI(\omega) = -\gamma(\omega)I(\omega)\,dx \tag{4.25}$$

$I(\omega)$ is the intensity of radiation (in energy units per unit area per second) incident on an absorbing medium of thickness dx. As dx is usually in cm, $\gamma(\omega)$ is usually given in cm^{-1}. Integration of (4.25) over the length, l, of the absorbing column yields the familiar Beer–Lambert relation

$$I(\omega) = I_0(\omega)\,e^{-\gamma(\omega)l} \tag{4.26}$$

where $I_0(\omega)$ is the initial incident intensity and $I(\omega)$ is the transmitted intensity. This expression holds as long as saturation effects can be neglected.

If we take a cylindrical section of the absorbing medium of length dx and unit cross-sectional area, then the energy absorbed per second will equal the absorbed intensity. Now neglecting spontaneous processes we need the net difference between the absorption and emission rates. As each photon absorbed or emitted corresponds to a loss or gain of one quantum of energy, $h\omega$, from the beam we see that

$$dI(\omega_{nm}) = -\left[\frac{N_m}{g_m} - \frac{N_n}{g_n}\right] R_{n\leftarrow m} h\omega_{nm}\,dx \tag{4.27}$$

where g_m and g_n are the degeneracies, and N_m and N_n are the concentrations of molecules in the states $|m\rangle$ and $|n\rangle$ respectively. If $N_m/g_m > N_n/g_n$, stimulated absorption occurs, and if $N_m/g_m < N_n/g_n$ we obtain stimulated emission—maser action. If $N_m/g_m = N_n/g_n$ the net beam intensity will not be modified, saturation conditions apply. This is an important point for we must always ensure that relaxation is fast enough that $N_m/g_m > N_n/g_n$ if absorption is desired. In the earliest attempts to carry out nuclear magnetic resonance experiments the relaxation process was too slow for resonance to be observed.

We can substitute for $R_{n\leftarrow m}$ in (4.27) using (4.20), and on comparison of the resulting relation with (4.25) we see that

$$\gamma(\omega_{nm}) = \left(\frac{N_m}{g_m} - \frac{N_n}{g_n}\right) B_{n\leftarrow m} \frac{h\omega_{nm}}{c} \tag{4.28}$$

Note that $I = \rho c$, where c is the velocity of light. We must also take into account the fact that several line-broadening processes occur which spread the intensity about the theoretical resonant frequency, ω_{nm}. To allow for this we can multiply the right-hand side of (4.28) by a normalized line shape function $S(\omega, \omega_0)$. ω_0 is the center of the line shape function (Section 4.6) and lies at ω_{nm}. The absorption coefficient thus becomes ω dependent and we obtain with the aid of (4.24)

$$\gamma(\omega) = \frac{8\pi^3}{3hc}\,\omega \left(\frac{N_m}{g_m} - \frac{N_n}{g_n}\right) |\langle n|\boldsymbol{\mu}|m\rangle|^2\, S(\omega, \omega_0) \tag{4.29a}$$

where

$$\gamma(\omega) = \gamma(\omega_{nm}) S(\omega, \omega_0) \tag{4.29b}$$

The general expression (4.29) is correct for all transitions in which spontaneous processes can be neglected. We shall now deal with the three factors $\langle n|\boldsymbol{\mu}|m\rangle$, $(N_m/g_m - N_n/g_n)$ and $S(\omega, \omega_0)$ in a little more detail with particular reference to rotational spectroscopy. The final evaluation of (4.29) is carried out in Section 4.7 for a linear molecule.

4.4 Dipole Moment Matrix Elements and Selection Rules

As we have seen, the dipole moment matrix element $\langle n|\boldsymbol{\mu}|m\rangle$ determines whether a transition can occur. If this factor is zero, as it is in the vast majority of cases, the transition is said to be forbidden. If on the other hand this quantity is non-zero then the transition is said to be allowed and its magnitude will play a part in determining the intensity of the associated absorption line as indicated by (4.29). It is also to be noted from the general discussion in Section 4.1 that these matrix elements govern all field-molecular dipole interactions whether the field oscillates or not. We thus need them when we come to derive the static field interactions which occur in the Stark effect. The general properties we seek are contained in the matrix elements of the space-fixed components of $\boldsymbol{\mu}$ in the representation which diagonalizes H_r. As we saw in the previous chapter the rotational eigenstates can always be described in terms of a linear combination of the eigenfunctions $|j\ k\ m\rangle$ and thus the matrix elements of μ in the *jkm* scheme are required. For a rigid molecule the molecule-fixed components of μ have constant values and we will take them to be experimentally determinable parameters. In certain simple cases where good electronic wave functions are known it may be possible to calculate them.

Using the **S** matrix as defined in (2.12) we see that

$$\mu_i = S_{i\alpha}^{-1}\mu_\alpha = S_{\alpha i}\mu_\alpha \tag{4.30}$$

and as we are taking the components μ_α to be constant parameters and $S_{i\alpha}^{-1} = S_{\alpha i}$ the relevant matrix relation is

$$\langle j'\ k'\ m'|\mu_i|\ j''\ k''\ m''\rangle = \langle j'\ k'\ m'|S_{\alpha i}|\ j''\ k''\ m''\rangle\mu_\alpha \tag{4.31}$$

The first important result, implicit in (4.31) is the familiar requirement that if transitions are to occur among *jkm* basis states, then *at least one molecule-fixed component of* μ must be non-zero. Thus such molecules as CO_2, C_6H_6, CH_4 and SF_6 do not show rigid-rotor electric dipole spectra.

The main problem is clearly the determination of the matrix properties of **S**. **S** is a rather special operator in that it can be considered as a vector in two spaces, (xyz) and (abc). Vectors such as **S**, $\boldsymbol{p}$ and $\boldsymbol{r}$ belong to an important class of vector operators known as T class operators. An operator is said to belong to class T if it obeys the commutation relation

$$[J_i, T_j] = ie_{ijk}T_k \tag{4.32a}$$

(see Condon and Shortley[49]). $\boldsymbol{J}$ is itself such an operator. An analogous relation holds for a T class operator in the $\alpha\beta\gamma$ space where now the 'anomalous sign of i' appears (3.30)

$$[J_\alpha, T_\beta] = -i\,e_{\alpha\beta\gamma}T_\gamma \tag{4.32b}$$

One can easily show that $\boldsymbol{r}$ also obeys these relations and as the direction cosine elements are linear functions of $\boldsymbol{r}$, **S** also belongs to class T. The relevant relations for **S** are

$$[J_i, S_{\alpha j}] = i\,e_{ijk}S_{\alpha k} \tag{4.33a}$$

$$[J_\alpha, S_{\beta i}] = -ie_{\alpha\beta\gamma}S_{\gamma i} \tag{4.33b}$$

In Chapter 3, the commutation relations which hold for the components of $\boldsymbol{J}$ were used to generate the angular momentum matrix elements. In the next sections we will use the relations (4.33) to develop the matrix elements of **S**. **S** is a kind of double T operator and as such is somewhat more complicated than the simple T operators.

4.4a The General Selection Rules on j, k and m

To determine under what circumstances the matrix elements of **S** are non-zero, we shall follow the general method given by Dirac[67] using notation and phase conventions in line with those of Condon and Shortley.[49]

From (4.33b) we can derive the relation

$$[J_c, S_i^{\pm}] = \mp S_i^{\pm} \tag{4.34}$$

where $S_i^{\pm} \equiv S_{ai} \pm iS_{bi}$. If we expand the commutator and write the relation in matrix format for the *jkm* scheme we have

$$\begin{aligned}\sum_{j''k''m''} \{&\langle j\ k\ m|J_c|j''\ k''\ m''\rangle\langle j''\ k''\ m''|S_i^{\pm}|j'\ k'\ m'\rangle \\ &-\langle j\ k\ m|S_i^{\pm}|j''\ k''\ m''\rangle\langle j''\ k''\ m''|J_c|j'\ k'\ m'\rangle\} \\ = \mp&\langle j\ k\ m|S_i^{\pm}|j'\ k'\ m'\rangle\end{aligned} \tag{4.35}$$

As we have chosen the representation in which J_c is diagonal (4.35) reduces and rearranges simply to

$$(k - k' \pm 1)\langle j\ k\ m|S_i^{\pm}|j'\ k'\ m'\rangle = 0 \tag{4.36}$$

This expression has an interesting form for it indicates that all matrix elements of S_i^+ must vanish unless $k = k' - 1$. Similarly those of S_i^- must vanish unless $k = k' + 1$.

We can process the commutator

$$[J_c, S_{ci}] = 0 \tag{4.37}$$

in a similar fashion to obtain the relation

$$(k - k')\langle j\ k\ m|S_{ci}|j'\ k'\ m'\rangle = 0 \tag{4.38}$$

from which we deduce that the matrix elements of S_{ci} vanish unless $k = k'$. In this way we have deduced the selection rules on *k*, i.e. *transitions may only occur if* $\Delta k = 0$ *or* ± 1.

This method of deriving selection rules has a pattern which has general application. The commutation relation (4.34) is a typical example, the power of $\boldsymbol{J}$ on the left-hand side is greater than the power of $\boldsymbol{J}$ on the right-hand side.

The derivation of the selection rule on *j* is more complicated. Dirac has shown that it can be derived using the following double commutator relation[67,49]

$$[\boldsymbol{J}^2, [\boldsymbol{J}^2, \mathbf{S}]] = \boldsymbol{J}^4\mathbf{S} - 2\boldsymbol{J}^2\mathbf{S}\boldsymbol{J}^2 + \mathbf{S}\boldsymbol{J}^4 = 2(\boldsymbol{J}^2\mathbf{S} + \mathbf{S}\boldsymbol{J}^2) - 4\boldsymbol{J}(\boldsymbol{J}\cdot\mathbf{S}) \tag{4.39}$$

The intermediate steps in the reduction are given in Section A3. If we first concentrate on the off-diagonal matrix elements (i.e. $j \neq j'$), then evaluation and rearrangement of the matrix form of this expression (A3.5) yields the result

$$[(j + j' + 1)^2 - 1][(j - j')^2 - 1]\langle j\, k\, m|\mathbf{S}|j'\, k'\, m'\rangle = 0 \qquad (4.40)$$

Of the three factors in (4.40) we note that if $j \neq j'$ the first can never be zero and the second is so only when $j' = j \pm 1$. The final result is that all matrix elements of **S** vanish unless $\Delta j = 0$ or ± 1. *The method is in actual fact non-committal about* $\Delta j = 0$ *matrix elements.*

The m selection rules are $\Delta m = 0, \pm 1$ and are obtained in a parallel fashion to that used for the k selection rules, this time using the commutation relation (4.33a). Thus the complete set of selection rules for the **S** matrix elements is

$$\Delta j = 0, \pm 1, \qquad \Delta k = 0, \pm 1, \qquad \Delta m = 0, \pm 1 \qquad (4.41)$$

Note the proviso $j = 0 \nrightarrow j' = 0$.

More specifically we see that the conditions, governing the quantum numbers, necessary for the matrix elements of the operator $\langle j'k'm'|S_{i\alpha}|jkm\rangle$ to be non-zero are:

$$\begin{array}{llll} S_{ai} + iS_{bi}, & k' = k - 1, & S_{\alpha x} + iS_{\alpha y}, & m' = m + 1 \\ S_{ai} - iS_{bi}, & k' = k + 1, & S_{\alpha x} - iS_{\alpha y}, & m' = m - 1 \\ S_{ci}, & k' = k, & S_{\alpha z}, & m' = m \end{array} \qquad (4.42)$$

These results can be understood in the context of photon induced transitions if we realize that electric dipole photons possess unit angular momentum. Angular momentum must be conserved in the absorption process and thus we immediately obtain the range over which j may change by the triangle rule:

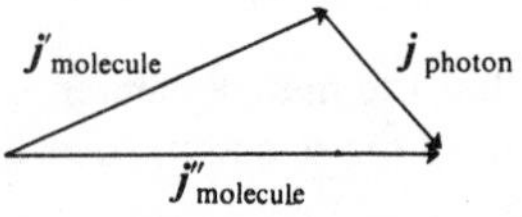

(4.43)

This procedure immediately yields all the results (4.41) by setting $j_{\text{photon}} = 1$ and in particular we see that $j' = 0 \nrightarrow j'' = 0$.

4.4b The Matrix Elements of S

We shall now determine how the matrix elements of **S** depend on j, k and m. Although some may prefer group theoretical methods we will use matrix operator methods based on the procedure given in Condon and Shortley.[49] Their phase convention is also adhered to. We will consider simple T vectors in xyz space and abc space separately and then graft the results together to derive those for **S** which is essentially a vector in both spaces. In this way we avoid some rather messy notation.

First let us consider the k dependence of a T vector in the abc space. As in previous derivations we start with commutation relations. Consider the commutator relation

$$[J^+, T^+] = 0 \tag{4.44}$$

which we can obtain from (4.32a). We can use the selection rules (4.42) and expand (4.44) in matrix notation for the jk basis as

$$\langle j\,k-1|J^+|j\,k\rangle\langle j\,k|T^+|j'\,k+1\rangle = \langle j\,k-1|T^+|j'\,k\rangle\langle j'\,k|J^+|j'\,k+1\rangle \tag{4.45}$$

As we already know the matrix elements of $\boldsymbol{J}$ (Table 3.1), we can simplify (4.45) as

$$\sqrt{(j+k)(j-k+1)}\,\langle j\,k|T^+|j\,k+1\rangle = \langle j\,k-1|T^+|j'\,k\rangle\sqrt{(j'+k+1)(j'-k)} \tag{4.46}$$

As $j' = j$ or $j \pm 1$, there are three separate cases to deal with. For $\Delta j = 0$ we can rearrange (4.46) as the ratio relation

$$\frac{\langle j\,k|T^+|j\,k+1\rangle}{\sqrt{(j+k+1)(j-k)}} = \frac{\langle j\,k-1|T^+|j\,k\rangle}{\sqrt{(j+k)(j-k+1)}} \tag{4.47}$$

This function is, according to (4.47), independent of k and is called a reduced matrix element which we will write as $\langle j\|\boldsymbol{T}\|j\rangle$. We can use this notation and write

$$\langle j\,k|T^+|j\,k+1\rangle = \langle j\|\boldsymbol{T}\|j\rangle\sqrt{(j+k+1)(j-k)} \tag{4.48}$$

We can write the multiplier in the equation in the form of a matrix element, $\langle j\,k|t^+|j\,k+1\rangle$ to obtain the relation

$$\langle j\,k|T^+|j\,k\rangle = \langle j\|\boldsymbol{T}\|j\rangle\langle j\,k|t^+|j\,k+1\rangle\cdot \tag{4.49}$$

We can carry out this same procedure for $\Delta j = \pm 1$. When $j' = j+1$ we obtain from (4.45) the relation

$$\sqrt{(j+k)(j-k+1)}\langle j\,k|T^+|j+1\,k+1\rangle = \langle j\,k-1|T^+|j+1\,k\rangle\sqrt{(j+k+2)(j-k+1)} \tag{4.50}$$

To be consistent with Condon and Shortley[49] and ensure that $f(k) = f(k+1)$ we will multiply both sides by the common factor $\sqrt{(j+k+1)/(j-k+1)}$, and thus obtain the ratio relation

$$\frac{\langle j\,k|T^+|j+1\,k+1\rangle}{\sqrt{(j+k+2)(j+k+1)}} = \frac{\langle j\,k-1|T^+|j+1\,k\rangle}{\sqrt{(j+k)(j+k+1)}} \tag{4.51}$$

This ratio is set equal to $(-1)\langle j\|\boldsymbol{T}\|j+1\rangle$ where the (-1) phase factor is introduced to be consistent with previous authors. Thus we can write

$$\langle j\,k|T^{+}|j+1\;k+1\rangle = -\langle j\|\boldsymbol{T}\|j+1\rangle\sqrt{(j+k+2)(j+k+1)} \tag{4.52}$$

By an analogous procedure we obtain the relation for $j' = j-1$ as

$$\langle j\,k|T^{+}|j+1\;k+1\rangle = \langle j\|\boldsymbol{T}\|j-1\rangle\sqrt{(j-k)(j-k-1)} \tag{4.53}$$

We can obtain the dependence of the c component T_c on k using the relation

$$[J^{-}, T^{+}] = [J^{+}, T^{+}] - 2i[J_b, T^{+}] = 2T_c \tag{4.54}$$

As before there are three cases for $\Delta j = 0, \pm 1$. For $j = j'$ we obtain the result

$$\langle j\|\boldsymbol{T}\|j\rangle[(j+k)(j-k+1) - (j-k)(j+k+1)] = 2\langle j\,k|T_c|j\,k\rangle \tag{4.55}$$

and thus we obtain

$$\langle j\,k|T_c|j\,k\rangle = \langle j\|\boldsymbol{T}\|j\rangle k \tag{4.56}$$

The rest of the relations for $\Delta j = \pm 1$ are similarly evaluated as

$$\langle j\,k|T_c|j+1\;k\rangle = \langle j\|\boldsymbol{T}\|j+1\rangle\sqrt{(j+k+1)(j-k+1)} \tag{4.57}$$

$$\langle j\,k|T_c|j-1\;k\rangle = \langle j\|\boldsymbol{T}\|j-1\rangle\sqrt{(j+k)(j-k)} \tag{4.58}$$

As T_c is real, the reduced matrix elements are Hermitian

$$\langle j'\|\boldsymbol{T}\|j\rangle = [\langle j\|\boldsymbol{T}\|j'\rangle]^{*} \tag{4.59}$$

As a consequence the matrix elements of T^{-} can be derived directly from those of T^{+}.

If we now consider a T vector in the xyz space we can develop a parallel set of relations by the following transformation

$$\begin{aligned} T^{\pm} &\rightarrow T_{\mp} \\ T_c &\rightarrow T_z \\ k &\rightarrow m \end{aligned} \tag{4.60}$$

These are derived by Condon and Shortley.[49] We can now graft these two sets of results together to determine the matrix properties of the ambivalent operator, **S**, with respect to both k and m. Both the k and the m dependence can now be separated out to yield the expression

$$\langle j\,k\,m|S_{\alpha i}|j'\,k'\,m'\rangle = \langle j\|\mathbf{S}\|j'\rangle\langle j\,k|s_\alpha|j'\,k'\rangle\langle j\,m|s_i|j'\,m'\rangle \tag{4.61}$$

where for the simple case S_{cz} we see from (4.56) that for $j = j'$

$$\begin{aligned} \langle j\,k|s_c|j\,k\rangle &\equiv \langle j\,k|t_c|j\,k\rangle = k \\ \langle j\,m|s_z|j\,m\rangle &\equiv \langle j\,m|t_z|j\,m\rangle = m \end{aligned} \tag{4.62}$$

These results are tabulated in Table 4.1. Note the phase convention differs from that of King, Hainer and Cross.[159]

The k and m dependence of **S** has been determined directly from the general T vector commutation relations (4.33). We have found that the k and m dependence is projected out in the form of multipliers such as $\langle j\, k|s_\alpha|j'\, k'\rangle$ which depend only on j and k. *The k and m dependence is thus the same for all operators which follow these commutation properties with **J**.* This statement is essentially the basis of the Wigner–Eckart Theorem.[298,259] To determine the functional behaviour of the reduced matrix element $\langle j\|\mathbf{S}\|j\rangle$ we must use the particular properties of **S**. The point that we are making is this; the orientation properties of a vector (which are associated with k and m) are independent of the type of vector, which is perhaps almost obvious.

A suitable relation from which we can develop the fully reduced matrix element $\langle j\|\mathbf{S}\|j\rangle$ (for which $j' = j$) is

$$J_c = S_{cx}J_x + S_{cy}J_y + S_{cz}J_z \tag{4.63}$$

This can be rewritten as

$$2J_c = S_{c+}J_- + S_{c-}J_+ + 2S_{cz}J_z \tag{4.64}$$

Let us consider the elements diagonal in j, k and m in matrix form; on reduction we obtain:

$$2k = \langle j\|\mathbf{S}\|j\rangle\langle j\,k|s_c|j\,k\rangle[(j+m)(j-m+1)+(j-m)(j+m+1)+2m^2] \tag{4.65}$$

We have already deduced the remaining factor in (4.65) which is given in Table 4.1. We thus obtain the result

$$\langle j\|\mathbf{S}\|j\rangle = \frac{1}{j(j+1)} \tag{4.66}$$

The remaining matrix elements for $j' = j \pm 1$ may be obtained using the relation

$$S_{cx}^2 + S_{cy}^2 + S_{cz}^2 = 1 \tag{4.67}$$

The complete set of results is given in the first row of Table 4.1.

4.4c *Matrix Elements of* $\boldsymbol{\mu}$ *and Molecular Selection Rules*

According to (4.30) we can rewrite the ith space-fixed component of $\boldsymbol{\mu}$ as

$$\mu_i = \tfrac{1}{2}(S_{ai} + iS_{bi})(\mu_a - i\mu_b) + \tfrac{1}{2}(S_{ai} - iS_{bi})(\mu_a + i\mu_b) + S_{ci}\mu_c \tag{4.68}$$

We immediately obtain a very simple result for symmetric top systems as we can take $\mu_a = \mu_b = 0$, and μ_c to lie along the axis of symmetry, and in this case we see that

$$\langle J\,K\,M|\mu_i|J'\,K'\,M'\rangle = \mu_c\langle J\,K\,M|S_{ci}|J'\,K'\,M'\rangle \tag{4.69}$$

Table 4.1 Direction Cosine Matrix Elements[a]

	$j' = j + 1$	$j' = j$	$j' = j - 1$
$\langle j \Vert \mathbf{S} \Vert j' \rangle$	$\dfrac{1}{(j+1)\sqrt{(2j+1)(2j+3)}}$	$\dfrac{1}{j(j+1)}$	$\dfrac{1}{j\sqrt{(2j+1)(2j-1)}}$
$\langle j\, k \vert s_c \vert j'\, k \rangle$	$\sqrt{(j+k+1)(j-k+1)}$	k	$\sqrt{(j+k)(j-k)}$
$\langle j\, k \vert s^{\pm} \vert j'\, k \pm 1 \rangle$	$\mp\sqrt{(j \pm k + 1)(j \pm k + 2)}$	$\sqrt{(j \mp k)(j \pm k + 1)}$	$\pm\sqrt{(j \mp k)(j \mp k - 1)}$
$\langle j\, m \vert s_z \vert j'\, m \rangle$	$\sqrt{(j+m+1)(j-m+1)}$	m	$\sqrt{(j+m)(j-m)}$
$\langle j\, m \vert s_{\mp} \vert j'\, m \pm 1 \rangle$	$\mp\sqrt{(j \pm m + 1)(j \pm m + 2)}$	$\sqrt{(j \mp m)(j \pm m + 1)}$	$\pm\sqrt{(j \mp m)(j \mp m - 1)}$

[a] $\langle j\, k\, m \vert S_{\alpha i} \vert j'\, k'\, m' \rangle$ is defined in (4.61), $s^{\pm} = s_a \pm i s_b$ and $s_{\mp} = s_x \mp i s_y$.

We thus obtain the following matrix elements for $\boldsymbol{\mu}$ which are diagonal in K and M.

$$\langle J\,K\,M|\mu_z|J\,K\,M\rangle = \mu_c\frac{KM}{J(J+1)} \tag{4.70a}$$

$$\langle J\,K\,M|\mu_z|J+1\,K\,M\rangle = \mu_c\left[\frac{(J+K+1)(J-K+1)(J+M+1)(J-M+1)}{(J+1)^2(2J+1)(2J+3)}\right]^{1/2} \tag{4.70b}$$

$$\langle J\,K\,M|\mu_z|J-1\,K\,M\rangle = \mu_c\left[\frac{(J+K)(J-K)(J+M)(J-M)}{J^2(2J-1)(2J+1)}\right]^{1/2} \tag{4.70c}$$

In the case of the $K = 0$ symmetric top stack we note that the matrix elements diagonal in J, K and M vanish. This applies to the linear molecule for which we obtain the relevant matrix elements by setting $K = 0$ in the relation (4.70b)

$$\langle J\,M|\mu_z|J+1\,M\rangle = \mu_c\left[\frac{(J+M+1)(J-M+1)}{(2J+1)(2J+3)}\right]^{1/2} \tag{4.71a}$$

$$\langle J\,M|\mu_x|J+1\,M\pm1\rangle = \mp\frac{\mu_c}{2}\left[\frac{(J\pm M+1)(J\pm M+2)}{(2J+1)(2J+3)}\right]^{1/2} \tag{4.71b}$$

$$\langle J\,M|\mu_y|J+1\,M\pm1\rangle = -\frac{i\mu_c}{2}\left[\frac{(J\pm M+1)(J\pm M+2)}{(2J+1)(2J+3)}\right]^{1/2} \tag{4.71c}$$

The last two relations are directly obtainable from Table 4.1 by evaluating $\frac{1}{2}(S_{c+} + S_{c-})\mu_c$ and $-(i/2)(S_{c+} - S_{c-})\mu_c$.

The intensity relation (4.29) involves the square of the matrix element of $\boldsymbol{\mu}$ between two states $|n\rangle$ and $|m\rangle$. If we consider, say, a linear molecule in homogeneous space, all states of the same value of J but differing in M are degenerate, and according to (4.71) there are three possible transitions for a given value of J. These are to the states $|J'\,M'\rangle$ with $M' = M$ and $M \pm 1$. We must thus sum over the accessible M' states according to

$$|\langle J|\boldsymbol{\mu}|J'\rangle|^2 = \sum_{M'=M-1}^{M+1} |\langle J\,M|\boldsymbol{\mu}|J'\,M'\rangle|^2 \tag{4.72}$$

to obtain the correct intensity relation when all transitions are of the same energy and the radiation is *unpolarized*. The result is

$$|\langle J|\boldsymbol{\mu}|J+1\rangle|_M^2 = \mu_c^2\frac{J+1}{2J+1} \tag{4.73}$$

The subscript M denotes that this matrix element applies to a particular M state. In the absence of a field various M transitions overlap and we must multiply (4.73) by $g_J = 2J + 1$. If the radiation is plane polarized we must divide by 3 and thus obtain

$$|\langle J|\mu_z|J + 1\rangle|^2 = \tfrac{1}{3}\mu_c^2(J + 1) \tag{4.74}$$

Another instructive way of evaluating this same matrix element (4.74) is to sum (4.71a) over all $(2J + 1)$, M states according to

$$|\langle J|\mu_z|J + 1\rangle|^2 = \sum_{M=-J}^{+J} |\langle J\ M|\mu_z|J + 1\ M\rangle|^2 = \tfrac{1}{3}\mu_c^2(J + 1) \tag{4.75}$$

where we have used the summation identity

$$\sum_{-J}^{J} M^2 = \tfrac{1}{3}(2J + 1)J(J + 1) \tag{4.76}$$

Note that to obtain (4.73) from (4.75) we must divide by $(2J + 1)$ to account for the M degeneracy, and multiply by 3 because of spatial homogeneity.

We can determine the appropriate relations for the symmetric top in a similar way. Assuming that the c-axis is the symmetry axis, $\mu_a = \mu_b = 0$ and thus from (4.68) we see that only the elements S_{ci} with $i = x$, y and z can contribute to the transition intensity. These elements are diagonal in K and thus the symmetric top selection rules are

$$\Delta J = 0, \pm 1, \qquad \Delta K = 0 \tag{4.77}$$

The $\Delta K = 0$ selection rule can quite easily be rationalized. It implies that the value of the angular momentum, about the symmetry axis, cannot be changed as there is no component of the molecule-fixed dipole moment perpendicular to the symmetry axis to generate a torque about this axis through interaction with the radiation field. We can obtain the appropriate matrix elements in a similar manner to that indicated in (4.75) using (4.70b) instead of (4.71a). We thus obtain

$$\begin{aligned} |\langle J|\mu_z|J + 1\rangle|_K^2 &= \sum_{M=-J}^{+J} |\langle J\ K\ M|\mu_z|J + 1\ K\ M\rangle|^2 \\ &= \frac{1}{3}\mu_c^2 \frac{(J + K + 1)(J - K + 1)}{(J + 1)} \end{aligned} \tag{4.78}$$

This is the matrix element appropriate for a single K state. Often as we shall see transitions originating in different K states with the same value of J coincide and thus we must sum (4.78) over the $(2J + 1)$ K states.

$$\sum_{K=-J}^{+J} |\langle J|\mu_z|J + 1\rangle|_K^2 = \tfrac{1}{9}\mu_c^2(2J + 1)(2J + 3) \tag{4.79}$$

In the case of the inversion transitions of ammonia $J \leftrightarrow J$ the appropriate matrix element is the square of (4.70a).

In the asymmetric top case we need to take into account the possibility that all three components μ_a, μ_b and μ_c may be non-zero. It is convenient to derive the selection rules from symmetry arguments. Let us consider the species of the molecule-fixed components of $\boldsymbol{\mu}$ with respect to the D_2 point group which was used to classify the asymmetric rotor eigenfunctions in Section 3.11b. The vector component μ_α must change sign on application of C_2^β and C_2^γ and we see from Table 3.7 that $\Gamma(\mu_A) = B_A$, $\Gamma(\mu_B) = B_B$ and $\Gamma(\mu_C) = B_C$. As the matrix element $\langle J''\,\tau''|\boldsymbol{\mu}|J'\,\tau'\rangle$ may only be non-zero if $\Gamma(\psi'') \times \Gamma(\boldsymbol{\mu}) \times \Gamma(\psi') = A$ we can set up a table of non-zero combinations (Table 4.2).

Table 4.2 Table of Species of ψ' for Allowed Transitions

$\Gamma(\psi'')$	μ_A	μ_B	μ_C
A	B_A	B_B	B_C
B_A	A	B_C	B_B
B_B	B_C	A	B_A
B_C	B_B	B_A	A

Using the results in this cross-product table we see that we can derive the following results.

$$\begin{array}{llll} \mu_A \neq 0 & \begin{matrix} A \leftrightarrow B_A \\ ee \quad eo \end{matrix} & \begin{matrix} B_B \leftrightarrow B_C \\ oo \quad oe \end{matrix} & \begin{matrix} \Delta K_A = 0, \pm 2, \cdots \\ \Delta K_C = \pm 1, \pm 3, \cdots \end{matrix} \\ \mu_B \neq 0 & \begin{matrix} A \leftrightarrow B_B \\ ee \quad oo \end{matrix} & \begin{matrix} B_A \leftrightarrow B_C \\ eo \quad oe \end{matrix} & \begin{matrix} \Delta K_A = \pm 1, \pm 3, \cdots \\ \Delta K_C = \pm 1, \pm 3, \cdots \end{matrix} \\ \mu_C \neq 0 & \begin{matrix} A \leftrightarrow B_C \\ ee \quad oe \end{matrix} & \begin{matrix} B_A \leftrightarrow B_B \\ eo \quad oo \end{matrix} & \begin{matrix} \Delta K_A = \pm 1, \pm 3, \cdots \\ \Delta K_C = 0, \pm 2, \cdots \end{matrix} \end{array} \tag{4.80}$$

The selection rules in (4.80) which apply to K_A and K_C are readily derived from (4.77) and (3.86) together with the summation restriction, that the linear combinations contain only symmetric rotor functions with K either all even or all odd. If we take, for instance, the mappings with $c \rightarrow A$, then we see that if the only non-zero component of $\boldsymbol{\mu}$ is μ_A then $\Delta K_A = 0$ by (4.77) *in the symmetric top limit.* Asymmetry, however, will cause the asymmetric rotor functions to be made up of linear combinations of basis functions of different values of K, *either all even or all odd.* Thus $\Delta K = 0, \pm 2, \pm 4, \cdots$, and if we note the *e–o* restrictions we see that ΔK_A and ΔK_C cannot both be even at the same time. ΔK_C selection rules are derived in a similar manner. The strength of a transition for which, say $\Delta K_A = \pm 2$, etc. depends on the degree of asymmetry. Near the symmetric rotor limit the linear combinations consist of but one term with a large coefficient and thus $\Delta K_A \neq 0$ transitions are very weak. As the asymmetry increases the relative magnitude of the other coefficients increases giving rise to significant intensity in those transitions for which ΔK_A or $\Delta K_C \neq 0$. The stronger asymmetric rotor transition selection rules are listed in Table 4.3. Some further transitions may also occur if the asymmetry is large, though they are usually very weak.

Table 4.3 Stronger Asymmetric Rotor Transitions

$\mu_A \neq 0$		$\mu_B \neq 0$		$\mu_C \neq 0$	
0 0 1 0 0 −1	1 0 1 −1 0 −1	0 −1 1 0 1 −1	1 1 1 −1 −1 −1	1 1 0 −1 −1 0	0 1 0 0 −1 0
0 2 −1 0 −2 1	1 2 −1 −1 −2 1	1 3 −1 −1 −3 1	1 −1 3 −1 1 −3	0 −1 2 0 1 −2	1 −1 2 −1 1 −2
Oblate only				Prolate only	

Each set of numbers represents ΔJ, ΔK_A and ΔK_C respectively. Each set can give rise to branches. For $\mu_A \neq 0$ the set 0 0 1 gives rise to ${}^AQ_{01}$ branches whereas the set −1 1 −2, $\mu_C \neq 0$ gives rise to ${}^CP_{1-2}$ branches. As the asymmetry increases, further transitions become allowed though they are usually very weak.[303]

In the calculation of intensities one method is to utilize the results of a computer calculation which diagonalizes H_r. Such programmes yield the eigenvectors which constitute the diagonalizing transformation of the Hamiltonian. Knowledge of the coefficients in the asymmetric rotor eigenfunctions together with the direction cosine matrix elements allows the appropriate intensities to be calculated.

We start with the relation

$$\mu_z = S_{\alpha z}\mu_\alpha \tag{4.81}$$

The general expression (4.61) for the factorization of the matrix elements must now be considered with regard to the states $|J\tau\rangle$. Thus we obtain

$$\langle J\ \tau\ M|S_{\alpha i}|J'\ \tau'\ M'\rangle = \langle J\|\mathbf{S}\|J'\rangle\langle J\ \tau|s_\alpha|J'\ \tau'\rangle\langle J\ M|s_i|J'\ M'\rangle \tag{4.82}$$

The evaluation over the asymmetric rotor states is a rather complicated procedure and the results have been tabulated in the literature.[59,346] We see that

$$|\langle J\ \tau\ M|S_{\alpha z}|J+1\ \tau'\ M\rangle|^2 = \frac{(J+M+1)(J-M+1)}{(J+1)^2(2J+1)(2J+3)}|\langle J\ \tau|s_\alpha|J+1\ \tau'\rangle|^2 \tag{4.83a}$$

$$|\langle J\,\tau\,M|S_{\alpha z}|J\,\tau'\,M\rangle|^2 = \frac{M^2}{J^2(J+1)^2}|\langle J\,\tau|s_\alpha|J\,\tau'\rangle|^2 \tag{4.83b}$$

$$|\langle J\,\tau\,M|S_{\alpha z}|J-1\,\tau\,M\rangle|^2 = \frac{(J+M)(J-M)}{J^2(2J+1)(2J-1)}|\langle J\,\tau|s_\alpha|J-1\,\tau'\rangle|^2 \tag{4.83c}$$

The line strength factors λ tabulated by Cross, Hainer and King[59,346] are defined by

$$\begin{aligned}\lambda_\alpha(J\,\tau,J'\,\tau') &= \sum_i\sum_M\sum_{M'}|\langle J\,\tau\,M|S_{\alpha i}|J'\,\tau'\,M'\rangle|^2 \\ &= 3|\langle J\|\mathbf{S}\|J'\rangle|^2|\langle J\,\tau|s_\alpha|J'\,\tau'\rangle|^2\sum_M|\langle J\,M|s_z|J'\,M\rangle|^2\end{aligned} \tag{4.84}$$

Using this definition we obtain with the aid of (4.75) and Table 4.1 the following relations

$$\lambda_\alpha(J\,\tau,J+1\,\tau') = |\langle J\,\tau|s_\alpha|J+1\,\tau'\rangle|^2/(J+1) \tag{4.85a}$$

$$\lambda_\alpha(J\,\tau,J\,\tau') = |\langle J\,\tau|s_\alpha|J\,\tau'\rangle|^2(2J+1)/J(J+1) \tag{4.85b}$$

$$\lambda_\alpha(J\,\tau,J-1\,\tau') = |\langle J\,\tau|s_\alpha|J-1\,\tau'\rangle|^2/J \tag{4.85c}$$

It is often the case that the relations (4.83) summed over the $(2J+1)$ values of M are required. In these cases we find that

$$\sum_{M=-J}^{J}|\langle J\,\tau\,M|S_{\alpha z}|J'\,\tau'\,M\rangle|^2 = \tfrac{1}{3}\lambda_\alpha(J\,\tau,J'\,\tau') \tag{4.86}$$

4.5 Population Factors

A second important factor which determines the intensity of a given line is, according to (4.29), the difference in population between two states, $N_m/g_m - N_n/g_n$. In the microwave region the collisional relaxation rate is usually fast enough to maintain a Boltzmann distribution though saturation may occur if high power is used.[58] If we assume equilibrium conditions, then the Boltzmann equation indicates that the population factors (concentrations) for the connected states are related according to

$$N_n/g_n = (N_m/g_m)\exp[-(E_n - E_m)/kT]$$

If the frequency associated with the transition is ω ($=\Delta E$ in the appropriate units, usually Hz or cm^{-1}) then

$$\frac{N_m}{g_m} - \frac{N_n}{g_n} = \frac{N_m}{g_m}(1 - e^{-\beta\omega}) \tag{4.87}$$

is the expression for the population difference between the two states.

$\beta = 1/kT$, h/kT or hc/kT depending on whether ω is in energy units, Hz or cm^{-1} respectively; when ω is in cm^{-1} and T in °K, $\beta = 1/0{\cdot}695\ T$. In the case of a typical microwave transition $\omega \sim 1\ cm^{-1}$ ($\equiv 29{\cdot}979$ GHz) and thus the exponential expansion can be truncated after the second term so that

$$\frac{N_m}{g_m} - \frac{N_n}{g_n} = \frac{Nf_m\beta\omega}{g_m}(1 - \tfrac{1}{2}\beta\omega + \tfrac{1}{6}\beta^2\omega^2 \cdots) \sim \frac{Nf_m\beta\omega}{g_m} \tag{4.88}$$

Here N is the number of molecules per unit volume and f_m is the fraction in the lower state $|m\rangle$. By substitution of (4.88) into (4.29) we obtain the expression applicable to low-frequency transitions such as occur in the microwave region.

$$\gamma(\omega) = \frac{8\pi^3}{3ckT}\frac{Nf_m\omega^2}{g_m}|\langle n|\mu|m|\rangle|^2 S(\omega\omega_0) \tag{4.89}$$

This expression applies whenever the transition energy is small relative to kT.

In general we need to evaluate the fractional population of the mth state

$$f_m = \frac{N_m}{N} = g_m\frac{e^{-\beta E_m}}{Q} \tag{4.90}$$

where E_m is the energy of the mth state above the ground state, g_m is the degeneracy of the state and Q is the partition function.

$$Q = \sum_i g_i\, e^{-\beta E_i} \tag{4.91}$$

If, as is usually the case, the Born–Oppenheimer approximation holds and the total wavefunction can be written as a product of the electronic, vibrational, rotational and nuclear spin functions $\Psi_m = \psi_e\psi_v\psi_r\psi_{ns}$ then

$$f_m = f_e f_v f_r f_{ns} \tag{4.92}$$

4.5a *Thermal Distribution over Vibrational States*

f_e, the fraction in a ground electronic state is usually ~ 1 except in rare cases such as NO where the excited $^2\Pi_{3/2}$ state is only 124 cm^{-1} higher than the ground $^2\Pi_{1/2}$ state.

The fraction in a given vibrational state can be written as

$$f_v = \frac{g_v e^{-\beta E_v}}{Q_v}, \qquad Q = \sum_v g_v\, e^{-\beta E_v} \tag{4.93}$$

and the sum carried over all vibrational states. For the simple case of a diatomic molecule with but one vibrational degree of freedom

$$f_v = \frac{e^{-\beta\omega v}}{Q_v}, \qquad Q_v = \sum_{v=0} e^{-\beta\omega v} \tag{4.94}$$

$$Q_v = 1 + e^{-\beta\omega} + e^{-2\beta\omega} + e^{-3\beta\omega} + \cdots \tag{4.95}$$

A useful tabulation of the numerical value of the term $\exp(-\beta E_v)$ for a range of values for E_v is given in Table A.1. This factor is $f(v = 1)/f(v = 0)$, the ratio of the number in the first excited state to the number in the ground vibrational state.

In the general case where there are $3N - 6$ normal coordinates, r

$$f_{[v]} = g_{[v]} \frac{e^{-\beta(E_{[v]} - E_0)}}{Q_v} \tag{4.96}$$

where

$$Q_v = \prod_r{}' Q_{v_r} \quad \text{and} \quad Q_{v_r} = \sum_{v_r} e^{-\beta\,\omega_r v_r} \tag{4.97}$$

The prime signifies that the product is carried out over distinct frequencies as we have to account for each of the $3N - 6$ modes separately. Note that the degeneracy of a vibrational level may be v_r dependent—in a doubly degenerate vibration such as the bending mode of a linear molecule, $g_{v_r} = v_r + 1$. As

$$\sum_n \exp(-nx) = [1 - \exp(-x)]^{-1}$$

Q_v can be written as

$$Q_v = \prod_r (1 - e^{-\beta\,\omega_r})^{-g_r} \tag{4.98}$$

4.5b *Thermal Distribution over Rotational States*

In the case of a rigid linear molecule

$$f_r = g_I(J)(2J + 1)\frac{e^{-\beta BJ(J+1)}}{Q_r} \tag{4.99}$$

is the fractional population of the Jth rotational state. In most cases $g_I(J) = 1$ for linear molecules which show microwave spectra because they do not possess a center of symmetry (Chapter 3). The factor $(2J + 1)$ arises because of the M spatial degeneracy of the orientation of $\mathbf{J}$. The summation sign in

$$Q_r = \sum_J g_I(J)(2J + 1)\, e^{-\beta BJ(J+1)} \tag{4.100}$$

can be replaced by an integral sign when, as is usually the case, the rotational states are closely stacked ($2B \ll kT$). If $g_I(J) = 1$, we see that

$$f_r = \beta B(2J + 1)\, e^{-\beta BJ(J+1)} \tag{4.101}$$

The homonuclear diatomic molecule O_2 shows a magnetic rotation spectrum in the microwave and far infra-red[186,91] and in this case $g_I(J)$ is 0 when J is even and unity when J is odd (Section 10.5a).

In the case of the symmetric top we have to take the nuclear spin degeneracy carefully into account as noted in Chapter 3. Thus

$$f_r = \frac{g_I(K)(2J+1)\,e^{-\beta E(J,K)}}{Q_r} \tag{4.102}$$

where

$$Q_r = \sum_J \sum_K g_I(K)(2J+1)\,e^{-\beta E(J,K)} \tag{4.102a}$$

When $E(J, K) \ll kT$ the simple partition function Q' *neglecting the spin statistics* becomes[123,303]

$$Q'_r = \sum_{J=0} \sum_{K=-J}^{+J} (2J+1)\,e^{-\beta E(J,K)} \sim \left[\frac{\pi}{B^2 A \beta^3}\right]^{1/2} \tag{4.103a}$$

A similar expression can be developed for the partition function for asymmetric tops[123]

$$Q'_r \sim \left[\frac{\pi}{ABC\beta^3}\right]^{1/2} \tag{4.103b}$$

This expression has been developed using approximate asymmetric rotor energy-level expressions. Even so the expression is in general quite accurate.[123]

All the simple partition functions should be corrected for the effects of nuclear spin degeneracy. This can most simply be done by defining the correct partition function Q as

$$Q = G_I Q' \tag{4.104}$$

Where the factor G_I allows for the extra degeneracy introduced by the spin factor $g_I(K)$ in (4.102) and neglected in developing the expressions (4.103a and b).

In the very common case of C_{3v} molecules with spin $\frac{1}{2}$ nuclei the appropriate statistical weights are given in Table 3.11. On average there are $\frac{8}{3}$ more states to sum over than are accounted for by (4.103a). In the general C_{3v} case[303] the result is

$$G_I = \tfrac{1}{3}(2I+1)(4I^2+4I+1) \tag{4.105}$$

which can be derived by averaging (3.105a) and (3.105b) over all states. Collecting the results (4.90), (4.103a) and (4.104) together we see that for a symmetric top in the high-temperature approximation where $E(J, K) \ll kT$

$$\frac{f_m}{g_m} = \left[\frac{h^3}{k^3\pi}\right]^{1/2} f_v \frac{g_I(K)}{G_I} \frac{BA^{1/2}}{T^{3/2}} \tag{4.106}$$

The most important result is that in those cases where the $|K|$ substructure of a particular $J + 1 \leftarrow J$ C_{3v} symmetric top transition is resolved, those lines for which $|K| = 3, 6, 9, \ldots$ are stronger than the rest by the ratio $(4I^2 + 4I + 3):(4I^2 + 4I)$. When $I = 0, \frac{1}{2}, 1$ and $\frac{3}{2}$ this ratio is 3:0, 2:1, 11:8 and 6:5 respectively. In symmetric top spectra the $K = 0$ line is single whereas all other $|K| \neq 0$ lines are doubled and not in general resolved as is discussed in Chapter 6. There are rare cases where the symmetry is C_{4v} as in SF_5Cl[152] or C_{5v} as in C_5H_5NiNO.[57] In such C_{nv} cases there will be a strong line when $|K| = n, 2n, 3n \ldots$ etc. A nice example of this type of intensity alternation has been observed in the spectrum of C_5H_5NiNO[57] where the lines with $|K| = 5$ and 10 are clearly stronger than the adjacent lines.

4.6 Line Shapes and Line Widths

The last factor that we must take into account before we can evaluate experimental intensities is the line width. From first principles we derived the relation (4.19) which contained a hypothetical delta function, $\delta(\omega_{nm} - \omega)$. Functions of this type are essentially mathematical devices which have the following curious properties: an infinite value at $\omega_{nm} - \omega = 0$, zero width and unit area. A more careful study of the radiation–molecule interaction carried out using quantum electrodynamics indicates that this delta function must be replaced by a line shape function whose line width is related to the spontaneous emission lifetime. This width is known as the *natural line width.* At microwave frequencies it is of the order of 10^{-6} Hz and can thus be neglected.[100]

There are however several other processes which set a limit on the experimental line width and therefore also on the resolution. The main ones are pressure broadening, modulation broadening, Doppler broadening and saturation broadening. One should also perhaps include the problem of source frequency stability often a limiting factor in the microwave region.

In a standard microwave spectrometer the main factor is pressure broadening which is due to binary collisions. Assuming collision limited lifetimes, Van Vleck and Weisskopf[311,287,303] developed the line shape function.

$$S(\omega, \omega_0) = \frac{\omega}{\pi\omega_0}\left[\frac{\Delta\omega}{(\omega_0 - \omega)^2 + (\Delta\omega)^2} + \frac{\Delta\omega}{(\omega_0 + \omega)^2 + (\Delta\omega)^2}\right] \tag{4.107}$$

which fits experiment fairly satisfactorily over quite a wide pressure range. In this expression $\Delta\omega = 1/2\pi\tau$, where τ is the average time between collisions, and ω_0 is the resonant frequency. Anderson[7] has discussed the mechanism of the collisions and how they relate to intermolecular forces. In the standard far infra-red or microwave experiment $\Delta\omega \ll \omega$ and $\omega \sim \omega_0$ which allows us to drop the second term in the brackets to yield the usual Lorentzian line shape function

$$S(\omega, \omega_0) = \frac{1}{\pi}\left[\frac{\Delta\omega}{(\omega_0 - \omega)^2 + (\Delta\omega)^2}\right] \tag{4.108}$$

As the pressure is increased $\rightarrow 1$ atmosphere, $\Delta\omega \rightarrow \omega_0$ and the second term must be retained, the line shape starts to depart from the symmetric distribution about ω_0 given by the Lorentzian function.[303] In this case $\Delta\omega$ is also the *half*-width at half peak height as can be seen from (4.108).

Substitution of the expression (4.108) for $S(\omega, \omega_0)$ into the intensity function (4.89) yields

$$\gamma(\omega) = \frac{8\pi^2}{3ckT}|\langle n|\mu|m\rangle|^2 N\frac{f_m}{g_m}\omega^2\left[\frac{\Delta\omega}{(\omega_0 - \omega)^2 + (\Delta\omega)^2}\right] \tag{4.109}$$

which is the standard general expression used in spectral intensity measurements for the microwave and far infra-red. The quantity which is most often measured and quoted is in fact the value of this function when $\omega = \omega_0$, $\gamma(\omega_0)$, and often called $\gamma_{\max}$. In this case we see that $S(\omega_0, \omega_0) = 1/\pi\Delta\omega_0$ which on substitution into (4.89) yields

$$\gamma(\omega_0) = \frac{8\pi^2}{3ckT}\frac{N}{\Delta\omega_0}\frac{f_m}{g_m}\omega_0^2|\langle n|\mu|m\rangle|^2 \tag{4.110}$$

This function is essentially the peak intensity. The number of molecules per cm^3, N is given by:

$$N = L^0\frac{P}{760}\cdot\frac{300}{T} = 9{\cdot}68 \times 10^{18}\left[\frac{P}{T}\right] \tag{4.111}$$

where L^0 is Loschmidt's number, P is the pressure in mm Hg and T the temperature in °K. If we also note that τ, the mean lifetime between collisions, is inversely proportional to the factor (P/T) then we see that we can write

$$\Delta\omega_0 = \Delta\omega_0^s 300\left[\frac{P}{T}\right] \tag{4.112}$$

When $\Delta\omega_0^s$ is the standard half width measured at 1 mm Hg pressure and 300°K. Typically $\Delta\omega_0^s$ is close to 10 MHz implying half widths of 1 MHz at 100 μ and 0·1 MHz at 10 μ. We can thus replace $N/\Delta\omega_0$ by

$$\frac{N}{\Delta\omega_0} = \frac{3{\cdot}219 \times 10^{16}}{\Delta\omega_0^s} \tag{4.113}$$

$\gamma(\omega_0)$ is thus independent of pressure over the range in which standard measurements are made; 1–10^5 μ (10^{-3}–100 mmHg).

Pressure broadening can be eliminated by lowering the pressure. Wall effects can become important also. Doppler-effect broadening which is due to the relative velocities of the radiation and the molecule having a range of values can also become a problem. This is eliminated to some extent in molecular-beam methods by crossing the molecular beam and the radiation beam at right angles. The Lamb dip technique has also been used by Costain[54] to obtain high resolution information (Chapter 11).

Saturation effects occur when the collisional relaxation rate cannot compete with the excited state population rate due to too high pumping power.[149] Oka has exploited high-power pumping techniques and the properties of molecular gases near the saturation point to obtain information about intermolecular collision mechanisms and intermolecular forces.[223]

4.7 Spectral Intensity

The evaluation of the absolute intensity expression which we derived as $\gamma(\omega)$ in (4.29) is clearly quite tortuous involving as it does several factors of very different natures. We are however, now in a position to evaluate $\gamma(\omega)$ or $\gamma_{\max} \equiv \gamma(\omega_0)$. We shall first consider the case of a linear molecule such as OCS as this molecule or one of its naturally occurring isotopic modifications is usually used as a test compound for microwave spectroscopy. We need to collect results from several previous sections and substitute them into (4.110). Let us consider the $J + 1 \leftarrow J$ transition. For a linear molecule the population factor is

$$\frac{f_m}{g_m} = f_v \frac{g_I(J)}{G_I} \frac{B}{T} \left[\frac{h}{k}\right] \tag{4.114}$$

which is obtained from (4.101) using the high-temperature approximation $E_r \ll kT$ in which case this exponential factor is close to unity. In this case $g_m = 2J + 1$. From relations (4.73), (4.74) and (4.75) we see that

$$|\langle J + 1|\boldsymbol{\mu}|J\rangle|^2 = \mu^2(J + 1) \tag{4.115}$$

for the dipole moment matrix element, summed over all M components. As a result we see that

$$\gamma(\omega_0) = \left[\frac{8\pi^2}{3ck}\right]\left[\frac{h}{k}\right] a f_v \frac{N}{\Delta\omega_0} \frac{\omega_0^2 B}{T^2} \frac{g_I(J)}{G_I} \mu^2(J + 1) \tag{4.116}$$

where a is an isotopic abundance factor. The frequency factors can conveniently be converted to MHz, the dipole moment to Debye and fundamental constant expressions evaluated. Thus we can use

$$\begin{aligned} 8\pi^2/3ck &= 6{\cdot}3588 \times 10^6 (\mathrm{cm}^{-1}\ \mathrm{sec}) \\ h/k &= 0{\cdot}4799 \times 10^{-10} (\mathrm{sec\ K}) \\ 10^{-18}\ \mathrm{esu} &= 1\ \mathrm{Debye} \\ 2B(J+1) &= \omega_0 \end{aligned} \tag{4.117}$$

and obtain $\gamma(\omega_0)$ (or $\gamma_{\max}$)

$$\gamma(\omega_0) = 1{\cdot}526 \times 10^{-28} a f_v \frac{N}{\Delta\omega_0} \frac{\omega_0^3}{T^2} \frac{g_I(J)}{G_I} \mu^2 \tag{4.118}$$

where ω_0 and $\Delta\omega_0$ are in MHz and μ is in Debyes. If the further substitution (4.113) for $N/\Delta\omega_0$ is made, we obtain

$$\gamma(\omega_0) = 4{\cdot}912 \times 10^{-12} a f_v \frac{\omega_0^3}{\Delta\omega_0^s T^2} \frac{g_I(J)}{G_I} \mu^2 \tag{4.119}$$

Note that this quantity is inversely proportional to T^2. For infra-red and submillimetre wave measurements it may be necessary to retain higher terms in the exponential factor (4.88) such as

$$(N_m/g_m - N_n/g_n) = N f_m \beta\omega(1 - \tfrac{1}{2}\beta\omega)/g_m \tag{4.120}$$

and also retain the exponential factor in (4.101) and not set it to unity.

The expression (4.119) can be evaluated for the particular case of $^{16}O^{12}C^{32}S$ using the following data

$$\begin{aligned} B_0 &= 6081{\cdot}5\ \mathrm{MHz} \quad (\text{Table 5.1}) \\ \Delta\omega_0^s &= 6{\cdot}27\ \mathrm{MHz}\ [81] \\ \mu &= 0{\cdot}71521\ \mathrm{Debye}\ [209] \end{aligned} \tag{4.121}$$

This yields the following expression for $\gamma(\omega_0)$

$$\gamma(\omega_0) = 0{\cdot}706(J+1)^3 T^{-2} f_v a \tag{4.122}$$

The vibrational frequencies ω_1, ω_2 and ω_3 are 2062, 520 and 859 cm^{-1} respectively, and thus at 300°K $f_v(000) = 0{\cdot}83$. The abundance factor, a, for $^{16}O^{12}C^{32}S$ is 0·94. Thus we find that at 300°K

$$\gamma(\omega_0) = 0{\cdot}624 \times 10^{-5}(J+1)^3 \quad (\mathrm{cm}^{-1}) \tag{4.123}$$

which yields a value of $4{\cdot}88 \times 10^{-5}\ \mathrm{cm}^{-1}$ for the $J = 2 \leftarrow 1$ line which is often used for calibration. This line thus will give rise to a 1 per cent absorption in a 2 metre cell.

We can carry out a similar calculation for *symmetric top molecules* and the somewhat involved procedure has been summarized in Table 4.4 to enable

Table 4.4 The Derivation of the Absorption Coefficient $\gamma(\omega_0)$ for a Symmetric Top

1. $\gamma(\omega) = \frac{8\pi^3}{3hc}\left[\frac{N_m}{g_m} - \frac{N_n}{g_n}\right]\omega\lvert\langle n\lvert\mu\rvert m\rangle\rvert^2 S(\omega\omega_0)$	General intensity expression (4.29a) which can be applied to all types of spectroscopic transitions
2. $\gamma(\omega_0) = \frac{8\pi^2}{3ck} a \frac{N}{T} \frac{\omega_0^2}{\Delta\omega_0} \frac{f_m}{g_m}\lvert\langle n\lvert\mu\rvert m\rangle\rvert^2$	For a rotational transition it is convenient to set (1) $S(\omega_0\omega_0) = 1/\pi\Delta\omega_0$ and (2) $N_m/g_m - N_n/g_n = (h/k)aNf_m\omega_0/g_m T$ and thus obtain the general expression for $\gamma(\omega_0)$ ($\equiv \gamma_{max}$) according to relations (4.89) and (4.110). a is the abundance factor.
3. $\gamma(\omega_0) = \left[\frac{8\pi^2}{3ck}\right]\left[\frac{h^3}{k^3\pi}\right]^{1/2} af_v \frac{N}{\Delta\omega_0}\omega_0^2 \frac{BA^{1/2}}{T^{5/2}} \frac{g_I(K)}{G_I}\mu^2\left[\frac{(J+1)^2 - K^2}{J+1}\right]$	For the particular case of a symmetric top we can make the substitutions (3) $f_m/g_m = (h^3/k^3\pi)^{1/2} f_v g_I(K) BA^{1/2}/G_I T^{3/2}$ (4) $\lvert\langle n\lvert\mu\rvert m\rangle\rvert^2 = \mu^2\{[(J+1)^2 - K^2]/(J+1)\}$ which follow from (4.103a) and (4.78) respectively.
4. $\gamma(\omega_0) = (5{\cdot}965 \times 10^{-31}) af_v \frac{N}{\Delta\omega_0}\omega_0^3 \frac{A^{1/2}}{T^{5/2}} \frac{g_I(K)}{G_I}\mu^2\left[1 - \frac{K^2}{(J+1)^2}\right]$ (ω_0, A in MHz, μ in Debye)	For this step the following substitutions have been made: (5) $2B(J+1) = \omega_0$ (6) $[h^3/k^3\pi]^{1/2} = 1{\cdot}876 \times 10^{-16}$ (7) $[8\pi^2/3ck] = 6{\cdot}359 \times 10^6$ (8) $\omega(\text{MHz}) = 10^6\omega(\text{Hz})$ (9) μ (Debye) $\equiv 10^{-18}$ μ (esu).
5. $\gamma(\omega_0) = (1{\cdot}920 \times 10^{-14}) af_v \frac{\omega_0^3}{\Delta\omega_0^s} \frac{A^{1/2}}{T^{5/2}} \frac{g_I(K)}{G_I}\mu^2\left[1 - \frac{K^2}{(J+1)^2}\right]$	Using the standardized half width $\Delta\omega_0^s$ at 1 mm Hg (4·113) we can set (10) $N/\Delta\omega_0 \equiv 3{\cdot}219 \times 10^{16}/\Delta\omega_0^s$.

the various steps and approximations to be appreciated most easily. Note that the second expression in Table 4.4 is applicable to all types of molecules.

In the case of the *asymmetric rotor* the procedure is closely related to that shown in Table 4.4. Instead of substitution step (3) we see from (4.90) and (4.103b) that we must use

$$f_m/g_m = (h^3/k^3\pi)^{1/2} f_v g_I(JK_A K_C)(ABC)^{1/2}/G_I T^{3/2} \tag{4.124}$$

The dipole moment matrix element cannot be replaced by a closed algebraic expression, however from (4.81) and (4.86) we see that we can use

$$\sum_{M=J}^{J} |\langle J'\, \tau'\, M'|\mu|J''\, \tau''\, M''\rangle|^2 = \mu_\alpha^2 \lambda_\alpha(J'\tau', J''\tau'') \tag{4.125}$$

where the λ_α factors have been tabulated.[346] The resulting expression for the asymmetric rotor is thus

$$\lambda(\omega_0) = (3{\cdot}840 \times 10^{-14}) a f_v \frac{\omega_0^2}{\Delta\omega_0^s} \frac{(ABC)^{1/2}}{T^{5/2}} \frac{g_I(JK_A K_B)}{G_I} \mu_\alpha^2 \lambda_\alpha(J'\tau', J''\tau'') \tag{4.126}$$

One could calculate the dipole moment of a molecule such as OCS by measuring the spectral intensity. Such a measurement is not, however, a trivial operation in a standard microwave spectrometer.[109] A good example of this procedure is the determination of the dipole moment of CH_3D as $5{\cdot}68 \times 10^{-3}$ Debye by Ozier, Ho and Birnbaum[233] from the far infra-red, pure rotational spectrum shown in Figure 5.3. The accuracy is quoted as $\pm 0{\cdot}3 \times 10^{-3}$ Debye or about 6%.

4.8 Raman Scattering

In 1923 Smekal[279] predicted on the basis of quantum arguments that a photon could interact with an atomic system inducing a non-resonant transition, the energy being conserved by transfer to or from the photon. In the quantum description, depicted in Figure 4.1 a photon interacts with a molecule or atom in an initial state $|m\rangle$. During the interaction the photon–molecule system can be considered as a *virtual* state which does not correspond to a stationary state of the molecule. Such a state can be described by a linear combination over the complete set of stationary states of the free molecule

$$\sum_r a_r |r\rangle$$

Finally a transition from the virtual level to a new stationary state $|n\rangle$ takes place with the emission of a photon. The difference in energy between the initial and final photons being equal in magnitude and opposite in sign to the difference in energy between the initial and final molecular states.

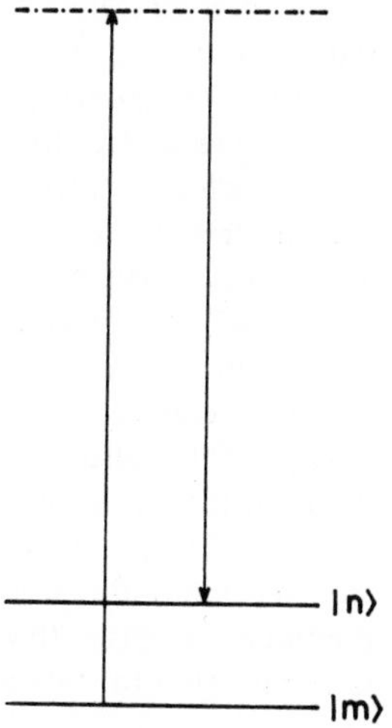

Figure 4.1 Quantum mechanical two-photon description of the Raman effect. This is a Stokes–Raman process. —·—·—· is the virtual state.

This process is called Raman Scattering when $|m\rangle$ and $|n\rangle$ are different and Rayleigh scattering if they are the same.

The classical picture of Raman and Rayleigh scattering provides a simple introductory explanation of the effect. Let us consider a rotating molecule and the behaviour of the electron distribution under the influence of the perturbing radiation. The *induced dipole moment* C_i is given by

$$C_i = \alpha_{ij} E_j^0 \cos \bar{\omega}_0 t \tag{4.127}$$

where α_{ij} is the ijth element of the space-fixed polarizability tensor and the electric vector of the radiation field is given by (4.4). In general we can relate the space-fixed components of the $\boldsymbol{\alpha}$ tensor to the molecule-fixed elements with the aid of the **S** matrix. To start off however we will consider a simple model situation of a homonuclear diatomic molecule such as H_2 rotating in a plane perpendicular to the y-axis with a rotational frequency ω. If the radiation is polarized along the z-axis and the direction of propagation is along the x-axis we can expand α_{xx} as a Fourier Series:

$$\alpha_{xx} = \alpha^{(0)} + \alpha_{xx}^{(2)} \cos 2\bar{\omega} t + \cdots \tag{4.128}$$

The first term in (4.128) is a symmetric polarizability component which is rotation invariant. The second term corresponds to a component which fluctuates at *twice* the rotational frequency. Using this simple classical description we see that if we truncate (4.128) after the second term and substitute into (4.127) we obtain the expression

$$C_x = \alpha^{(0)} E_x^0 \cos \bar{\omega}_0 t + \tfrac{1}{2}\alpha_{xx}^{(2)} E_x^0 [\cos(\bar{\omega}_0 + 2\bar{\omega})t + \cos(\bar{\omega}_0 - 2\bar{\omega})t] \tag{4.129}$$

for the induced dipole moment. This expression indicates that the perturbing radiation induces electric dipole oscillations in the molecule which can behave as secondary sources of radiation. The molecule thus behaves like a scattering centre. In this simple example we see that the Rayleigh component, which corresponds to the first term in (4.129), oscillating at the exciting frequency $\bar{\omega}_0$ is due to the isotropic polarizability coefficient $\alpha^{(0)}$. The second term in (4.129) corresponds to the anti-Stokes–Raman component $(\bar{\omega}_0 + 2\bar{\omega})$ and the third term to the Stokes–Raman component $(\bar{\omega}_0 - 2\bar{\omega})$.

A very extensive review of the basic general theory has been given by Placzek[240] (English translation in Reference 240a). The theory is also to be found in the book by Koningstein[164] and in rather less detail elsewhere.[83] A second-order time-dependent perturbation treatment which is essentially an extension of the procedure in Section 4.3 yields the following relation for the intensity in the ith component of the scattered radiation corresponding to a transition to state $|n\rangle$ from state $|m\rangle$.

$$I_i^{nm} = \frac{64\pi^4}{3c^3}(\omega_0 + \Delta\omega)^4 |\langle n|C_i|m\rangle|^2 \tag{4.130}$$

where $\langle m|C_i|n\rangle$ is the matrix element of the induced electric dipole transition moment and is a *vector* quantity given by

$$\langle n|C_i|m\rangle = \frac{1}{2\hbar}\sum_j\sum_r \left\{\langle n|\mu_i|r\rangle\left[\frac{\langle r|E_j^0\mu_j|m\rangle}{\bar{\omega}_{rm} - \bar{\omega}_0}\right] + \left[\frac{\langle n|E_j^0\mu_j|r\rangle}{\bar{\omega}_{rn} + \bar{\omega}_0}\right]\langle r|\mu_i|m\rangle\right\} \tag{4.131}$$

When arranged in this way we can get a good idea of the process which gives rise to the most important contribution to the scattering intensity. The first term within the curved brackets can be written as a linear combination

$$\frac{1}{2\hbar}\sum_j\sum_r \langle n|\mu_i|r\rangle a_{rm}^j(\bar{\omega}_0) \tag{4.132}$$

This has the form of a set of emission transition moments from the set of states $|r\rangle$ to the state $|n\rangle$ where the coefficient $a_{rm}^j(\bar{\omega}_0)$ is given by the first square-bracketed term. One can consider this as a type of population factor for the states $|r\rangle$ arising via *induced* absorption from state $|m\rangle$ by radiation of frequency $\bar{\omega}_0$. The resonant type denominator $\bar{\omega}_{rm} - \bar{\omega}_0$ strongly governs the contribution that a particular stationary state $|r\rangle$ can make to the virtual state ensuring the greatest weighting to states close to the virtual state where $\bar{\omega}_{rm} - \bar{\omega}_0$ is small. The term blows up as resonance is approached and this treatment is no longer valid. The second term corresponds to the Inverse Raman Effect observed by Jones and Stoicheff[145] in which the emission of a photon is followed by absorption.

We can factorize out the electric field amplitude E_j^0 from this expression and write

$$\langle n|C_i|m\rangle = \sum_j \langle n|c_{ij}|m\rangle E_j^0 \tag{4.133}$$

where

$$\langle n|c_{ij}|m\rangle = \frac{1}{2\hbar}\sum_r \left[\frac{\langle n|\mu_i|r\rangle\langle r|\mu_j|m\rangle}{\bar{\omega}_{rm} - \bar{\omega}_0} + \frac{\langle n|\mu_j|r\rangle\langle r|\mu_i|m\rangle}{\bar{\omega}_{rn} + \bar{\omega}_0}\right] \tag{4.134}$$

where $\langle n|c_{ij}|m\rangle$ is the *nm*th matrix element of the space-fixed scattering tensor c_{ij}. In general c_{ij} is an antisymmetric second-rank cartesian tensor. In Chapter 8 such tensors are discussed in more detail and it is shown that they can be decomposed into three components

$$c_{ij} = \tfrac{1}{3}\delta_{ij}c^t + c_k^a + c_{ij}^s \tag{4.135}$$

We thus obtain c^t which is a rotationally invariant trace tensor, a traceless antisymmetric second-rank tensor, $c_k^a = \frac{1}{2}(c_{ij} - c_{ji})$, and a symmetric second-rank tensor with zero trace, $c_{ij}^s = \frac{1}{2}(c_{ij} + c_{ji}) - \frac{1}{3}\delta_{ij}c^t$. If the combining states are non-degenerate and the perturbing radiation is far from resonance, Placzek[240] has shown that the antisymmetric part vanishes and the resulting symmetric tensor is often called the polarizability tensor $\boldsymbol{\alpha}$. It has only six independent components whereas the scattering tensor has nine. The polarizability theory thus is of more limited applicability and, in those cases where c_k^a does not vanish, the more rigorous theory must be considered as extra transitions will in general be allowed. If we restrict ourselves to the simple polarizability treatment, we see that

$$\alpha_{ij} = S_{\alpha i}S_{\beta j}\alpha_{\alpha\beta} \tag{4.136}$$

As we are also restricting ourselves to pure rotational transitions of molecules in totally symmetric vibronic ground states, the molecule-fixed components of the polarizability tensor, $\alpha_{\alpha\beta}$, can be considered as constant parameters in the same way that the molecule-fixed components of the dipole moments are. As a result, we see that the selection rules which apply in this case are obtained directly from the matrix elements of $\mathbf{S}^2$. These govern the scattering intensity relations and are most easily evaluated using spherical tensor techniques (Chapter 8). For symmetric top molecules the scattered intensity can be written as

$$I = A(\omega_0 + \Delta\omega)^4 f_r b_{J''K''}^{J'K'} \tag{4.137}$$

where A is a constant and $b_{J''K''}^{J'K'}$ are the matrix elements of $\mathbf{S}^2$ in the notation of Placzek and Teller.[241] A complete list of these elements is given by Stoicheff[285] and an extended listing by Lepard.[174] The main ones for our purposes are

$$b_{JK}^{J+2K} = \frac{3[(J+1)^2 - K^2][(J+2)^2 - K^2]}{2(J+1)(J+2)(2J+1)(2J+3)} \tag{4.138a}$$

$$b_{JK}^{J+1K} = \frac{3K^2[(J+1)^2 - K^2]}{J(J+1)(J+2)(2J+1)} \tag{4.138b}$$

$$b_{JK}^{JK} = \frac{[J(J+1) - 3K^2]^2}{J(J+1)(2J-1)(2J+3)} \tag{4.138c}$$

These factors are adequate for linear molecules and symmetric top molecules in their totally symmetric ground vibronic states. The resulting selection rules are:

Linear molecules and symmetric tops with $|K| = 0$ $\quad J = 0, \pm 2. \ \Delta K = 0$

Symmetric tops with $|K| \neq 0$ $\quad J = 0, \pm 1, \pm 2. \ \Delta K = 0$

More complicated selection rules apply in non-totally symmetric vibronic states when $\Delta K = \pm 2$, when non-degenerate and $\Delta K = \pm 1, \pm 2$ when degenerate. Stoicheff[285] has listed the rules which apply in other important cases.

For the diatomic molecule the intensity relation can be written as

$$I(J + 2 \leftrightarrow J) = A(\omega_0 \pm \Delta\omega)^4 f_r b_{J0}^{J+2\,0} \tag{4.139}$$

Transitions for which $\Delta J = +2$ are called S-branch transitions and when $\Delta J = -2$ they are called O-branch transitions.

Chapter 5

Basic Rotational Spectra and Analysis

In this chapter we bring all the results developed in previous chapters together in order to explain observed basic spectral patterns. We will follow the following format as much as possible:

(a) Draw up the basic rigid-rotor *energy-level pattern* using the procedures of Chapter 3.

(b) Using the selection rules discussed in Chapter 4 determine *which transitions are allowed.*

(c) By combining (b) with (a) we can develop the *resulting spectrum.*

(d) *Analysis* requires a recognition of the type of spectral pattern together with an assignment of the observed lines to transitions between particular levels. The next step is to determine the set of parameters which allows the calculated transition frequencies to fit those observed most closely. The criterion as to how good this fitting is, is often the least squares requirement. In this way, A, B and C can be determined. The relationship between these constants and the molecular structural parameters requires careful consideration (Chapter 6) as it is not straight-forward.

(e) Under the highest resolution it is often found that the spectral pattern differs in many ways from that predicted by the rigid-rotor theory we have developed. The origins of these departures from the simple theory are the main considerations of the following chapters where it is shown that a quantitative analysis of the modified patterns can yield a wealth of detailed molecular information.

5.1 The Absorption Spectra of Linear Molecules

From the expression (3.49) we can set up the basic energy level diagram for a diatomic or linear polyatomic molecule, Figure 3.1. If the molecule has a dipole moment, absorption can occur according to the selection rule $\Delta J = +1$ with intensity governed by (4.118). The frequencies of the allowed transitions are governed by the expression

$$\Delta E(J) = E(J + 1) - E(J) = B[(J + 1)(J + 2) - J(J + 1)] = 2B(J + 1) \tag{5.1}$$

According to this result, a set of R branch transitions, ($\Delta J = +1$) should occur in absorption at frequencies $2B, 4B, 6B, \ldots$, etc. They can be assigned to the transitions $J = 1 \leftarrow 0$, $J = 2 \leftarrow 1$, $J = 3 \leftarrow 2, \ldots$ etc. respectively. In Figure 5.1 the far infra-red spectrum of CO observed by Dowling[73] is shown.

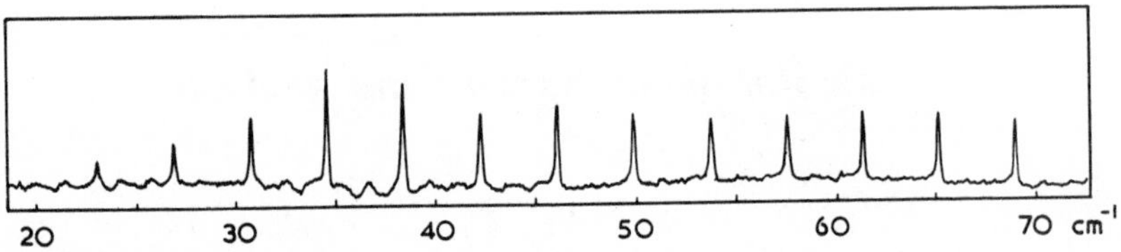

Figure 5.1 Part of the far infra-red absorption spectrum of CO observed by Dowling.[73] The background varies considerably over this range distorting the intensity contour of the branch

The characteristic pattern of lines, separated by $2B$ is observed. In Table 5.1 the observed frequencies of microwave and millimetre wave absorptions of

Table 5.1 Analysis of the Rotational Data for OCS[346]

J''	ΔE_{obs}	$\Delta E_{obs} - 2\tilde{B}(J+1)$	$\dfrac{\Delta E_{obs}}{(J+1)}$[a]
0	12 162·97	12·97	12 162·970
1	24 325·921	25·92	12 162·960
2	36 488·82	38·82	12 162·940
3	48 651·40	51·40	12 162·910
4	60 814·08	64·08	12 162·816
5	72 976·80	76·80	12 162·800
7	97 301·19	101·19	12 162·648
9	121 624·63	124·63	12 162·463
11	145 946·79	146·79	12 162·232
13	170 267·49	167·49	12 161·963
15	194 586·66	186·66	12 161·666
17	218 903·41	203·41	12 161·300
19	243 218·09	218·09	12 160·904
21	267 529·56	229·56	12 160·434
23	291 839·22	239·22	12 159·967
25	316 144·7	244·70	12 159·416
27	340 449·2	249·2	12 158·900
29	364 747·5	247·5	12 158·250
31	389 041	241·0	12 157·531
39	486 184·2	184·2	12 154·605
41	510 457·3	157·3	12 153·745

[a] This function is useful for an accurate analysis of the OCS data as shown in Section 6.2. The data yield $B_0 = 6081{\cdot}49$ and $D_0 = 0{\cdot}00130$ MHz.

OCS are given. The procedure for analysis of this data depends on the precision with which the frequencies are known and the information we wish to derive. To start off, let us assume that the simple rigid-rotor expression holds, and plot ΔE_{obs} against $(J + 1)$ as in Figure 5.2. We obtain a very good straight line; *too good*, because we are throwing information away. From the gradient we obtain a value of $2B$ to the accuracy that our scale allows, which is much less than the quoted precision implies we could obtain. If we now plot $[\Delta E_{obs} - 2\tilde{B}(J + 1)]$ against $(J + 1)$, where $\tilde{B}$ is this first estimate, we obtain the result shown in Figure 5.2(b). We not only obtain a correction to $\tilde{B}$ from the gradient but also we see the range of applicability of the simple

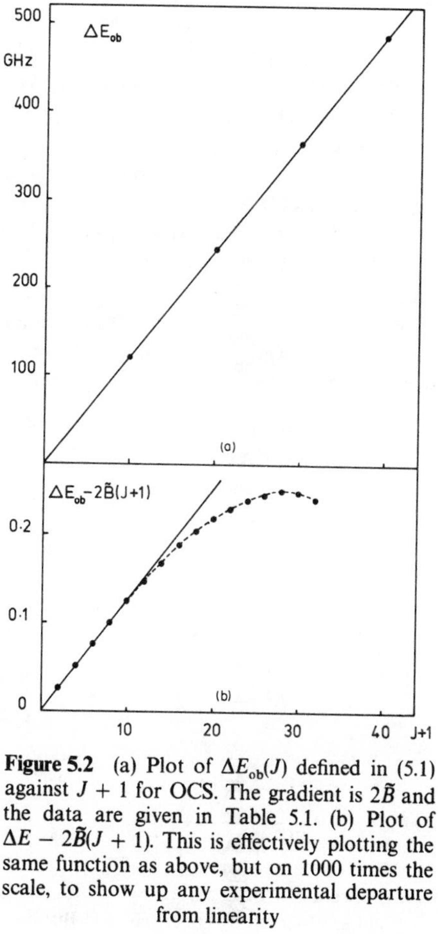

Figure 5.2 (a) Plot of $\Delta E_{ob}(J)$ defined in (5.1) against $J + 1$ for OCS. The gradient is $2\tilde{B}$ and the data are given in Table 5.1. (b) Plot of $\Delta E - 2\tilde{B}(J + 1)$. This is effectively plotting the same function as above, but on 1000 times the scale, to show up any experimental departure from linearity

rigid-rotor formalism. There is a noticeable departure from a linear plot, which becomes more and more severe as J increases, indicating that the effective B value is *decreasing*. The implication is that I, the moment of inertia, and the associated internuclear distances are increasing as the molecular rotational frequency increases. This is the effect we expect if the molecule is not completely rigid and centrifugal distortion occurs. This phenomenon is discussed more fully in Sections 6.2 and 6.5, where a more effective analytical procedure is applied to this type of data.

5.2 The Absorption Spectra of Symmetric Tops

In Figure 3.2 are shown typical energy-level schemes for a prolate and an oblate symmetric top. Transitions are allowed with $\Delta J = \pm 1$ and $\Delta K = 0$ as shown in Section 4.4c, and may occur only within the vertical K stacks of Figure 3.2. Application of these selection rules using (3.51) and (3.52) yields an expression identical with (5.1) for the transition frequency. The transition frequencies, in the rigid-rotor approximation, are independent of K and the resulting spectra consist of lines separated by $2B$ as in the linear molecule case. Each $J + 1 \leftarrow J$ line consists of $J + 1\,|K|$ components which in the rigid-rotor approximation would not be resolved. In Figure 5.3 a fine example, the far infra-red absorption of CH_3D observed by Ozier, Ho and Birnbaum,[233] is shown. A good discussion of intensities is also given in this paper. Note that because transitions between the $|K|$ stacks are forbidden the A constant cannot be determined. This information is sometimes obtainable from high-resolution vibrational spectra. As in the linear molecule case centrifugal

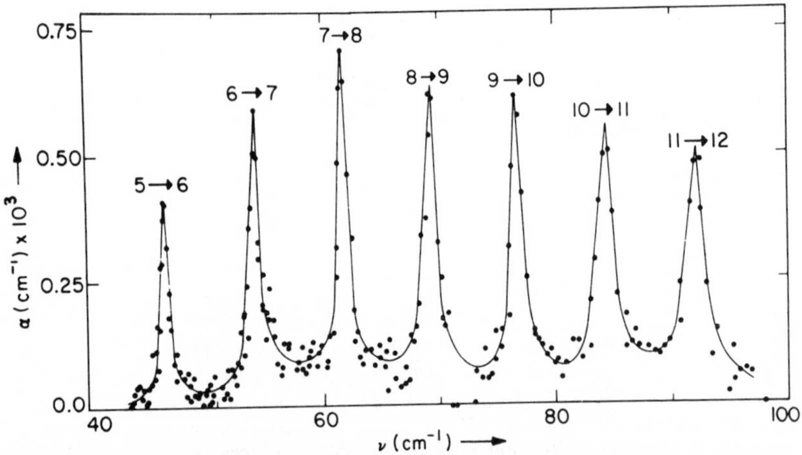

Figure 5.3 The far IR spectrum of CH_3D caused by zero point shortening of the CD bond as observed by Ozier, Ho and Birnbaum.[233]

effects show up, in that there is a departure from the simple linear relationship (5.1). As well as this quantitative departure from (5.1) a qualitative effect is detectable at high resolution in that the components of different $|K|$, for a given $J + 1 \leftarrow J$ transition, may be resolved (Section 6.6), yielding a sub-branch with $K + 1$ members.

5.3 The Absorption Spectra of Asymmetric Rotors

In Figure 3.3 the asymmetric rotor energy levels are correlated schematically between the prolate and oblate symmetric top limiting cases. Close to the limits the patterns of the levels are still fairly organized in that each level, which was degenerate in the symmetric rotor limit ($K \neq 0$), has split into two. In Figure 5.4 the levels of cyanogen azide, NCN_3,[56] which is a near prolate symmetric rotor are shown and in Figure 5.5 the levels of carbonyl fluoride, F_2CO[173,199] which is a near oblate case are shown schematically. In the highly asymmetric ($\kappa \sim 0$) case, as you might guess, the levels may not

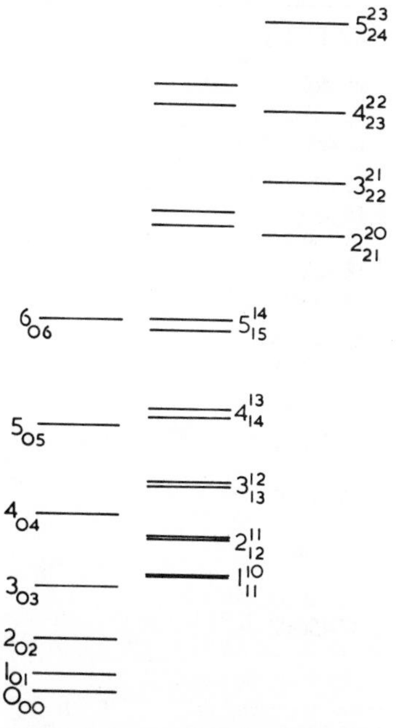

Figure 5.4 Energy-level diagram for the near prolate asymmetric rotor NCN_3 drawn to scale. Superscripts apply to the upper levels, subscripts to the lower ones

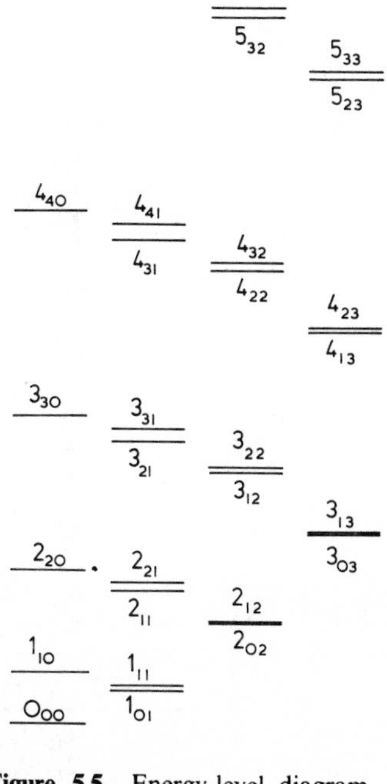

Figure 5.5 Energy-level diagram for the near oblate asymmetric rotor F_2CO. The asymmetry splittings are *not* drawn to a scale

form any obvious simple pattern and the resulting spectrum can be quite a mess.

Many different types of transition may now occur as discussed in Section 4.4 and in Table 4.3, the stronger general asymmetric rotor transitions are summarized and correlated with the dipole moment components which cause them to appear. Let us see how these results apply to NCN_3 and F_2CO. NCN_3 is a planar V-shaped molecule and has both μ_A and μ_B components.[56] The μ_A component gives rise to transitions with $\Delta K_A = 0, \pm 2, \pm 4, \ldots$ However, this molecule is very near the prolate limit and the wave functions (3.86) with a particular value of K_A are only slightly contaminated by $K_A \pm 2n (n = 1, 2, \ldots)$ symmetric rotor functions. Thus the only strong type A transitions, are those for which $\Delta K_A = 0$. These transitions give rise to bunches of lines at intervals of approximately $B + C$. The lines are called R branch lines and specified as members of the ${}^AR_{01}$ branch (Table 5.2).

Table 5.2 Some of the Observed Frequencies[a] in the Microwave Spectrum of Cyanogen Azide, NCN_3[56]

${}^AR_{01}$ Branch	Frequency
2_{12}–1_{11}	11 985·7
2_{02}–1_{01}	12 235·7
2_{11}–1_{10}	12 489·4
3_{13}–2_{12}	17 978·3
3_{01}–2_{02}	18 350·2
3_{22}–2_{21}	18 359·3
3_{21}–2_{20}	18 362·9
3_{12}–2_{11}	18 733·0
6_{16}–5_{15}	35 946·76
6_{06}–5_{05}	36 663·08
6_{25}–5_{24}	36 711·28
6_{34}–5_{33}, 6_{33}–5_{32}	36 733·29
6_{43}–5_{42}, 6_{42}–5_{41}	36 742·22
6_{52}–5_{51}, 6_{51}–5_{50}	36 756·2
6_{24}–5_{23}	36 758·65
6_{15}–5_{14}	37 456·02
${}^BQ_{1\,-1}$ Branch	
1_{10}–1_{01}	35 133·6
2_{11}–2_{02}	35 387·0
3_{12}–3_{03}	35 770·0
4_{13}–4_{04}	36 285·2

[a] $A_0 = 38066{\cdot}8 \pm 0{\cdot}15$, $B_0 = 3185{\cdot}15 \pm 0{\cdot}03$, $C_0 = 2933{\cdot}30 \pm 0{\cdot}03$ MHz.

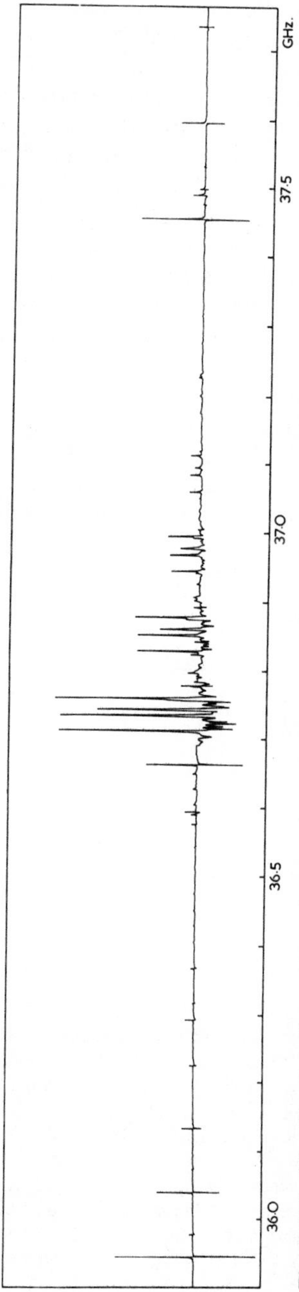

Figure 5.6 The $J = 6 \leftarrow 5$ transition of cyanogen azide, NCN_3, observed by Costain and Kroto.[56] The line frequencies are given in Table 5.2. A prominent feature of the spectrum is the vibrational satellite structure which causes the basic pattern to be repeated at intervals of approximately +150 MHz with decreasing intensity, see also Figure 5.9

In Figure 5.6 the $J = 6 \leftarrow 5$ set of transitions of NCN_3 are shown. The asymmetry is not however so large that all $2J + 1$ transitions are resolved. Only those with $K_A = 1$ and 2 are well separated, Table 5.2. For a given value of J the $K_A = 1$ (or $K_C = 1$) levels are the most strongly split and this behaviour manifests itself in the spectrum in that the $6_{16} \leftarrow 5_{15}$ and $6_{15} \leftarrow 5_{14}$ components in Figure 5.6 are the most widely separated. A simple calculation using the result (3.78) indicates that, to first order, the $K_A = 1$ doublets are separated by $(B - C)(J + 1)$. The splitting factor $(A - B)(J + 1)$ applies to the $K_C = 1$ doublets in the oblate limit. From the list of line frequencies for NCN_3 given in Table 5.2 we see that the separation of the $K_A = 1$ doublet shown in Figure 5.6 is 1509·6 MHz which yields a value of 251·6 MHz for $B - C$. A least squares fitting of all the observed lines yields a value of 251·59 MHz. This accuracy is somewhat fortuitous and is not a gauge of the validity of the simple first-order calculation *in general*. It does however highlight a phenomenon of cancellation of terms to second order of perturbation. In this case if we take the expansion (3.80) to terms in C_2b^2 we find that these last terms, in C_2, cancel. Such behaviour is quite a common feature of perturbation theory calculations and is clarified in Figure 5.7.

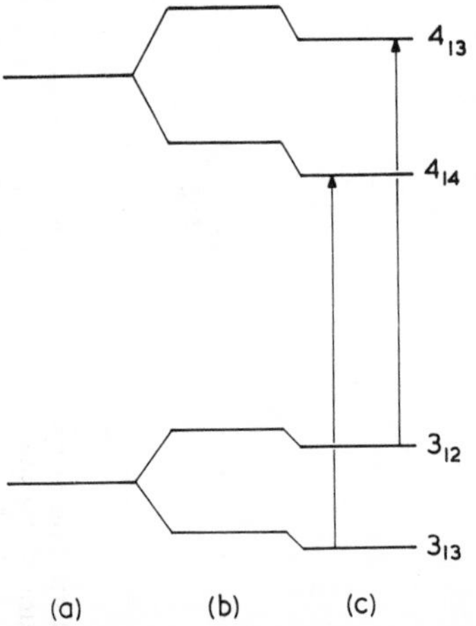

Figure 5.7 First- and second-order perturbation contributions to the asymmetric rotor energy levels and transition frequencies for the $J = 3$ and 4 levels with $K_A = 1$. $E \sim \bar{B}J(J + 1) + \bar{A}(K_A^2 + C_1b + C_2b^2)$ according to (3.80) for a near prolate case, where $\bar{B} = \frac{1}{2}(B + C)$ and $\bar{A} = A - \frac{1}{2}(B + C)$. (a) The pseudosymmetric rotor contribution is $(20\bar{B} + \bar{A})$ for the $J = 4$ level and $(12\bar{B} + \bar{A})$ for the $J = 3$ level. (b) The first-order contribution is $\bar{A}C_1b$. As $C_1 = \pm 10$ for the $J = 4$ level and ± 6 for the $J = 3$ level[303], the states are split apart symmetrically. (c) The second-order contribution is $\bar{A}C_2b^2$ where $C_2 = -7{\cdot}875$ for $J = 4$ and $-1{\cdot}875$ for $J = 3$. As a result the 4_{13} and 4_{14} levels are pushed down by equal amounts as are the 3_{12} and 3_{13} levels. As a result the *separation* of the $4_{13} \leftarrow 3_{12}$ and $4_{14} \leftarrow 3_{13}$ lines is independent of C_2

On first attempting an analysis of a spectrum one in general searches for the lines of lowest J value such as $J = 1, 2$ and 3, as they include lines for which simple algebraic relations exist. They also tend to show the clearest Stark patterns which also aid assignment and determination of dipole moment information as discussed in Chapter 7. We see immediately from Table 3.2 that the following simple relations hold

$$\Delta E(1_{01} \leftarrow 0_{00}) = B + C \tag{5.2a}$$

$$\Delta E(2_{12} \leftarrow 1_{11}) = B + 3C \tag{5.2b}$$

$$\Delta E(2_{11} \leftarrow 1_{10}) = 3B + C \tag{5.2c}$$

allowing B and C to be directly determined. Of course centrifugal distortion corrections should be applied but they are usually very small for these low-frequency lines. In fact it is the effects of quadrupole interactions which can cause the most serious problems at low J, Chapter 8.[22]

Although the higher J, μ_A, lines are slightly sensitive to the value of A, they are not sensitive enough in this molecule to allow a reliable value of the A constant to be determined. This molecule has, however, a μ_B dipole moment component and this allows the electromagnetic radiation field to perturb the motion about the A-axis and cause perpendicular transitions with $\Delta K_A = \pm 1, \pm 3, \ldots$ to occur, Table 4.3. From the energy-level diagram we see that the Q branch ($^{B}Q_{1\,-1}$) should have its origin at about $(A - B) \sim$ 35 MHz. Measurement of these lines would allow us to make an accurate estimate of the A constant. Such transitions may sometimes be difficult to assign, because of their sensitivity to the value of A. In NCN_3 they are also weak because μ_B, the dipole moment component which gives rise to them, is not large ($\mu_B \sim 0{\cdot}44$ Debye as compared to $\mu_A \sim 2{\cdot}9$ Debye.[56] The search for such lines is often facilitated by using a correlation diagram such as that shown in Figure 5.8. This uses the inertial defect $\Delta^0 = I_C^0 - I_A^0 - I_B^0$ which is discussed in Section 6.7. Δ^0 should, in the case of a planar molecule such as NCN_3 is expected to be, have a small positive value. In the case of NCN_3 one can estimate that $\Delta^0 \sim 0{\cdot}3 \pm 0{\cdot}1$ amu Å^2 by making a rough comparison with molecules with similar vibrational modes. As I_B and I_C are known accurately, one can determine I_A^0 as a function of Δ^0 over the predicted range and thus calculate the resulting positions of the first few $^{B}Q_{1\,-1}$ branch lines. Once a possible contender has been measured, the assignment is confirmed if the remaining lines are observed as predicted by a horizontal line through Figure 5.8 at the associated value of Δ^0. Some of these lines appear in Figure 5.9 and can be found by using this procedure.

Another type of Q branch which can occur involves what are sometimes called K-doubling transitions because they take place between the K doublet components. When $K_A = 1$ in a near prolate case such transitions occur at a frequency $\sim\frac{1}{2}(B - C)J(J + 1)$, and those observed by Johnson and

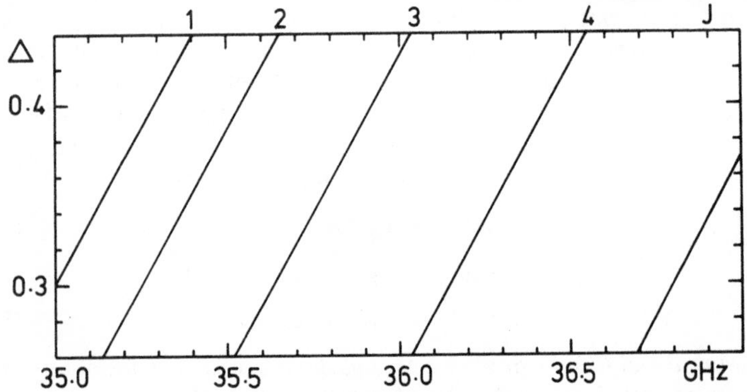

Figure 5.8 Plot of the positions of the Q_{1-1} branch lines with $K_A = 1 \leftarrow 0$ for NCN_3 as a function of the inertial defect Δ^0. Once a contending line has been measured and a possible assignment made, the diagram predicts a value of Δ^0. The assignment can then be confirmed by detecting the other members which must lie at the frequencies given by the intersection of a horizontal line at this value of Δ^0

Powell[143,143a] in the ${}^{A}Q_{0-1}$ branches with $K_A = 1$ and 2 for H_2CS are given in Table 5.3.

The F_2CO molecule is near oblate,[173,199] and the dipole moment lies along the A-axis. From Table 4.3 we see that as well as the R branch series ${}^{A}R_{01}$, branches such as ${}^{A}Q_{0-1}$ and ${}^{A}Q_{2-1}$ should also occur. In fact, we might expect that as the dipole moment is perpendicular to the C-axis the spectrum will be dominated by transitions for which $\Delta K_C = \pm 1$. The Q branch transitions ${}^{A}Q_{0-1}$ and ${}^{A}Q_{2-1}$ for $K_C = 2 \leftarrow 3$ have their origin at about 29·5 GHz as shown in Figure 5.10. Consider this spectrum in conjunction with the energy-level diagram Figure 5.5. Note how the two branches come to a head at about 29 510 MHz. In the symmetric rotor limit such a head should lie at $(B - C)(2K'_C + 1)$ or $(A - C)(2K'_C + 1)$. In this case $K'_C = 2$ and the mean of $5(B - C) \equiv 29\,361$ MHz and $5(A - C) \equiv 29\,664$ MHz is 29 512 MHz, extremely close to the origin of the two branches.[173,199]

There is a clear intensity alternation evident in these branches. The operation C_2^A interchanges the fluorine atoms and by using the results of Section 3.11d given in Table 3.16, we see that states with K_A even have a statistical weight of 1, and when K_A is odd it is 3.

For some molecules such as malonitrile, $CH_2(CN)_2$, the spectrum consists only of B-type transitions[126] and can be quite difficult to assign. In highly asymmetric molecules the lines can be quite difficult to locate. In F_2CS which is fairly asymmetric ($\kappa = -0{\cdot}62640$) and has only A type transitions (Table 5.4) the lines are sparsely distributed throughout the microwave region rather

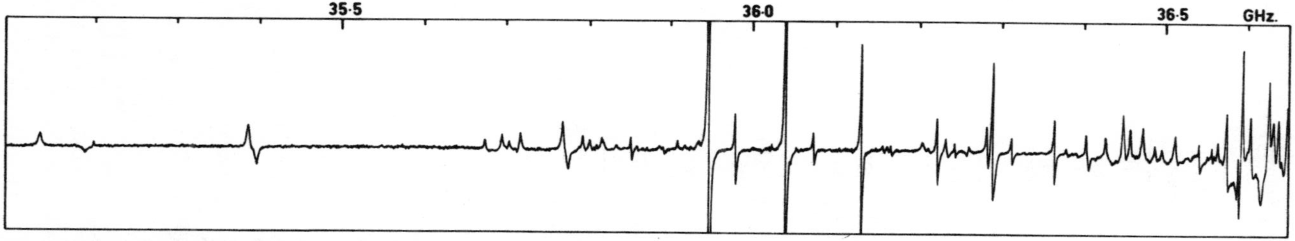

Figure 5.9 A scan at high sensitivity of the region between 35·1 to 36·7 GHz in the spectrum of NCN_3 observed by Costain and Kroto.[56] Part of this spectrum appears in Figure 5.6 under much less sensitive conditions. One can detect at least 5 quanta of the low frequency bending vibration in this scan. Several members of the ${}^{B}Q_{1-1}$ branch with $K_A = 1 \leftarrow 0$ can be identified on this scan with the aid of the correlation diagram Figure 5.8

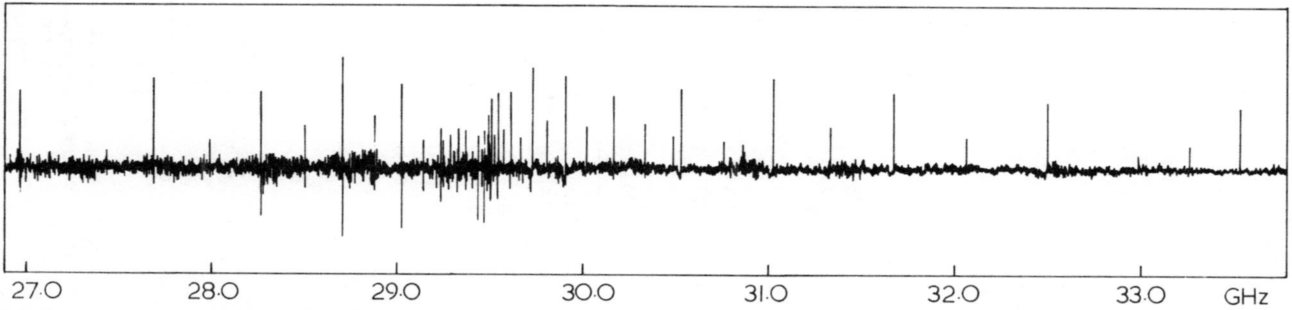

Figure 5.10 The ${}^{A}Q_{0-1}$ and ${}^{A}Q_{2-1}$ branches of F_2CO with $K_C = 2 \leftarrow 3$ degrade to lower and higher frequencies respectively starting at an origin at approximately 29·4 GHz. Compare these branches with the energy-level pattern given in Figure 5.5. The intensity alternation is due to the fluorine spin statistics. Some of these lines have been measured previously[173,199]

Table 5.3 Table of observed ${}^{A}Q_{0-1}$ *K*-Doubling Transitions in the Spectrum of H_2CS[143]

Assignment	Observed frequency[a]	Centrifugal distortion[b]
2_{11}–2_{12}	3 139·38	1·618
3_{12}–3_{13}	6 278·65	3·061
4_{13}–4_{14}	10 463·97	4·713
5_{14}–5_{15}	15 695·12	6·342
6_{15}–6_{16}	21 971·71	7·657
7_{16}–7_{17}	29 293·21	8·309
8_{17}–8_{18}	37 658·83	7·895
9_{18}–9_{19}	47 067·73	5·951
10_{19}–10_{110}	57 518·80	1·958
11_{110}–11_{111}	69 010·57	−4·657
15_{213}–15_{214}	7 052·07	−19·999
16_{214}–16_{215}	9 050·16	−27·802
17_{215}–17_{216}	11 438·53	−38·004
18_{216}–18_{217}	14 261·90	−51·158
19_{217}–19_{218}	17 565·92	−67·902
20_{218}–20_{219}	21 397·44	−88·962
21_{219}–21_{220}	25 803·14	−115·159
22_{220}–22_{221}	30 830·35	−147·407
23_{221}–23_{222}	36 525·32	−186·719
24_{222}–24_{223}	42 933·80	−234·201

[a] Frequency in MHz.
[b] Combined effect of all τ contributions, see Chapter 6.

Table 5.4 The ${}^{A}R_{01}$ Branch Line Frequencies of $F_2C^{32}S^{34}$ [a]

3_{21}–2_{20}	27 057·53
3_{12}–2_{11}	28 307·24
4_{14}–3_{13}	31 333·51
4_{04}–3_{03}	32 747·61
4_{23}–3_{22}	34 667·06
4_{32}–3_{31}	35 267·69
4_{31}–3_{30}	35 404·35
4_{22}–3_{21}	36 771·16
4_{13}–3_{12}	37 407·72
5_{16}–4_{14}	38 854·78
5_{05}–4_{04}	39 903·28

[a] $A_0 = 11\,892{\cdot}6 \pm 0{\cdot}5$,
$B_0 = 5\,133{\cdot}03 \pm 0{\cdot}03$,
$C_0 = 3\,580{\cdot}32 \pm 0{\cdot}03$.

than in convenient bunches as in NCN_3. However, because of the asymmetry in this case, the R_{01} lines are now quite sensitive to the A constant which thus can be determined quite accurately ($\pm 0{\cdot}5$ MHz). The higher K_A lines such as those with $K_A = 3$ and $J = 4 \leftarrow 3$ however still have residual symmetric rotor character lying fairly close to $(B + C)(J + 1)$. They are also useful in identification as they should show a mirror image relationship (Figure 5.11) with Stark lobes which converge. The approximate determination of $(B - C)$ directly from the $K_A = 1$ lines is not now so valid, as in the case of NCN_3. From Table 5.4 we see that $(B - C) \sim 1519$ MHz and the correct value is 1552·7 MHz, an error of $\sim 2\%$. The A constant is largely determined by the asymmetry splitting of $K_A = 2$ levels and resulting transitions, as can be seen from (3.80). We can truncate at terms higher than b_A^3 and note that all odd-order coefficients vanish for $K_A = 2$ levels. On simplification we see that the $K_A = 2$ doublet frequencies depend essentially on sums and differences of the type $C_2(B - C)^2/(2A - B - C)$. We can thus estimate that, in a case like this, the accuracy with which A can be determined is of the same order as the accuracy with which the $K_A = 2$ doublet *separation* can be measured. This of course assumes that $B - C$ is well determined, as is often the case.

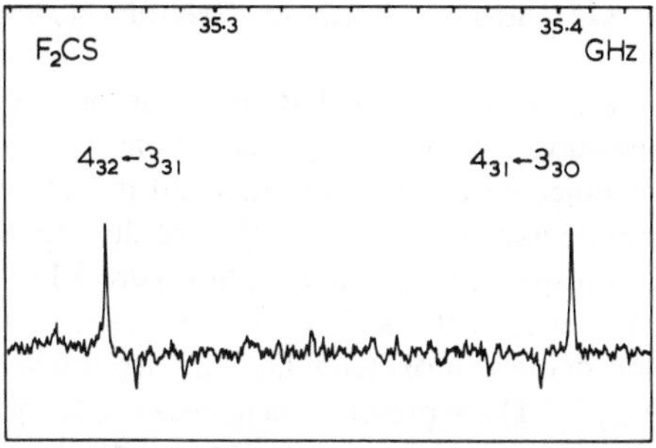

Figure 5.11 The $K_A = 3$ doublets of F_2CS for $J = 4 \leftarrow 3$ observed by Careless, Kroto and Landsberg.[34] The mean of these two lines lies within 500 MHz of $(B + C)(J + 1)$ which in this case is 34.85 GHz, Table 5.4

In the high J transitions of complex molecules there may be just too many lines to assign and indeed assignment may not be a worthwhile exercise because the accessible data may only yield fragmentary and not particularly interesting information. High resolution spectra are often so complicated that one cannot see the wood for the trees. The development of automatic wide-band microwave sweepers has enabled low-resolution wide-range spectra to be obtained. In these spectra only spectral contours are evident. The effects of asymmetry are strongest at low K_A or K_C as is indicated in Figures 3.5 and 5.6. As a consequence, if the molecule is not too asymmetric the majority of the lines clump together close to $(B + C)(J + 1)$ or $(A + C)(J + 1)$ as is shown in Figures 5.6 and 9.6. In a low resolution scan one observes the contours associated with the positions where the lines tend to pile up. In Figure 5.12 the low-resolution spectrum of *p*-iodotoluene is shown. The bands are separated by essentially $(B + C) = 1{\cdot}08$ GHz.

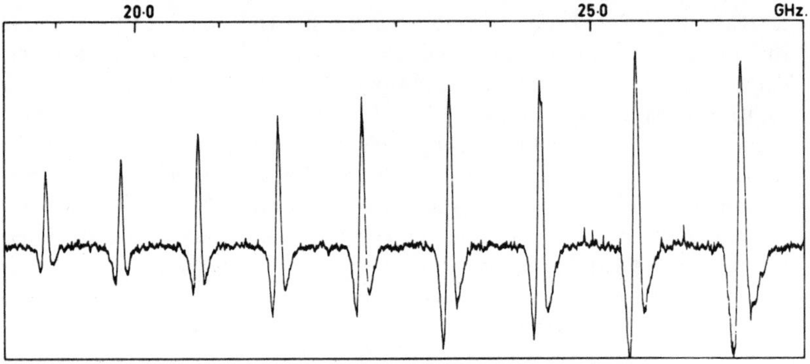

Figure 5.12 Broadband microwave spectrum of *p*-iodotoluene

A similar type of spectrum to that of *p*-iodotoluene was observed by Costain and Srivastava [56a] for the hydrogen bonded dimer $CF_3COOH \cdots HCOOH$. This molecule, which has an eight-membered elongated ring structure, is a prolate asymmetric rotor. Part of the broad band microwave spectrum of this unusual species is shown in Figure 5.13.

The spectra associated with the $J = 3 \leftarrow 2$ *R*-branch transitions of thioketene, H_2CCS, and deuterothioketene, D_2CCS, are shown in Figures 5.14(a) and (b) respectively.[92a] These present excellent examples of how spin statistics govern the intensity patterns of spectra. According to the results obtained in Table 3.16, as H_2CCS is a prolate asymmetric rotor, the $K_A = 1$ lines have statistical weights three times those of the $K_A = 0$ and 2 lines, consistent with the observed spectrum (Figure 5.14a). In the case of D_2CCS we see that the central $K_A = 0$ and 2 lines have statistical weights twice those of the $K_A = 1$ lines, which is also consistent with the observed spectrum shown in Figure 5.14(b).

Some of the vibrational satellites show reversals of the ground state spin statistics where the vibrational wavefunctions are antisymmetric under C_2. In the case of H_2CCS the even K_A lines of a bending vibrational satellite are the stronger and the odd K_A lines are too weak to be detected in Figure 5.14(a). In D_2CCS a similar reversal occurs and in this case it is the $K_A = 1$ lines which are the stronger and most easily detected.

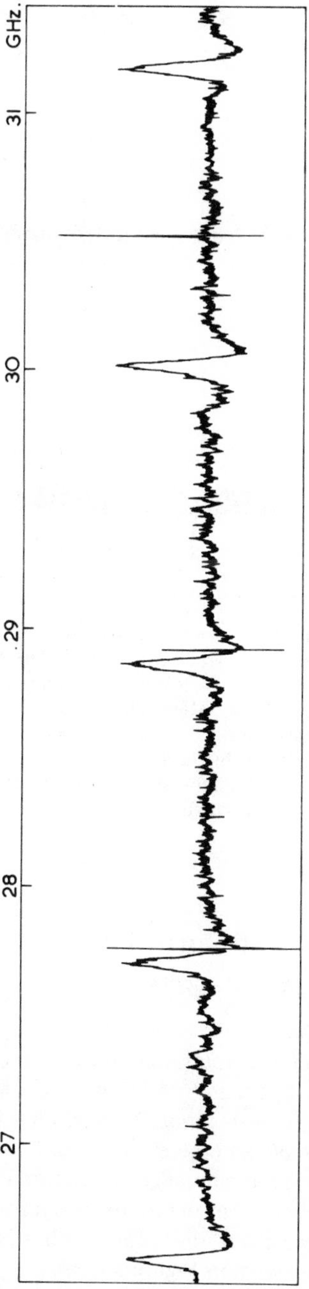

Figure 5.13 Part of the broadband microwave spectrum of the hydrogen bonded dimer $CF_3COOH \cdots HCOOH$. The molecule has an eight-membered elongated ring structure and is a prolate asymmetric rotor. Costain and Srivastava[56a] obtained the following rotational constants $A = 2900{\cdot}9$, $B = 604{\cdot}7$ and $C = 549{\cdot}4$ MHz. The above *R*-branch complexes are separated by $B + C = 1154$ MHz

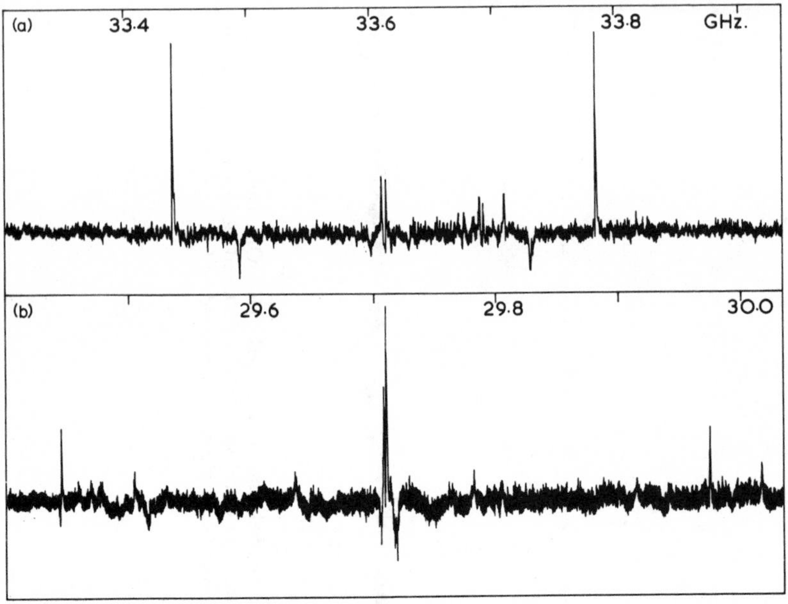

Figure 5.14 (a) The spectrum of the $J = 3 \leftarrow 2$ R-branch transitions of thioketene, H_2CCS, observed by Georgiou, Kroto and Landsberg.[92a] The $K_A = 0$ and 2 lines lie close together (at 33,611·70 and 33,607·84 MHz respectively) and are flanked by the $K_A = 1$ lines at 33,438·37 and 33,783·23 MHz. The statistical weights of the odd K_A lines are three times those of the even K_A lines. The $K_A = 0$ and 2 vibrational satellite lines lie close to 33,692 MHz (see text). (b) The analogous spectrum of deuterothioketene D_2CCS. In this case the $K_A = 0$ and 2 lines have statistical weights twice those of the $K_A = 1$ lines. The $K_A = 1$ vibrational satellite lines lie at 29,503 and 30,016 MHz

5.4 Raman Spectra

If we apply the selection rule $\Delta J = \pm 2$ to the energy-level expression (3.47) for a linear molecule we obtain the expression

$$|\Delta E(J)| = 4B(J + 3/2) \tag{5.3}$$

for the displacement of the lines from the exciting line. In Figure 5.15 is shown a recorder trace of the rotational Raman spectrum of N_2. The lines, as seen from (5.3), are separated by $4B$ from each other, and the spacings between the $J = 0$ lines and the exciting (Rayleigh) line is $6B$. In the case of N_2, the nuclei possess unit spin and therefore follow Bose–Einstein statistics. As discussed in Section 3.11d the generalized Exclusion Principle applies and in this case we obtain the result that the transitions for which J is even must have twice the statistical weight of those for which J is odd. The intensity alternation is quite clearly shown in Figure 5.15.

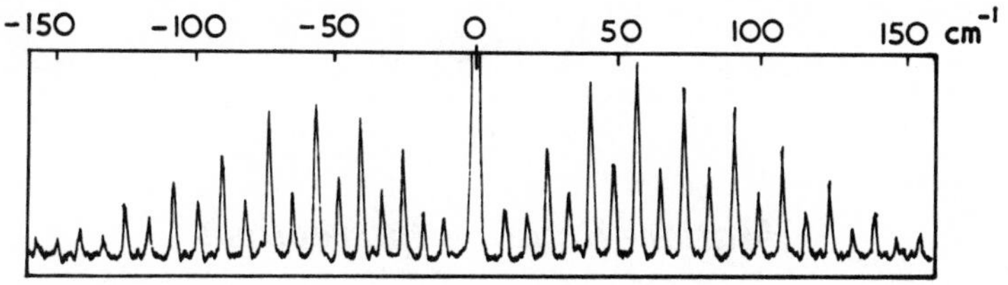

Figure 5.15 The Raman spectrum of N_2 obtained using the 6328 Å He–Ne laser line and photoelectric recording. The 2–1 intensity alternation is due to the spin statistics of the equivalent N atoms ($I = 1$)

The resolution is much inferior to microwave spectroscopy and as a result the technique is limited to symmetric or near-symmetric top systems with fairly simple energy-level patterns. Stoicheff has studied a large number of the feasible systems and has determined many of their rotational constants and the accessible structural data.[285] The technique is most important for molecules which do not possess permanent dipole moments for which standard microwave techniques are not applicable.

Chapter 6

Vibration–Rotation Interactions

6.1 The Vibration–Rotation Hamiltonian for Semi-rigid Molecules

In Chapters 3, 4 and 5 the effects of motions relative to the molecule-fixed axes were assumed to be negligible and in this approximation it was possible to treat the molecule as a rigid structure. That this is a relatively good approximation is witnessed by the fact that the representative spectra given in Chapter 5 are well accounted for on the basis of this simple model. However, we saw that when one studies the spectra more carefully, especially under the high resolution obtainable using microwave techniques, departures from the behaviour expected for a rigid structure are observed. These departures manifest themselves in correction terms which must be added to the rigid-rotor expressions in order to accurately account for the energy-level pattern *as determined experimentally* from the spectrum. If the assumption of rigidity is now relaxed, the modified derivation yields the form of the correction factors. We shall deal in this chapter with molecules in which the rigid-rotor model still remains a good basis approximation and we can treat the effects of non-rigidity by perturbation methods, i.e. the semi-rigid class of molecules as defined in Section 9.1. Molecules in which large amplitude displacements, such as internal rotation and inversion, are feasible may be classed as flexible molecules and are discussed separately in Chapter 9.

One of the main problems which will be discussed in this context is how to determine the structure of a molecule from rotational information.

In Chapter 2 the classical expression for the kinetic energy of a vibrating–rotating molecule was shown to consist of three terms and for a complete treatment which includes non-rigidity we must combine the three terms together. We can write the total energy (2.29) as

$$T = \tfrac{1}{2}I_{\alpha\beta}\omega_\alpha\omega_\beta + \sum_n m_n\omega_\alpha e_{\alpha\beta\gamma}d_{n\beta}\dot{d}_{n\gamma} + \tfrac{1}{2}\sum_n m_n\dot{d}_{n\alpha}\dot{d}_{n\alpha} \tag{6.1}$$

and then from this expression derive a complete quantum mechanical Hamiltonian. The derivation which is non-trivial was first carried out by Wilson and Howard,[336,335] modified by Darling and Dennison[62] and simplified by Watson.[318]

The first step is to transform (6.1) into the *classical momentum* form as follows. From vibrational theory we know that for a molecule in which the vibrational motion can be considered to be harmonic (often a good approximation) the vibrations can, as indicated in Section 6.4b, be described in terms of the *normal modes* of vibration.

We can thus transform from the cartesian atomic displacements $d_{n\alpha}$ to the normal coordinates Q_r by

$$\sqrt{m_n}d_{n\alpha} = \sum_r l_{nr}^{(\alpha)} Q_r \tag{6.2}$$

where $l_{nr}^{(\alpha)}$ are elements of the transformation matrix which is subject to the orthogonality condition

$$\sum_n l_{nr}^{(\alpha)} l_{nr'}^{(\alpha)} = \delta_{rr'} \tag{6.3}$$

(6.2) essentially determines how much displacement (mass-weighted) along a cartesian axis, α, a particular normal coordinate, Q_r, contributes. This is clearly an important quantity if one is interested, as we are, in any changes in the moment-of-inertia tensor during vibration. On substituting (6.2) into (6.1) and taking note of (6.3) we obtain

$$T = \tfrac{1}{2} I_{\alpha\beta}\omega_\alpha\omega_\beta + \sum_n \sum_r \sum_{r'} \omega_\alpha e_{\alpha\beta\gamma} l_{nr}^{(\beta)} l_{nr'}^{(\gamma)} Q_r \dot{Q}_{r'} + \tfrac{1}{2} \sum_r \dot{Q}_r^2 \tag{6.4}$$

This expression can be further simplified by realizing that the coriolis ζ constant is defined by

$$\zeta_{rr'}^{(\alpha)} = \sum_n e_{\alpha\beta\gamma} l_{nr}^{(\beta)} l_{nr'}^{(\gamma)} \tag{6.5}$$

$\zeta_{rr'}^{(\alpha)}$ is the coriolis coupling constant between the two normal coordinates Q_r and $Q_{r'}$.[25] We can get a feel for its physical significance by noting that it scales the amount of angular momentum about the α-axis generated by vibrational motion.[202] We thus obtain

$$T = \tfrac{1}{2} I_{\alpha\beta}\omega_\alpha\omega_\beta + \sum_r \sum_{r'} \omega_\alpha \zeta_{rr'}^{(\alpha)} Q_r \dot{Q}_{r'} + \tfrac{1}{2} \sum_r \dot{Q}_r^2 \tag{6.6}$$

Using this expression we can now define the following *classical momenta*.

$$J_\alpha = \frac{\partial T}{\partial \omega_\alpha} = I_{\alpha\beta}\omega_\beta + \sum_r \sum_{r'} \zeta_{rr'}^{(\alpha)} Q_r \dot{Q}_{r'} \tag{6.7}$$

$$P_r = \frac{\partial T}{\partial \dot{Q}_r} = \sum_{r'} \omega_\alpha \zeta_{r'r}^{(\alpha)} Q_{r'} + \dot{Q}_r \tag{6.8}$$

Multiplying (6.7) by ω_α and (6.8) by $\dot{Q}_r$ and adding, we obtain

$$T = \tfrac{1}{2} J_\alpha \omega_\alpha + \tfrac{1}{2} \sum_r P_r \dot{Q}_r \tag{6.9}$$

By substituting for $\dot{Q}_r$ using (6.8) we obtain

$$T = \tfrac{1}{2}J_\alpha\omega_\alpha + \tfrac{1}{2}\sum_r P_r^2 - \tfrac{1}{2}\sum_r P_r \sum_{r'} \omega_\alpha \zeta_{rr'}^{(\alpha)} Q_{r'} \tag{6.10}$$

which may be written as

$$T = \tfrac{1}{2}(J_\alpha - \pi_\alpha)\omega_\alpha + \tfrac{1}{2}\sum_r P_r^2 \tag{6.11}$$

where a *modified* 'vibrational' induced internal angular momentum has been introduced, defined by

$$\pi_\alpha = \sum_r \sum_{r'} \zeta_{rr'}^{(\alpha)} Q_r P_{r'} \tag{6.12}$$

If we now substitute (6.8) for P_r we see that

$$\pi_\alpha = \sum_r \sum_{r'} \zeta_{rr'}^{(\alpha)} Q_r \sum_{r''} \omega_\beta \zeta_{r''r'}^{(\beta)} Q_{r''} + \sum_r \sum_{r'} \zeta_{rr'}^{(\alpha)} Q_r \dot{Q}_{r'} \tag{6.13}$$

(note the implied summation over the double index β). From (6.7) and (6.13) we thus obtain the result

$$J_\alpha - \pi_\alpha = I'_{\alpha\beta}\omega_\beta \tag{6.14}$$

where

$$I'_{\alpha\beta} = I_{\alpha\beta} - \sum_r \sum_{r'} \sum_{r''} \zeta_{rr'}^{(\alpha)} \zeta_{r''r'}^{(\beta)} Q_r Q_{r''} \tag{6.15}$$

We can write (6.14) in matrix form

$$(\boldsymbol{J} - \boldsymbol{\pi}) = \mathbf{I}'\boldsymbol{\omega} \tag{6.16}$$

Now, as $\mathbf{I}' = \mathbf{I}'^{\dagger}$ and $\boldsymbol{\mu} = \mathbf{I}'^{-1}$ we can rewrite the first term in (6.11) as

$$\tfrac{1}{2}\boldsymbol{\omega}^{\dagger}\mathbf{I}'\boldsymbol{\omega} = \tfrac{1}{2}(\boldsymbol{J} - \boldsymbol{\pi})^{\dagger}\boldsymbol{\mu}(\boldsymbol{J} - \boldsymbol{\pi}) \tag{6.17}$$

Thus the *classical Hamiltonian* form which can be derived from (6.1) is

$$H = \tfrac{1}{2}(J_\alpha - \pi_\alpha)\mu_{\alpha\beta}(J_\beta - \pi_\beta) + \tfrac{1}{2}\sum_r P_r^2 + V \tag{6.18}$$

In arriving at this expression we introduced an arbitrarily defined internal momentum $\boldsymbol{\pi}$ which strictly should not be called vibrational angular momentum as this would be given by

$$P_\alpha = \sum_r \sum_{r'} \zeta_{rr'}^{(\alpha)} Q_r \dot{Q}_{r'} \tag{6.19}$$

(the second term of (6.13)). The quantity $\boldsymbol{\pi}$ causes the transformation to the quantum mechanical Hamiltonian to require a special procedure. This is because $\boldsymbol{\pi}$ is not conjugate to any particular coordinate and in fact includes a contribution from rotational motion, as is clear from (6.13). Using the

Podolsky method[335,343] (6.18) can be transformed to the quantum mechanical operator form:

$$H = \tfrac{1}{2}\mu^{1/4}(J_\alpha - \pi_\alpha)\mu_{\alpha\beta}\mu^{-1/2}(J_\beta - \pi_\beta)\mu^{1/4} + \tfrac{1}{2}\sum_r \mu^{1/4}P_r\mu^{-1/2}P_r\mu^{1/4} + V \tag{6.20}$$

where μ is the determinant of $\boldsymbol{\mu}$.

This rather complicated expression has been simplified by Watson[318] who showed that (6.20) can be rearranged to the simpler form

$$H = \tfrac{1}{2}(J_\alpha - \pi_\alpha)\mu_{\alpha\beta}(J_\beta - \pi_\beta) + \tfrac{1}{2}\sum_r P_r^2 + U + V \tag{6.21}$$

The extra term U which this rearrangement yields turns out to be

$$U = -\tfrac{1}{8}\mu_{\alpha\alpha} \tag{6.22}$$

which is a function only of the coordinates and not momenta and can be considered as part of the potential energy term.[318] We shall now show how (6.21) can be used as the basis for the calculation of energy-level expressions for a rotating molecule in which non-rigidity has been taken into account. Finally Watson[318] has shown that

$$[\pi_\alpha, \mu_{\alpha\beta}] = 0 \tag{6.23}$$

and as a consequence the Hamiltonian (6.21) is independent of the order of the operators in the first term. Note that (6.23) implies summation over α when expanded. The field-free Hamiltonian may thus be written as

$$H = \tfrac{1}{2}\mu_{\alpha\beta}(J_\alpha - \pi_\alpha)(J_\beta - \pi_\beta) + \tfrac{1}{2}\sum_r P_r^2 + V(Q_r) \tag{6.24}$$

6.2 The Semi-rigid Diatomic Molecule

It is useful at this point to consider how the Hamiltonian (6.24) may be used to account for non-rigidity in diatomic molecules. In this case one can study, in a fairly simple way, centrifugal distortion effects as well as zero-point vibrational contributions to the *observed* rotational constants. The derivation can be formulated such that the expressions have the same form as those given by the much more complicated general polyatomic treatment. It does however have a disadvantage in that the effects of vibrational angular momentum π are not considered.

A few semi-quantitative points are worth considering first, which highlight in a simple way the dependence of B, the rotational constant, on vibrational motion. In a classical description, as a molecule rotates through one cycle it will have vibrated perhaps a hundred times. Typical rotational frequencies are of the order of 10^9–10^{12} Hz and vibrational frequencies 10^{13}–10^{14} Hz. We thus require the value of B *averaged* over this vibrational motion.

Quantum mechanically this is determined by evaluating $\langle v|B|v\rangle$ where B is related to the *instantaneous* moment of inertia I, by $B = 1/2I$. For a diatomic molecule the definitions (2.26) indicate that $I = mr^2$ where r is the internuclear distance and $m = m_1m_2/(m_1 + m_2)$. We thus see that we need the value of r^{-2} averaged over the vibrational motion

$$\langle v|B|v\rangle = \langle v|\tfrac{1}{2}I^{-1}|v\rangle = \frac{1}{2m}\langle v|r^{-2}|v\rangle \tag{6.25}$$

In Figure 6.1 is depicted a simple oscillator model for a diatomic molecule with a fairly representative structure and vibrational amplitude.[122] For the purposes of this example we assume that the system oscillates such that it spends half of its time at each turning point. In Table 6.1, algebraic expressions are developed for the average values of various powers of r for a symmetric oscillation using the binomial expansion. In the last row the average value of r^{-2} for a non-symmetric oscillator is given.

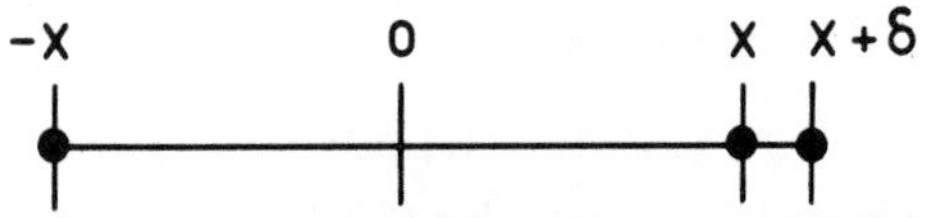

Figure 6.1 Schematic diagram of a hypothetical oscillator which spends 50% of its time at two turning points. In the symmetric mode they are at $-x$ and $+x$ and in the unsymmetric mode at $-x$ and $+(x + \delta)$

Table 6.1 Table of Terms Which Contribute to Errors in the Determined Value of r in a Simple Oscillator, Figure 6.1

r^n	$\langle r^n\rangle$	$\langle r^n\rangle^{1/n}$	δr_{calc} [a]
r	1	1	0
r^2	$1 + x^2$	$1 + \frac{1}{2}x^2 + \cdots$	+0·0013
r^{-1}	$1 + x^2 + \cdots$	$1 - x^2 + \cdots$	−0·0025
r^{-2}	$1 + 3x^2 + \cdots$	$1 - \frac{3}{2}x^2 + \cdots$	−0·0038
r_a^{-2}	$1 - \delta + 3x^2 + \cdots$	$1 + \frac{1}{2}\delta - \frac{3}{2}x^2 + \cdots$	+0·0012

[a] In the last column δr has been evaluated with $r = 1{\cdot}0$, $x = 0{\cdot}05$ and $\delta = 0{\cdot}01$. x and δ are defined in Figure 6.1.

The first interesting result is that in the case of the symmetric (and therefore also a simple harmonic) oscillator the bond length, as determined by taking the inverse square root of the average value of r^{-2}, yields a bond length *shorter* than the equilibrium value. If we now take anharmonicity into account the anharmonic term δ contributes in 'first order' and thus even if δ is small relative to x it will, in general, counteract the harmonic shortening effect and yield an apparently *long* bond length. In fact the anharmonic contribution is generally two to four times the magnitude of the harmonic part[118,119] and thus in the case of diatomic molecules one almost always obtains a long bond length. These points highlight the basic problem which exists when one comes to the question of determining molecular geometries from rotational data, particularly the part played by anharmonicity.

6.2a Perturbation Theory Approach to Vibration–Rotation Interactions in Diatomic Molecules

In Chapter 1 the Hamiltonian for the simple harmonic oscillator was discussed and written first in terms of x, a linear displacement (1.7) and then transformed to (1.9a) where it is written in terms of $Q = \sqrt{m}\,x$, a mass-weighted coordinate. It was finally solved in the form (1.9b) where $q = \sqrt{m\omega}\,x$. It is convenient for our purposes to inspect the form of I, the *instantaneous* moment of inertia of a diatomic molecule under these same coordinate transformations. We see that we can write

$$I = mr^2 = m(r^e + x)^2 = (\sqrt{I^e} + Q)^2 = I^e + 2\sqrt{I^e}Q + Q^2 \tag{6.26}$$

For reasons that will become clear when we discuss the general polyatomic case in the next section, we will rewrite I in terms of $a = (\partial I/\partial Q)_e (= 2\sqrt{I^e}$ by inspection of (6.26)) as

$$I = I^e + aQ + \frac{a^2}{4I^e}Q^2 = I^e\left[1 + \frac{aQ}{I^e} + \frac{a^2Q^2}{4I^{e\,2}}\right] \tag{6.27}$$

If we now introduce the scaled coordinate β defined by ‡

$$\beta = \frac{aQ}{I^e} = \frac{\partial I}{\partial Q}\Big/\frac{I^e}{Q} \tag{6.28}$$

We see that we can factorize (6.27) as

$$I = I^e(1 + \tfrac{1}{2}\beta)^2 \tag{6.29}$$

As we have seen, we need the average value of I^{-1} so from (6.29) we obtain an expansion

$$I^{-1} = I^{e-1}(1 - \beta + \tfrac{3}{4}\beta^2 + \cdots) \tag{6.30}$$

β can also be written in terms of q as $\beta = \gamma q$ where $\gamma = a/I^e\sqrt{\omega} = (8B_e/\omega)^{1/2}$. We see that the first term in the expansion of I^{-1} is just I^{e-1}, and the effect of the extra terms can be determined by carrying out a perturbation calculation. However, we should note that the discussion in Section 6.2 implied

‡ $\frac{1}{2}\beta \equiv \xi$ where ξ is the coordinate usually used in diatomic molecule calculations $\xi = (r - r^e)/r^e$.

that the most important correction terms to I^e may have their source in the anharmonicity of the potential curve. We thus consider the Hamiltonian.

$$H = B_e(1 - \beta + \tfrac{3}{4}\beta^2)J^2 + \tfrac{1}{2}\omega(p^2 + q^2) + \frac{\varphi_3}{3!}q^3 + \cdots \tag{6.31}$$

where we have truncated the expansion of I^{-1} after the term in q^2, i.e. β^2 and the vibrational potential energy after the cubic, q^3, term. We shall take our zeroth-order Hamiltonian to be

$$H^0 = B_eJ^2 + \tfrac{1}{2}\omega(p^2 + q^2) \tag{6.32}$$

which defines our basis functions as $|J\,v\rangle = |J\rangle|v\rangle$. The solutions of this have already been studied in Chapter 1 and Chapter 3. The perturbation terms are thus

$$H' = [-B_e\gamma]qJ^2 + [\tfrac{3}{4}B_e\gamma^2]q^2J^2 + [\tfrac{1}{6}\varphi_3]q^3 \tag{6.33}$$

J^2 is diagonal in our basis and thus we can evaluate the first- and second-order perturbation corrections from (6.33). They are according to the results discussed in Chapter 1 and Section A4 the following three terms.

$$[-B_e\gamma]^2[-1/2\omega]J^2(J + 1)^2 \tag{6.34a}$$

$$[\tfrac{3}{4}B_e\gamma^2][v + \tfrac{1}{2}]J(J + 1) \tag{6.34b}$$

$$2[-B_e\gamma][\tfrac{1}{6}\varphi_3][-\tfrac{3}{2}(v + \tfrac{1}{2})/\omega]J(J + 1) \tag{6.34c}$$

We have neglected terms which are independent of J as these will normally be considered as part of the vibrational energy. We can collect all these results and add them to the zeroth-order energy to obtain the following expressions for the rotational energy or *term value*:

$$E(vJ) = B_vJ(J + 1) - DJ^2(J + 1)^2 \tag{6.35}$$

where

$$B_v = B_e - \alpha(v + \tfrac{1}{2}), \qquad D = 4B_e^3/\omega^2 \tag{6.36a}$$

$$\alpha = -\frac{6B_e^2}{\omega}(1 + a_1), \qquad a_1 = \frac{\varphi_3}{6}\sqrt{\frac{2}{\omega B_e}} = \frac{\varphi_3}{6}\frac{a}{\sqrt{\omega}} \tag{6.36b}$$

It is worth noting how the various correction terms arose. The term in $DJ^2(J + 1)^2$ is independent of v and is usually called the centrifugal distortion correction. It originates from the term in $-B_e\beta J^2$ of (6.31) via the second-order perturbation correction (6.34a). Terms (6.34b) and (6.34c) both yield corrections with a $(v + \tfrac{1}{2})J(J + 1)$ dependence, the first is the *harmonic* correction to B, the second the *anharmonic* correction and it is illuminating to relate them to the correction terms in δ and x^2 respectively discussed in the previous section. From Figure 6.3 we see that a realistic potential curve will in general require φ_3 and thus also a_1 in (6.36b) to be negative. a_1 is usually in the range -2 to -4[118,119] and thus α is almost always positive causing the value of B_v to decrease with v and the *effective* value of r to increase. There is perhaps a warning lesson for compulsive standardizers here in that the convention usually used requires the negative sign in (6.35) and (6.36a) and causes nothing but confusion. The sign of α and D should be allowed to arise naturally.

Note that in this approach one obtains at the end of the day only as much as one has put into the calculation at the beginning. Injection of more terms, such as $(1/4!)\varphi_4 q^4, \cdots$ etc, into the starting Hamiltonian yields increased flexibility in the final energy-level expression in that terms in v^2 and J^6 are disgorged. The inclusion of higher-order corrections also tends to readjust the expressions for the lower constants.

Dunham has calculated these expressions[78] by the WKB (Wentzel–Kramers–Brillouin) method and the results are given in terms of the expression

$$E(vJ) = \sum_{lj} Y_{lj}(v + \tfrac{1}{2})^l[J(J + 1)]^j \tag{6.37}$$

The Dunham Y_{lj} coefficients are related to the spectroscopic parameters according to $Y_{01} \to B_e$, $Y_{11} \to -\alpha_e$, $Y_{02} \to -D_e$, etc. The potential energy expression used by Dunham is

$$U = a_0\xi^2(1 + a_1\xi + a_2\xi^2 + \cdots) \tag{6.38}$$

where $\xi = x/r_e = (r - r_e)/r_e$. The relations between the Y coefficients and the a coefficients in (6.38) are listed by Townes and Schawlow.[303] Note that a_1 in (6.36) and a_1 in (6.38) are the same. Herschbach and Laurie[118,119] have published a useful tabulation of vibrational force constants.

6.2b *High-resolution Rotational Spectra of Diatomic Molecules*

The final step which we wish to make is to fit our more flexible energy-level expression (6.35) to the observed spectrum. We note two main results of this treatment: (1) The transition frequencies for different vibrational states will not coincide under high resolution, and (2) the separations between successive $\Delta J = \pm 1$ transitions will not be constant and will depart slightly from the rigid-rotor result, $2B$, deduced in Section 5.1.

For a typical $\Delta J = +1$ transition we obtain from (6.35) the expression

$$\Delta E(vJ) = 2B_v(J + 1) - 4D(J + 1)^3 \tag{6.39}$$

for the transition frequencies. In Table 6.2 are given some very accurate experimental data obtained by Wyse, Gordy and Pearson[344] for AlF. They observed several sets of rotational transitions belonging to the molecules in

Table 6.2 Observed Frequencies of AlF in MHz[344]

	v				
J	0	1	2	3	4
5	197 833·15	196 052·54	194 284·39	192 528·97	
6	230 793·89	228 716·54	226 653·84	224 605·75	222 572·54
7	263 749·35	261 375·27	259 017·92	256 677·28	254 353·69
8	296 698·87	294 028·10	291 376·02		
9	329 641·64	326 674·01	323 727·41		
10	362 576·79	359 312·50	356 071·28		

various vibrationally excited states. We can analyse the data conveniently using the relation

$$\frac{\Delta E(vJ)}{J+1} = 2B_v - 4D_v(J+1)^2 \tag{6.39a}$$

which we get directly from (6.39). We can plot this function as implied by the right-hand side against $(J + 1)^2$. The intercept yields $2B_v$ and the gradient $-4D_v$. The plot for the ground state ($v = 0$) is shown in Figure 6.2. The result is that $B_0 = 16\,488{\cdot}325$ MHz and $D_0 = 0{\cdot}0314$ MHz. One can make similar plots of (6.39a) for the excited states to obtain B_v with $v = 1$ to 5. This data is so accurate that Wyse, Gordy and Pearson were able to fit B_v, by a least squares procedure, to a relation such as (6.70) up to terms in $(v + \frac{1}{2})^3$. The relation (6.70) is an extended version of the relation given in (6.36a). They were also able to detect a vibrational dependence in the centrifugal distortion correction. The above simple graphical procedure using (6.70) up to terms in

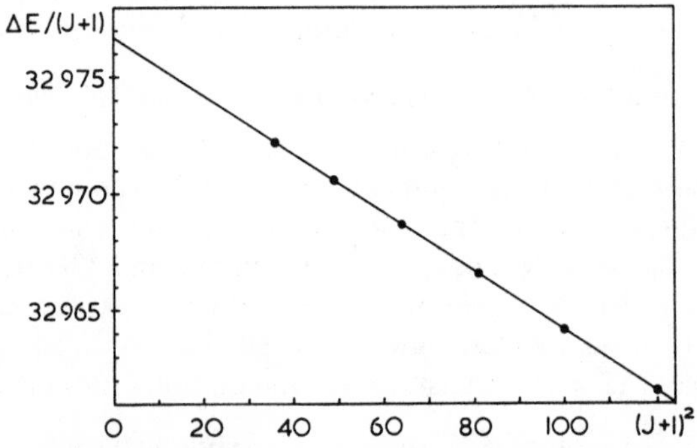

Figure 6.2 Graphical determination of B_0 and D_0 for AlF using (6.39a). The gradient $-4D_0 = -0{\cdot}1256$ MHz and the intercept $2B_0 = 32\,976{\cdot}65$ MHz. The relevant experimental data is that of Wyse, Gordy and Pearson[344] given in Table 6.2

$\gamma(v + \frac{1}{2})^2$ is adequate for our purposes and yields the following spectroscopic parameters (MHz)

$$B_0 = 16\,488{\cdot}325$$
$$D_0 = 0{\cdot}0314 \quad (D_e = 0{\cdot}03137 \pm 0{\cdot}00002)$$
$$\alpha = 149{\cdot}424 \quad (149{\cdot}420 \pm 0{\cdot}009)$$
$$\gamma = 0{\cdot}519 \quad (0{\cdot}515 \pm 0{\cdot}004)$$
$$B_e = 16\,562{\cdot}910 \quad (16\,562{\cdot}930 \pm 0{\cdot}006)$$

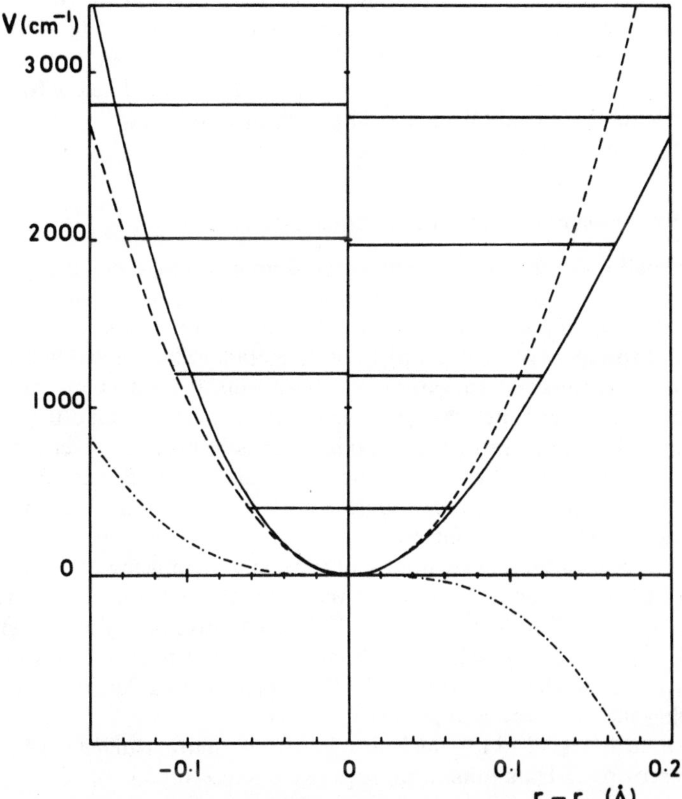

Figure 6.3 The potential function for AlF. --- represents the harmonic parts of the potential, —·—·— represents the cubic part and ——— represents the sum of the harmonic and cubic terms. The curves have been drawn using the results of Sections 6.2a, 6.2b and A.7. On the left the vibrational levels are harmonically (equidistantly) spaced and on the right they have closed up slightly because of the anharmonicity

The results of Wyse, Gordy and Pearson[344] are given in brackets. A similar analysis of the data given in Table 5.1 for the ground vibrational state of OCS yields the results $B_0 = 6081{\cdot}49$ MHz and $D_0 = 0{\cdot}00131$ MHz.

We can use this data to obtain information about the molecular force field. Details of the harmonic force field are obtainable directly from the centrifugal distortion constant D. According to (6.36a) we see that $\omega_e = (4B_e^3/D)^{1/2}$ and using our above values of the appropriate constants we find that $\omega = 802{\cdot}3\ \mathrm{cm}^{-1}$. The value observed from the electronic spectrum is $801{\cdot}95\ \mathrm{cm}^{-1}$.[213,118] As we also know α, (6.36b) indicates that we can use this value of ω to determine the cubic constant φ_3. In section A7 we use this data to determine φ_3 and thus draw the potential diagram shown in Figure 6.3. If in addition we determine the value of β in (6.70b) we could also obtain φ_4.[344,78,303] Note that we need both D and α to obtain the cubic constant. On the other hand we could have obtained ω by some other more direct method such as infra-red, or in this case analysis of the electronic spectrum.[118]

6.3 Expansion of the Vibration–Rotation Hamiltonian

We shall now see how the simple procedure used in the previous section for a diatomic molecule may be generalized to polyatomic molecules. The method essentially separates the exact Hamiltonian into a sum of various terms of successively smaller orders of magnitude. These individual terms in general correspond to various physically understood effects and are usually treated separately as and when it is necessary to take them into account. The treatment of the individual terms in order to determine the corresponding energy-level dependence can be quite complicated, and as an example in Section 6.5 the centrifugal distortion corrections are derived for a system possessing C_{3v} symmetry.

One approach to the expansion of the Hamiltonian using the VanVleck transformation which also includes a very general discussion of the ensuing terms has been given by Nielsen.[217,218] An alternative procedure suggested by Oka[224] will be utilized here which essentially consists of assigning orders of magnitude to the various terms in the expansion and determining their resulting effects by standard perturbation theory.

We note that it is $\mathbf{I}'$ (6.15) which must be inverted and expanded to generate the approximate Hamiltonian we require from (6.21) in an analogous way to the procedure in Section 6.2a. It is possible to simplify $\mathbf{I}'$ and the subsequent expansion procedure. Amat and Henry have shown that if (2.23) is substituted into (2.25) and account taken of the zeta sum rule[195] then $\mathbf{I}'$ as defined in (6.15) can be rewritten as[5]

$$I'_{\alpha\beta} = I^e_{\alpha\beta} + \sum_r a_r^{(\alpha\beta)} Q_r + \tfrac{1}{4} \sum_{rr'} a_r^{(\alpha\gamma)} I^{e\,-1}_{\gamma\delta}\, a_{r'}^{(\delta\beta)} Q_r Q_{r'} \qquad (6.40)$$

In the literature[227] this relation also is written as

$$I'_{\alpha\beta} = I^e_{\alpha\beta} + \sum_r a_r^{(\alpha\beta)} + \sum_{rr'} (A_{rr'}^{(\alpha\beta)} - \sum_{r''} \zeta_{rr''}^{(\alpha)} \zeta_{r'r''}^{(\beta)}) Q_r Q_{r'} \tag{6.40a}$$

where

$$a_r^{(\alpha\beta)} = (\partial I_{\alpha\beta}/\partial Q_r)_e \text{ and } (\partial^2 I_{\alpha\beta}/\partial Q_r \partial Q_{r'})_e = A_{rr'}^{(\alpha\beta)} + A_{r'r}^{(\alpha\beta)}$$

If we now define yet another inertia type quantity $\mathbf{I}''$ by the relation[318]

$$I''_{\alpha\beta} = I^e_{\alpha\beta} + \tfrac{1}{2} \sum_r a_r^{(\alpha\beta)} Q_r \tag{6.41}$$

then we can factorize (6.40) as

$$\mathbf{I}' = \mathbf{I}'' \mathbf{I}^{e-1} \mathbf{I}'' \tag{6.42}$$

which can be neatly inverted[318]

$$\boldsymbol{\mu} = \mathbf{I}''^{-1} \mathbf{I}^e \mathbf{I}''^{-1} \tag{6.43}$$

The matrix notation simplifies the final steps.

If $\mathbf{a}_r$ has elements $a_r^{(\alpha\beta)}$ and $\mathbf{b}_r = \mathbf{I}^{e-\frac{1}{2}} \mathbf{a}_r \mathbf{I}^{e-\frac{1}{2}}$ then we can introduce a useful quantity $\boldsymbol{\beta}$

$$\boldsymbol{\beta} = \sum_r \mathbf{b}_r Q_r \tag{6.44}$$

Using this definition, the inverse of $\mathbf{I}''$ defined in (6.41) may be written

$$\mathbf{I}''^{-1} = \mathbf{I}^{e-\frac{1}{2}} (1 + \tfrac{1}{2}\boldsymbol{\beta})^{-1} \mathbf{I}^{e-\frac{1}{2}} \tag{6.45}$$

Using this expression in (6.43) and substituting into the Hamiltonian (6.21) allows us to write

$$\boldsymbol{\mu} = \mathbf{I}^{e-\frac{1}{2}} (1 + \tfrac{1}{2}\boldsymbol{\beta})^{-2} \mathbf{I}^{e-\frac{1}{2}} \tag{6.46}$$

$$H_r = \tfrac{1}{2} \boldsymbol{\Phi}^\dagger (1 + \tfrac{1}{2}\boldsymbol{\beta})^{-2} \boldsymbol{\Phi} \tag{6.47}$$

where

$$\boldsymbol{\Phi} = \mathbf{I}^{e-\frac{1}{2}} (J - \boldsymbol{\pi}) \tag{6.47a}$$

and expanding the central matrix by the binomial expansion we obtain

$$H_r = \tfrac{1}{2} \boldsymbol{\Phi}^\dagger \boldsymbol{\Phi} - \tfrac{1}{2} \boldsymbol{\Phi}^\dagger \boldsymbol{\beta} \boldsymbol{\Phi} + \tfrac{3}{8} \boldsymbol{\Phi}^\dagger \boldsymbol{\beta}^2 \boldsymbol{\Phi} \ldots \tag{6.48}$$

as the expanded Hamiltonian in matrix form.

The first term of (6.48) contributes to a basis vibration–rotation Hamiltonian and subsequent terms are corrections to this zeroth-order expression. This is a convenient form because the orders of magnitude of the contributions are neatly labelled by the power of $\boldsymbol{\beta}$ which is of the same order of magnitude as the Born–Oppenheimer constant κ defined in (2.1). As δr_{el}, electronic displacements, are of the order of r_e, an equilibrium bond length, we can extend the approximation relation (2.1) to β by

$$\kappa \sim \frac{\delta r}{r_e} \sim \frac{aQ}{I^e} \sim \frac{\delta I}{I^e} \sim \beta \tag{6.49}$$

which follows from a comparison of (6.28) with (6.44). If we note that $\kappa \sim 0{\cdot}1$ we can tabulate the terms which are comprised in the expanded Hamiltonian (6.48) together with those of H_v according to their order of magnitude as has been done in Table 6.3. The most convenient form of H_v for our purposes is (see Sections 1.4a and 6.4b).

$$H = \tfrac{1}{2}\sum_r \omega_r(p_r^2 + q_r^2) + \tfrac{1}{6}\sum_{rr'r''} \varphi_{rr'r''}q_r q_{r'} q_{r''} + \tfrac{1}{24}\sum_{rr'r''r'''} \varphi_{rr'r''r'''}q_r q_{r'} q_{r''} q_{r'''} + \cdots \tag{6.50}$$

Table 6.3 Explicit Expressions for Some of the Terms in the Hamiltonian Expansion[a]

T_v	$\frac{1}{2}\sum_r \omega_r(p_r^2 + q_r^2)$
κT_v	$\frac{1}{6}\sum_{rr'r''} \varphi_{rr'r''}q_r q_{r'} q_{r''}$
$\kappa^2 T_v$	$\frac{1}{2}(J_\alpha - \pi_\alpha)^2/I^e_{\alpha\alpha} + \frac{1}{24}\sum_{rr'r''r'''} \varphi_{rr'r''r'''}q_r q_{r'} q_{r''} q_{r'''}$
$\kappa^3 T_v$	$-\frac{1}{2}[\sum_r a_r^{(\alpha\beta)}Q_r/I^e_{\alpha\alpha}I^e_{\beta\beta}](J_\alpha - \pi_\alpha)(J_\beta - \pi_\beta)$
$\kappa^4 T_v$	$\frac{3}{8}[(\sum_r a_r^{(\alpha\gamma)}Q_r)(\sum_r a_r^{(\gamma\beta)}Q_r)/I^e_{\alpha\alpha}I^e_{\gamma\gamma}I^e_{\beta\beta}](J_\alpha - \pi_\alpha)(J_\beta - \pi_\beta)$

[a] Quintic and higher-order vibrational terms have been omitted.

In this expression the first term will be used as a basis Hamiltonian, H_v^0, and the higher terms define the anharmonic force constants. Note that essentially $\varphi_{rr} \equiv \omega_r$.

It is convenient to set up a symbolic table for the terms as given in Table 6.4. Oka[224] indicates how this ordering procedure can be used as a guide in deciding which terms should be retained in the perturbation treatment for a particular correction. As an example, in the diatomic molecule case discussed in the previous section the terms retained in the perturbing Hamiltonian (6.33) are, according to Table 6.4, of order κ^3, κ^4 and κ respectively. The rigid-rotor term is of order κ^2 whereas the vibrational (harmonic) energy is of order κ^0; all other terms were neglected as they are of higher order. The general method has been applied by Oka to the problem of l-type doubling

Table 6.4 Symbolic Table of Contributions to the Vibration–Rotation Hamiltonian[a]

J^2	$-2J\pi$	π^2	H_v	$O(\kappa^n)$
			p^2+q^2	κ^0
			q^3	κ^1
J^2	Jpq	p^2q^2	q^4	κ^2
J^2q	Jpq^2	p^2q^3	q^5	κ^3
J^2q^2	Jpq^3	p^2q^4	q^6	κ^4

[a] At the top is given the origin of the terms, at the right-hand side the order in terms of κ the Born–Oppenheimer constant.

in symmetric top molecules.[224] In the next section we will follow this procedure and determine the centrifugal distortion corrections to the symmetric top energy levels.

6.4 General Vibration–Rotation Theory Procedure

The situation in the case of polyatomic molecules is rather more complicated than in the diatomic case. There are not only many more terms in the expanded Hamiltonian to be considered, but also, the possibility of degeneracy, or near degeneracy, causes resonances to occur leading to a breakdown of the perturbation procedure (Chapter 1). It is convenient to partition the complete Hamiltonian, H, which contains all the terms implied in Table 6.3 as

$$H = H_v^0 + H_r^0 + \sum_m H'_m \tag{6.51}$$

where $[H_v^0 + H_r^0] = H^0$ is the basis Hamiltonian and defines the representation $|[v][J]\rangle$ in which the matrix elements of H are to be written. Here $[v]$ and $[J]$ represent complete sets of vibrational and rotational specifications respectively.

$$\sum_m H'_m$$

contains all the terms which remain. Many are found to vanish because each term must transform as does the totally symmetric representation of the point group to which the molecule belongs. On evaluating the eigenvalues of H by perturbation theory we obtain a result which can be written as

$$E = \sum_{k=0}^{\infty} E^{(k)} + E^{(*)} \tag{6.52}$$

The term $E^{(*)}$ contains all contributions to E that, due to resonance conditions, must be considered separately. Terms of this type are considered in Section 6.8.

$$E^{(0)} = \langle[v]|H_v^0|[v]\rangle + \langle[J]|H_r^0|[J]\rangle \tag{6.53a}$$

$$E^{(1)} = \sum_m E_m^{(1)} = \sum_m \langle[v\,J]|H'_m|[v\,J]\rangle \tag{6.53b}$$

$$E^{(2)} = \sum_{mn} E_{mn}^{(2)} = \sum_{mn} \sum_{[v'J']} \frac{\langle[v\,J]|H_m|[v'\,J']\rangle\langle[v'\,J']|H_n|[v\,J]\rangle}{E_{[v\,J]}^{(0)} - E_{[v'J']}^{(0)}} \tag{6.53c}$$

The sum of terms in (6.52) can be rearranged into three sets.

$$E = E([\tilde{J}]) + E([\tilde{v}]) + E([\tilde{v}\,\tilde{J}]) \tag{6.54}$$

where $[\tilde{J}]$ and $[\tilde{v}]$ represent the sets of rotational and vibrational specifications of the corrected functions and eigenstates. E is now in the form of an analytical term formula, one part of which contains only rotational specifications, a second contains only vibrational ones and the third contains corrections which depend on both. With the aid of the selection rules of Chapter 3 we can generate from E, (6.54), an analytical frequency relation which can be fitted to empirically observed spectra. In this way we can determine the *spectroscopic parameters* such as B_v, α, ω and D etc., and from these work back through the results (6.54) → (6.53) → (6.52) → (6.51) to determine the *molecular parameters* such as r_e, φ_2 and φ_3 which have definite physical significance.

There is a final point that must be made, that as it stands the summation in (6.52) is infinite and must be truncated using some criteria or other. Here we shall use the order of magnitude guide suggested by Oka.[224] Thus we can write

$$O(E_{mn\ldots}^{(k)}) = \prod_{mn\ldots}^{k} O(H'_m)\{O(\Delta E_{mn})\}^{-1} \tag{6.55}$$

and neglect all terms whose contributions are small compared with the experimental accuracy of our measurement. It is also not always necessary to evaluate all the resulting terms because of the partitioning (6.54). In this book we are in general only interested in two terms in (6.54) and do not in general need to evaluate $E([\tilde{v}])$.‡

We have essentially applied this procedure in Section 6.2 in the analysis of the diatomic molecule: H^0 was given by (6.32) and defined the basis set.

H' was given by (6.33) and only those terms were included which yielded contributions to E of $O(\kappa^4)$ or greater. *We could have proceeded to higher-order terms had we wished.* This is done when highly accurate information is required and is experimentally accessible.

‡ Some care should be exercised here as indicated in Section 6.9.

Terms such as $|\langle v\ J|\frac{1}{6}\varphi_3\boldsymbol{q}^3|v'\ J'\rangle|^2/\Delta E$ were not evaluated as these contribute to $E([v])$ only in the vibrational correction. This term shifts one rotational manifold *en bloc* relative to the rest and must therefore be considered when vibrational transitions with $\Delta v \neq 0$ occur.

The order of magnitude of (6.34c) is given according to (6.55) as $O(E_{mn}^{(2)}) = \kappa \times \kappa^3 \times (\kappa^0)^{-1} = \kappa^4$. The individual selected contributions to E (6.52) are given by the three terms (6.34a, b and c).

The energy-level expression which resulted, (6.35), is in the required form (6.54). Using the selection rule $\Delta J = +1$ we generated the analytical frequency formula (6.39) which was fitted to the observed spectrum. As a result we were able to determine the *spectroscopic parameters* and use one of them, D, the centrifugal distortion constant to obtain the vibrational frequency ω which is directly related to φ_2 or k_2 the *molecular parameter.*

We should finally point out that this formal procedure is equivalent to the approach outlined by Wilson and Howard,[336] Nielsen[217,218] and others[211] who use the Van Vleck transformation (Section A8) or the analogous contact transformation. In this procedure the complete Hamiltonian is transformed to a new Hamiltonian, $\tilde{H}$, which is block diagonal in the vibrational quantum number. An analogous procedure is discussed in Section 10.5c. Such a partial diagonalization is feasible because the energy separations in the rotational state manifold differ from those in the vibrational manifold by at least an order of magnitude in general. The result is that one obtains an effective Hamiltonian for each vibrational state.

6.4a Vibrational Hamiltonian, H_v^0

We have already considered the rigid-rotor part of H^0 and thus it remains for us to consider H_v^0. The general solution has been discussed by many authors and in particular by Wilson, Decius and Cross.[335] Let us restrict the potential governing the vibrational motion to quadratic terms only. We can thus write the total energy as

$$E_v = \tfrac{1}{2}\sum_i m_i\dot{x}_i^2 + \tfrac{1}{2}\sum_{ij} k_{ij}x_ix_j \tag{6.56}$$

where the x_i are some set of generalized displacement coordinates. Then we can transform to a new set of coordinates, Q_r in which this system is diagonal, i.e. involves no cross terms. This is the definition of the normal coordinates, Q_r, for which

$$E_v = \tfrac{1}{2}\sum_r \dot{Q}_r + \tfrac{1}{2}\sum_r \lambda_r Q_r^2 \tag{6.56a}$$

and the resulting Hamiltonian is

$$H_v^0 = \tfrac{1}{2}\sum_r (P_r^2 + \omega_r^2 Q_r^2) \tag{6.57}$$

where $\omega_r^2 = \lambda_r$ and P_r are the momenta conjugate to Q_r. We have already discussed the evaluation of the individual terms in (6.57) and thus the solution of (6.57) can be written as

$$E([v_r]) = \sum_r \omega_r(v_r + \tfrac{1}{2}) \tag{6.58}$$

The resulting eigenfunctions can be written as

$$|[v_r]\rangle = \prod_r |v_r\rangle \tag{6.59}$$

This product is not always the most convenient representation in which to work. In some cases when we are interested in the properties of degenerate vibrational states it is useful to modify this expression.

Consider the doubly degenerate isotropic harmonic oscillator Hamiltonian (as implied in (1.9) we can transform to $p_r = P_r/\sqrt{\omega_r}$ and $q_r = \sqrt{\omega_r}Q_r$)

$$H = \tfrac{1}{2}\omega_t[(p_{t_1}^2 + p_{t_2}^2) + (q_{t_1}^2 + q_{t_2}^2)] \tag{6.60}$$

where the subscript t denotes a degenerate vibration. The solution of this type of equation is discussed by Moffitt and Liehr[201] and summarized in Section A5 where the appropriate matrix elements are tabulated. An important result is that an angular momentum operator, L, arises in a natural way out of the theory with the form

$$L = q_{t_1}p_{t_2} - q_{t_2}p_{t_1} \tag{6.61}$$

L commutes with H in (6.60) and thus the eigenfunctions of H are simultaneously eigenfunctions of L. We can thus specify the eigenfunctions as $|vl\rangle$ according to the result

$$L|v\,l\rangle = l|v\,l\rangle \tag{6.62}$$

The so called 'pure vibrational' function can thus possess angular momentum and clearly can yield a contribution to the overall angular momentum via π (see Section 6.1). Thus it is sometimes (though not always, Section 6.5) convenient to specify the wave functions by

$$|[v_r]\rangle = [\prod_s |v_s\rangle][\prod_t |v_t\,l_t\rangle] \tag{6.63}$$

where v_s represents the non-degenerate and $v_t l_t$ the doubly degenerate normal modes of vibration.

This point is important when we come to discuss linear and symmetric top molecules in degenerate vibrational states. In Figure 6.4 the bending vibrational states of a linear molecule are represented.

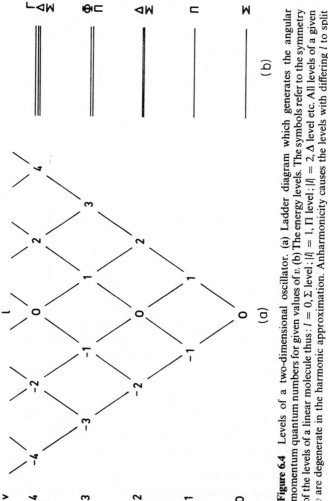

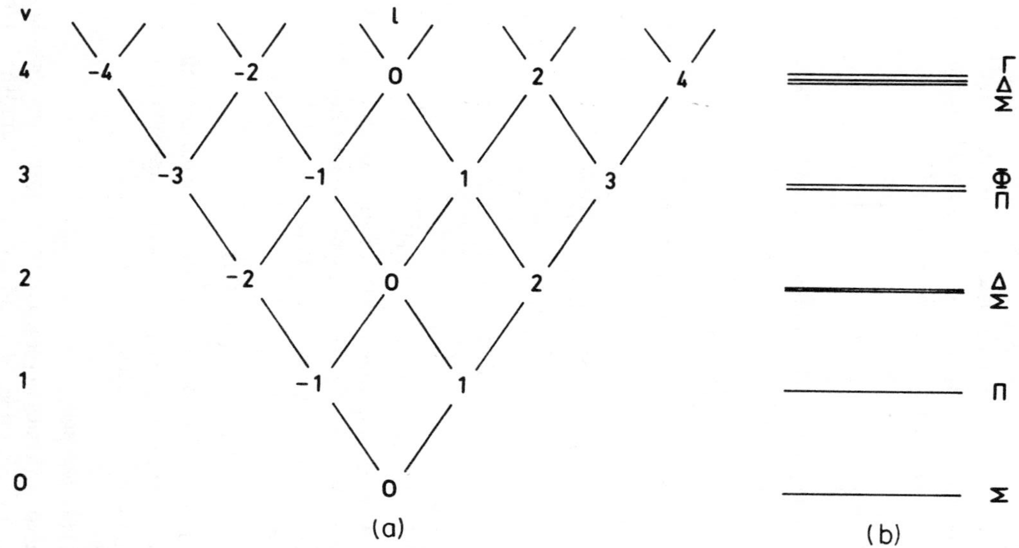

Figure 6.4 Levels of a two-dimensional oscillator. (a) Ladder diagram which generates the angular momentum quantum numbers for given values of v. (b) The energy levels. The symbols refer to the symmetry of the levels of a linear molecule thus: $l = 0$, Σ level; $|l| = 1$, Π level; $|l| = 2$, Δ level etc. All levels of a given v are degenerate in the harmonic approximation. Anharmonicity causes the levels with differing l to split apart as shown

6.4b *Vibrational Angular Momentum*

As a result of the discussion leading to (6.62) we see that vibrations can possess intrinsic rotational character. For instance a linear triatomic molecule in the $v_2 = 1$ state has $l = \pm 1$ and $J \geqslant |l|$ and thus in this 'pure vibrational' state the molecule cannot be described as 'not-rotating'! In the Hamiltonian Table (6.4) we see that the general term which represents this motion is $\boldsymbol{\pi}$ which was defined in (6.12). We see that if $q_r = \sqrt{\omega_r} Q_r$, then

$$\pi_\alpha = \sum_{rr'} \zeta_{rr'}^{(\alpha)} \left\{ \left[\frac{\omega_{r'}}{\omega_r} \right]^{1/2} q_r p_{r'} - \left[\frac{\omega_r}{\omega_{r'}} \right]^{1/2} q_{r'} p_r \right\} \tag{6.64}$$

and in the case of a degenerate vibration $\omega_r = \omega_{r'}$ and thus we see that the contribution to π_α from degenerate vibrations, π_α^*, is[218]

$$\pi_\alpha^* = \sum_t \zeta_t^{(\alpha)} L_t \tag{6.65}$$

where L_t is given by (6.61). In the case of a symmetric top we obtain a term of order κ^2

$$H' = -2(\tfrac{1}{2}\mu_{\alpha\alpha}^e J_\alpha \sum_t \zeta_t^{(\alpha)} L_t) \tag{6.66}$$

If the basis Hamiltonian is set up in the representation $|J^{2'}\, J_c'\, v_t\, l_t \ldots\rangle$ then we obtain a first-order coriolis contribution to the rotation energy and we can write

$$E(JKv_t l_t) = B_e J(J+1) + (A_e - B_e)K^2 - 2A_e \zeta_t K l_t \tag{6.67}$$

as the energy-level expression to first order of approximation for a prolate symmetric top in a state with vibrational angular momentum. Note that in the case of a linear molecule the terms in A do not occur in the Hamiltonian and can be dropped from (6.67). As $K = l_t$ and we obtain the result

$$E(Jl_t) = B_e[J(J+1) - l_t^2] \tag{6.68}$$

In the asymmetric rotor there is no vibrational degeneracy and thus only terms of the type (6.64) with $\omega_r \neq \omega_{r'}$ can occur. These give rise to second-order coriolis interactions and coriolis resonances.

6.4c *Energy-level Expressions*

It is convenient to work backwards from the energy-level expressions which have proved useful in describing molecular spectra. Again we shall neglect the resonance contributions, $E^{(*)}$. In the diatomic molecule case we saw that

$$E(vJ) = B_v\langle J^2\rangle - D_v\langle J^4\rangle + \cdots \tag{6.69}$$

where B_v and D_v are essentially given by (6.36a) and we have retained the v subscript on D here for completeness. One can extend the power series in

$(v + \frac{1}{2})$ to higher terms which may be necessary for a satisfactory fit to accurate experimental data (as is the case with AlF[344]), i.e.

$$B_v = B_e - \alpha(v + \tfrac{1}{2}) + \gamma(v + \tfrac{1}{2})^2 + \cdots \tag{6.70a}$$

$$D_v = D_e - \beta(v + \tfrac{1}{2}) + \cdots \tag{6.70b}$$

Terms in $H_v\langle \boldsymbol{J}^6 \rangle$ may be added to (6.69) also. In general the accuracy of microwave data, especially in the case of polyatomic molecules, is such that we are justified in dropping all terms other than B_e, α and D where the v subscript can be dropped.

In the asymmetric rotor case we cannot deduce simple analytical expressions to account for terms in $\boldsymbol{J}^4$ any more than we could in the case of the basic $\boldsymbol{J}^2$ terms. It is thus convenient to write the energy-level expression in terms of average values

$$\begin{aligned} E([v]J\tau) = {} & a_{[v]}\langle J_a^2 \rangle + b_{[v]}\langle J_b^2 \rangle + c_{[v]}\langle J_c^2 \rangle - d_J\langle \boldsymbol{J}^4 \rangle - d_{JK}\langle \boldsymbol{J}^2 J_c^2 \rangle \\ & - d_K\langle J_c^4 \rangle - d_{EJ}\langle H^0 \boldsymbol{J}^2 \rangle - d_{EK}\langle H^0 J_c^2 \rangle \end{aligned} \tag{6.71}$$

where $H^0 = a_{[v]}J_a^2 + b_{[v]}J_b^2 + c_{[v]}J_c^2$. The terms containing the d parameters are all of order $\boldsymbol{J}^4$ and are a set of five terms which Watson has shown[315] account satisfactorily for centrifugal distortion effects in asymmetric rotors (Section 6.5b). We can write

$$b_{[v]} = b_e - \sum_r \alpha_r^b(v_r + \tfrac{1}{2}) + \sum_{r \geqslant r'} \gamma_{rr'}^b(v_r + \tfrac{1}{2})(v_{r'} + \tfrac{1}{2}) + \cdots \tag{6.72}$$

Because of the possibility of degeneracy the symmetric rotor energy levels are rather different. The basis Hamiltonian is also as we have seen diagonal in K. In terms of average values we obtain

$$\begin{aligned} E([v]JK) = {} & b_{[v]}\langle \boldsymbol{J}^2 \rangle + (c_{[v]} - b_{[v]})\langle J_c^2 \rangle \\ & - D_J\langle \boldsymbol{J}^4 \rangle - D_{JK}\langle \boldsymbol{J}^2 J_c^2 \rangle - D_K\langle J_c^4 \rangle \\ & - \sum_t [2c\zeta_t\langle J_c L_c \rangle - \eta_{tJ}\langle \boldsymbol{J}^2 J_c L_c \rangle \\ & - \eta_{tK}\langle J_c^3 L_c \rangle] + E^{(*)} + \cdots \end{aligned} \tag{6.73}$$

where

$$b_{[v]} = b_e - \sum_r \alpha_r^b\left(v_r + \frac{d_r}{2}\right) + \cdots \tag{6.73a}$$

The terms in square brackets occur only when degenerate vibrations are excited and $L_c \neq 0$. The η terms which can be thought of as centrifugal distortion corrections to the first-order coriolis term were first discussed by Maes.[189] The three centrifugal distortion correction terms involving D coefficients were first derived by Slawsky and Dennison[278] and are discussed in detail in Section 6.5a. The Hamiltonian, apart from the terms corresponding to $E^{(*)}$ is diagonal in J and K, and also in L_c as discussed in Section 6.4c, and thus (6.73) is easily evaluated in terms of these quantum numbers. The $E^{(*)}$ terms are usually most important when $l_c = K = \pm 1$ and are the l-doubling terms to be discussed in Section 6.8.

In the linear molecule case we find that (6.73) reduces to

$$E([v]Jl) = B_{[v]}[\langle J^2 \rangle - \langle J_c^2 \rangle] - D_J[\langle J^2 \rangle - \langle J_c^2 \rangle]^2 + E^{(*)} + \cdots \quad (6.74)$$

6.5 Centrifugal Distortion in Polyatomic Molecules

We are now in a position to determine the ways in which the energy-level expressions for the rigid rotor are modified when the correction terms for non-rigidity given in Table 6.3 are taken into account. We take as an example here the terms which can be related to centrifugal distortion and yield correction terms which depend only on the rotational quantum numbers and not on the vibrational quantum numbers. The $DJ^2(J + 1)^2$ term in (6.34) is of this type. We shall first determine the general centrifugal distortion term in the Hamiltonian and then proceed to calculate the resulting energy-level expression for a C_{3v} symmetric top molecule. This problem was first studied by Slawsky and Dennison.[278]

We can use the experience gained in the previous diatomic procedure to guide us in isolating the term we need. In this case we saw that a second-order perturbation contribution of the form $(-\frac{1}{2}\mu \boldsymbol{J}^2 \boldsymbol{q})^2/\Delta E$ (in the nomenclature of Table 6.4) gave rise to a term of the type $DJ^2(J + 1)^2$ of order κ^6. Terms of this type occur in the general polyatomic case, and as indicated in Tables 6.3 and 6.4 they are

$$-\tfrac{1}{2}\Big(\sum_r a_r^{(\alpha\beta)} Q_r / I_{\alpha\alpha}^e I_{\beta\beta}^e\Big) J_\alpha J_\beta \quad (6.75)$$

If we expand the expression for $\boldsymbol{\mu}$ as given in (6.46) we can obtain the expression for the derivatives of $\boldsymbol{\mu}$[318] and in particular we see that

$$u_r^{(\alpha\beta)} = \left[\frac{\partial \mu^{(\alpha\beta)}}{\partial Q_r}\right]_e = -\frac{a_r^{(\alpha\beta)}}{I_{\alpha\alpha}^e I_{\beta\beta}^e} \quad (6.76)$$

This allows us to simplify the notation in (6.75)

$$\tfrac{1}{2}\sum_r u_r^{(\alpha\beta)} Q_r J_\alpha J_\beta \quad (6.77)$$

If we substitute $Q_r = q_r/\sqrt{\omega_r}$ into (6.77) and take advantage of the fact that rotational energy differences are in general much smaller than vibrational intervals (i.e. set $E_{vJ} - E_{v'J'} = E_v - E_{v'}$) then the second-order perturbation correction term in the Hamiltonian can be written as

$$H' = \tfrac{1}{4}\sum_r \sum_{v_r'} \left[\frac{u_r^{(\alpha\beta)} u_r^{(\gamma\delta)}}{\omega_r}\right]\left[\frac{\langle v_r|q_r|v_r'\rangle\langle v_r'|q_r|v_r\rangle}{E_{v_r} - E_{v_r'}}\right] J_\alpha J_\beta J_\gamma J_\delta \quad (6.78)$$

essentially carrying out the vibrational diagonalization first. The bracketed term on evaluation becomes $-1/(2\omega_r)$ (Section A4) and thus we obtain the expression originally given by Wilson and Howard.[336]

$$H' = \tfrac{1}{4}\tau_{\alpha\beta\gamma\delta}J_\alpha J_\beta J_\gamma J_\delta \tag{6.79}$$

where

$$\tau_{\alpha\beta\gamma\delta} = -\tfrac{1}{2}\sum_r u_r^{(\alpha\beta)}u_r^{(\gamma\delta)}/\omega_r^2 \tag{6.80}$$

Compare these results (6.75) to (6.80) to the results for the diatomic molecule (6.33) and (6.34a). From (6.80) we see that the following relations hold among the τ coefficients:

$$\tau_{\alpha\beta\gamma\delta} = \tau_{\alpha\beta\delta\gamma} = \tau_{\beta\alpha\gamma\delta} = \tau_{\beta\alpha\delta\gamma} = \tau_{\gamma\delta\alpha\beta} = \tau_{\gamma\delta\beta\alpha} = \tau_{\delta\gamma\alpha\beta} = \tau_{\delta\gamma\beta\alpha} \tag{6.81}$$

In the case of a polyatomic molecule the evaluation of the matrix elements (6.79) can involve summing over a large number of terms. The procedure is facilitated by taking advantage of symmetry restrictions which cause most of them to vanish.

In the previous section we saw how we could determine the vibrational frequency for AlF from the centrifugal distortion correction. In general one can determine information about the harmonic force field from centrifugal distortion analyses. In the case of such molecules as OF_2, it may be completely determined.[238,100] For this purpose it is sometimes convenient to write the τs in terms of some set of coordinates, R, other than normal coordinates. Usually these are internal valence coordinates for which we have a feel for the force constants. In this case τ can be written as

$$\tau_{\alpha\beta\gamma\delta} = -\frac{1}{2I_{\alpha\alpha}^e I_{\beta\beta}^e I_{\gamma\gamma}^e I_{\delta\delta}^e}\sum_{mn}\frac{\partial I_{\alpha\beta}}{\partial R_m}\frac{\partial I_{\gamma\delta}}{\partial R_n}(f^{-1})_{mn} \tag{6.82}$$

where R_m and R_n are two internal valence coordinates and f_{mn} is a force constant in the *harmonic* potential energy defined by

$$V = \tfrac{1}{2}\sum_{mn} f_{mn}R_m R_n$$

as in (6.56). This expression for τ is obtained by first substituting (6.76) into (6.80) to obtain τ in terms of the derivatives of $\mathbf{I}$ rather than $\boldsymbol{\mu}$. If we define the coordinate $\boldsymbol{R}$ by $R_m = L_{mr}Q_r$ where $\mathbf{L}$ is a linear transformation matrix, then we see that its elements are $L_{mr} = \partial R_m/\partial Q_r$. This allows us to substitute for the terms $a_r^{(\alpha\beta)}$ as

$$a_r^{(\alpha\beta)} = \sum_m \frac{\partial I_{\alpha\beta}}{\partial R_m}\frac{\partial R_m}{\partial Q_r} = \sum_m \frac{\partial I_{\alpha\beta}}{\partial R_m}L_{mr} \tag{6.83}$$

The resulting expression for τ thus can be written

$$\tau_{\alpha\beta\gamma\delta} = \frac{-1}{2I^e_{\alpha\alpha}I^e_{\beta\beta}I^e_{\gamma\gamma}I^e_{\delta\delta}} \sum_{mn} \frac{\partial I_{\alpha\beta}}{\partial R_m} \frac{\partial I_{\gamma\delta}}{\partial R_n} \sum_r L_{mr} \frac{1}{\omega_r^2} L_{nr} \tag{6.84}$$

From the definition of normal coordinates as given by relation (6.57) we see that

$$(f^{-1})_{mn} = \sum_r L_{mr} \frac{1}{\omega_r^2} L_{nr} \tag{6.85}$$

thus yielding required form of τ (6.82).

6.5a *Centrifugal Distortion in Symmetric Top Molecules*

Symmetry allows us to show that many of the τs are zero and that others are related. In fact we shall see that for C_{3v} molecules there are only six independent τ coefficients. As an example it is clear that during a totally symmetric (a_1) vibration of CH_3F the axes of the instantaneous inertia tensor cannot rotate. There will thus be no contributions to the off-diagonal elements of $\mathbf{I}$ from a_1 vibrations, i.e., $a_s^{(\alpha\beta)} = (\partial I_{\alpha\beta}/\partial Q_s)_e = 0$, if $\alpha \neq \beta$ (s represents an index for summation over symmetric vibrations). Clearly $u_r^{(\alpha\beta)}$ under these conditions is also zero. Secondly it is clear that as C_{3v} symmetry is preserved during these vibrations that $u_s^{(aa)} = u_s^{(bb)}$ (the c-axis being the symmetry axis).

A systematic symmetry procedure utilizes the general rule that the Hamiltonian must be invariant to any symmetry operation of the point group to which the molecule in its equilibrium geometry belongs (i.e. transform as A_1). In this case this rule allows us to go back and determine which of the numerous coefficients $u_r^{(\alpha\beta)}$ in the correction Hamiltonian matrix (6.79) are non-zero by determining which of the many combinations $Q_r J_\alpha J_\beta$ are or are not of A_1 species. Those which do not possess the correct symmetry may thus be systematically removed.

One could apply C_{3v} symmetry operations to (6.77) directly as outlined by Amat and Henry.[6] This procedure however involves triple products of the type $E \times E \times E$ in cases where J_a, J_b and a degenerate vibration Q_t are involved. Oka[224] uses rather more convenient operators which can be classified under C_3. The ones used here for degenerate operators are

$$Q^{\pm} = Q_1 \pm iQ_2 \tag{6.86a}$$

$$J^{\pm} = J_a \pm iJ_b \tag{6.86b}$$

where Q_1 and Q_2, the degenerate normal coordinates behave like T_b and T_a respectively (note that this choice differs from that of Oka). If we define the symmetry operation C_3 according to the discussion in section (3.11b) then with $e^{2\pi i/3} = \varepsilon$ we see that

$$C_3Q^{\mp} = \varepsilon^{\mp 1}Q^{\mp} \tag{6.87a}$$

$$C_3J^{\pm} = \varepsilon^{\mp 1}J^{\pm} \tag{6.87b}$$

If we also single out the (ca) plane as typical we find on application of $\sigma_v(ca)$ that

$$\sigma_v Q^{\pm} = -Q^{\mp} \tag{6.88a}$$

$$\sigma_v J^{\pm} = -J^{\mp} \tag{6.88b}$$

The simplification afforded by using the above symmetry operators may now be displayed. For instance several of the terms required in the summation (6.77) are contained in the product $Q^+J^+J^+$. On application of C_3 we see that

$$C_3Q^+J^+J^+ \rightarrow \varepsilon^{-1}Q^+J^+J^+ \tag{6.89}$$

These terms may not therefore contribute to H' whereas the term $Q^-J^+J^+$ can because

$$C_3Q^-J^+J^+ \rightarrow Q^-J^+J^+ \tag{6.90}$$

as $\varepsilon^{-3} = 1$. A convenient systematization of the application of C_3 is given in Table 6.5.

Table 6.5 Table of the Species of Operator Products under C_3

Q_s	J^+	J^-	J_c	Q^+	J^+	J^-	J_c	Q^-	J^+	J^-	J_c
J^+	E	A	E	J^+	E	E	A	J^+	A	E	E
J^-	A	E	E	J^-	E	A	E	J^-	E	E	A
J_c	E	E	A	J_c	A	E	E	J_c	E	A	E

Having determined those operator products which have A species under C_3 it only remains to form those linear combinations which possess A_1 symmetry on application of $\sigma_v(ca)$. Thus for instance

$$\sigma_v(Q^-J^+J^+ - Q^+J^-J^-) \rightarrow (-Q^+J^-J^- + Q^-J^+J^+) \tag{6.91}$$

Only four sets survive this procedure and they are tabulated in Table 6.6.

We can expand the linear combinations to find the relationships between the coefficients $u_r^{(\alpha\beta)}$. For instance if we expand the third term in Table 6.6 we obtain

$$\tfrac{1}{2}(Q^+J^-J^- - Q^-J^+J^+) = iQ_{t_2}(J_a^2 - J_b^2) - iQ_{t_1}(J_aJ_b + J_bJ_a) \tag{6.92}$$

where Q_{t_1} and Q_{t_2} are the two components of the degenerate vibration t.

Table 6.6 A_1 Linear Combinations of the Operator Products of Table 6.5 with A_1 Symmetry

$u_s^{(aa)}$	$\frac{1}{2}Q_s(J^+J^- + J^-J^+) = Q_s(J_a^2 + J_b^2) = Q_s(J^2 - J_c^2)$
$u_s^{(cc)}$	$Q_sJ_c^2$
$u_{t_1}^{(aa)}$	$\frac{1}{2}(Q^+J^-J^- - Q^-J^+J^+) = iQ_{t_2}(J_a^2 - J_b^2) - iQ_{t_1}(J_aJ_b + J_bJ_a)$
$u_{t_1}^{(ac)}$	$\frac{1}{2}(Q^+J^+J_c - Q^-J^-J_c + Q^+J_cJ^+ - Q^-J_cJ^-) = iQ_{t_1}(J_bJ_c + J_cJ_b)$
	$+ iQ_{t_2}(J_aJ_c + J_cJ_a)$

This indicates that the following relationship holds between the u coefficients

$$-u_{t_2}^{(aa)} = u_{t_2}^{(bb)} = u_{t_1}^{(ab)} = u_{t_1}^{(ba)} \tag{6.93}$$

The complete set of relations is given in Table 6.7.

Table 6.7 Non-vanishing Coefficients, $u_r^{(\alpha\beta)}$, and Relations

$u_s^{(aa)} = u_s^{(bb)}$	$u_s^{(cc)}$
$-u_{t_2}^{(aa)} = u_{t_2}^{(bb)} = u_{t_1}^{(ab)} = u_{t_1}^{(ba)}$	
$u_{t_1}^{(bc)} = u_{t_2}^{(ac)} = u_{t_1}^{(cb)} = u_{t_2}^{(ca)}$	

We can note that the first relation $u_s^{(aa)} = u_s^{(bb)}$ is one of the results we determined originally on the basis of simple qualitative symmetry considerations.

We have now essentially determined which terms of the type (6.77) are non-vanishing in our expansion (6.48). The next step is to evaluate the second-order perturbation terms (6.78) which come from all possible cross-products among the four non-vanishing sets in Table 6.6. The summation over symmetric vibrations are particularly simple as the individual terms are already diagonal in J and K. For instance, the second term from Table 6.6 yields, according to (6.80)

$$\tfrac{1}{4}\tau_{cccc}J_c^4 \tag{6.94}$$

The others are given in Table 6.8. The summation over the degenerate vibrational contributions are more cumbersome to evaluate.

Table 6.8 Complete Set of Diagonal Operator products[a]

$u_s^{(aa)}u_s^{(aa)}$	$(J^2 - J_c^2)^2$	J^4	$-2J^2J_c^2$	$+J_c^4$		
$u_s^{(cc)}u_s^{(cc)}$	J_c^4			$+J_c^4$		
$u_s^{(aa)}u_s^{(cc)}$	$2(J^2 - J_c^2)J_c^2$		$+2J^2J_c^2$	$-2J_c^4$		
$u_{t_1}^{(aa)}u_{t_1}^{(aa)}$	$(J_a^2 - J_b^2)^2 + (J_aJ_b + J_bJ_a)^2$	J^4	$-2J^2J_c^2$	$+J_c^4$	$-2J^2$	$+5J_c^2$
$u_{t_2}^{(ac)}u_{t_2}^{(ac)}$	$(J_bJ_c + J_cJ_b)^2 + (J_aJ_c + J_cJ_a)^2$		$+4J^2J_c^2$	$-4J_c^4$	$+J^2$	$-5J_c^2$

[a] The off-diagonal term in τ_{abbc} has been omitted. Note that the first three terms in $(u_{t_1}^{(aa)})^2$ contribute to τ_{aaaa} and the last two to τ_{abab}. In evaluating (6.78) note that we must take matrix element products such as $H_{mn}H_{nm}$.

We should note that the products between the third and fourth terms in Table 6.6 will involve products in components which are *off-diagonal* in K of the type $J^+J^+J^+J_c$; they will yield matrix elements between K and $K \pm 3$. We can neglect these off-diagonal terms in the summation, as they will yield much smaller contributions than the others. These terms were in fact discussed by Slawsky and Dennison[278] in their original treatment. The fact that the ground-state symmetric rotor Hamiltonian is *not quite* diagonal up to terms of $O(J^4)$ appears to have been forgotten until it was shown that such off-diagonal terms can split the $K = \pm 2$ degeneracy in C_{4v} molecules[6] and in C_{3v} systems give rise to transitions with $\Delta K = 3\,(\Delta|K| = 1)$[231] and have been observed by Chu and Oka.[47a]

We can sum over Q_{t_1} and Q_{t_2} separately for the third term to obtain

$$-\tfrac{1}{8}\sum_t (u_{t_1}^{(aa)}/\omega_t)^2[(J_a^2 - J_b^2)^2 + (J_aJ_b + J_bJ_a)^2] \qquad (6.95)$$

This term can be reduced by careful application of the commutation relations (3.30) to yield the diagonal matrix element

$$-\tfrac{1}{8}\sum_t (u_{t_1}^{(aa)}/\omega_t)^2\langle J\,K|\mathbf{J}^4 - 2\mathbf{J}^2J_c^2 + J_c^4 + 5J_c^2 - 2\mathbf{J}^2|J\,K\rangle \qquad (6.96)$$

A similar reduction must be carried out for the fourth term in τ_{acac}.

By equating coefficients we find that the rotational term values of a particular vibrational state can be written as

$$\begin{aligned} E(JK) = {} & B'J(J+1) + (A' - B')K^2 \\ & - D_JJ^2(J+1)^2 - D_{JK}J(J+1)K^2 - D_KK^4 \end{aligned} \qquad (6.97)$$

where

$$\begin{aligned} A' &= A + \tfrac{3}{4}\tau_{abab} - \tau_{acac} \\ B' &= B - \tfrac{1}{2}\tau_{abab} + \tfrac{1}{4}\tau_{acac} \\ D_J &= -\tfrac{1}{4}\tau_{aaaa} \\ D_{JK} &= +\tfrac{1}{2}\tau_{aaaa} - \tfrac{1}{2}\tau_{aacc} - \tau_{acac} \\ D_K &= -\tfrac{1}{4}\tau_{aaaa} - \tfrac{1}{4}\tau_{cccc} + \tfrac{1}{2}\tau_{aacc} + \tau_{acac} \end{aligned} \qquad (6.98)$$

This expression for the rotational energy is corrected for the first-order centrifugal distortion contribution. From Table 6.6 we see that terms in τ_{aabc} can contribute (in *second* order) but as they are off-diagonal in K by $\Delta K = \pm 3$ they can be neglected. Amat and Henry[6] have drawn attention to the particularly interesting case of C_{4v} molecules where the off-diagonal terms include matrix elements with $\Delta K = \pm 4$. In this case the degenerate pair with $|K| = 2$ are connected. This degeneracy should thus be split even in the ground vibrational state and has been observed in BrF_5.[22a]

There are some important points that we should note about (6.97): (i) calculation yields a correction to A and B which for accurate work must be taken into account, (ii) the energy-level expression to this degree of approximation depends on only five parameters which are linear combinations of seven terms (6.98), thus in general not all the τs may be determined from the rotational data alone.

In the case of planar symmetric top molecules $(\partial I_{\alpha\beta}/\partial Q_r)_e = 0$ when $\alpha = a$ or b and $\beta = c$ causing τ_{acac} to vanish. Also $\tau_{cccc} = \frac{1}{2}\tau_{aacc}$. With these simplifications we find that

$$\begin{aligned} D_J &= -\tfrac{1}{4}\tau_{aaaa} \\ D_{JK} &= \tfrac{1}{2}\tau_{aaaa} - \tau_{cccc} \\ D_K &= \tfrac{3}{4}\tau_{cccc} - \tfrac{1}{4}\tau_{aaaa} \end{aligned} \tag{6.99}$$

which yields the relation[71]

$$D_{JK} = -\tfrac{2}{3}(D_J + 2D_K) \tag{6.100}$$

between the centrifugal distortion coefficients.

The relation for the diatomic molecule has already been derived in Section 6.2. For a linear molecule the same energy-level expression applies. D is however rather more complicated to derive. For an XYZ type linear triatomic molecule such as HCN the centrifugal distortion constant[214] can be written as

$$D = 4B_e^3\left[\frac{\zeta_{21}^2}{\omega_3^2} + \frac{\zeta_{23}^2}{\omega_1^2}\right] \tag{6.101}$$

Watson[320] has discussed the consequences of centrifugal distortion in spherical top molecules on the rotational eigenstates. This mechanism causes a large enough distortion from spherical symmetry that the so-induced dipole moment in methane is sufficient to allow electric dipole transitions to be detectable. Rosenberg, Ozier and Kudian have observed the pure rotational infra-red spectra of CH_4 type molecules[260,261,262,232] and Curl and Oka have observed such a transition using an infra-red microwave double resonance technique.[60a]

6.5b *Centrifugal Distortion in Asymmetric Rotors*

The centrifugal distortion correction procedure in asymmetric rotor molecules turns out to be unusually complicated. The problem was first discussed by Kivelson and Wilson.[162] Complications arise however when the Kivelson and Wilson energy-level expressions are applied to the experimentally observed spectroscopic data.[76,77] The source of the complication has been identified by Watson[314] and the situation has now been clarified.[315,316]

Wilson and Howard[336] showed that the general centrifugal distortion correction (to order $\boldsymbol{J}^4$) could be written as in (6.79). This is called a *general form.*[315] Application of D_2 symmetry operations in a similar way to the procedure in Section 6.4a indicates that all but twenty-one τs in (6.79) vanish.[162] These are $\tau_{\alpha\alpha\alpha\alpha}$, $\tau_{\alpha\alpha\beta\beta}$, $\tau_{\alpha\beta\alpha\beta}$ and $\tau_{\alpha\beta\beta\alpha}$ ($\alpha \neq \beta$). The energy contributions of these τs are not all linearly independent as application of the commutation relations (3.30) allows us to rearrange (6.79) and write the rotational Hamiltonian as:[162]

$$H_r = \tfrac{1}{2}\mu'_\alpha J_\alpha^2 + \tfrac{1}{4}\tau'_{\alpha\alpha\beta\beta} J_\alpha^2 J_\beta^2 \tag{6.102}$$

where

$$\begin{aligned} \tau'_{\alpha\alpha\alpha\alpha} &= \tau_{\alpha\alpha\alpha\alpha} \qquad & \tfrac{1}{2}\mu'_\alpha &= \tfrac{1}{2}\mu_\alpha - \tfrac{1}{2}\tau_{\alpha\beta\alpha\beta} - \tfrac{1}{2}\tau_{\alpha\gamma\alpha\gamma} + \tfrac{3}{4}\tau_{\beta\gamma\beta\gamma} \\ \tfrac{1}{2}\mu_\alpha &= \frac{1}{2I_\alpha} \qquad & \tau'_{\alpha\alpha\beta\beta} &= \tau_{\alpha\alpha\beta\beta} + 2\tau_{\alpha\beta\alpha\beta} \quad (\alpha \neq \beta) \end{aligned} \tag{6.103}$$

Note how this simple procedure has injected τ contributions into the 'rigid-rotor' part of H. After this rearrangement only six τs remain *apparently* independent: τ'_{aaaa}, τ'_{bbbb}, τ'_{cccc}, τ'_{aabb}, τ'_{aacc} and τ'_{bbcc}. We now need to evaluate the diagonal matrix elements of (6.102) in the $|J\tau\rangle$ scheme to obtain an expression which can be fitted to the experimentally observed energy-level pattern.

The first problem with which we should concern ourselves is to determine the number of independent parameters on which the energy-level *expression* depends. In the case of the non-planar symmetric top $E(J, K)$ according to (6.98) depends on five. In the case of a non-planar asymmetric rotor Watson has shown that the following general rule applies: If $H = H(\boldsymbol{J}^2) + H(\boldsymbol{J}^4) + H(\boldsymbol{J}^6) + \cdots$ then the number of parameters (coefficients of J terms) on which *the energy levels depend* is $3 + 5 + 7 + \cdots$ respectively. Thus the rigid-rotor expressions depend on three, which we already knew, and if we include centrifugal distortion up to $\boldsymbol{J}^4$ then there are only eight,[315] not nine as is implied by (6.102). This problem showed up as an ill-conditioning when the Kivelson and Wilson expressions were applied to the spectra of $(CH_3)_2S$ and $(CH_3)_2O$ by Dreizler, Dendl and Rudolph.[76,77] Previous attempts to fit these expressions had been limited to planar molecules for which planarity conditions reduced the number of centrifugal distortion parameters from six to four[71] and thereby glossed over the difficulty.

The general procedure starts from a general Hermitian expression for the rotational Hamiltonian of the type

$$H = \sum_{pqr} h_{pqr}(J_a^p J_b^q J_c^r + J_c^r J_b^q J_a^p) \tag{6.104}$$

called a *standard form.*[315] Equation (6.102) is in such standard form whereas (6.79) is not. The h coefficients have the following properties: (i) they vanish when $p + q + r = n$ is odd, (ii) in general $h(n = 2) \gg h(n = 4) \gg h(n = 6) \gg \cdots$ causing (6.104) to converge rapidly.

The rigid-rotor Hamiltonian (2.33) is a *general form* and if we note that μ is symmetric it is also in standard form. By application of a rotation of reference axes the off-diagonal elements of μ vanish leaving a Hamiltonian with only three independent parameters μ_A, μ_B and μ_C, and called by Watson a *reduced form*. The reduced form contains only as many independent parameters as the correct energy-level expressions. The reduced rigid-rotor Hamiltonian was derived from the standard form by a unitary transformation which in this case is a simple rotation of axes. This procedure has been generalized to higher-order terms by Watson by finding unitary transformations which will reduce (6.104). There is no unique way of doing this but one which complements the rigid-rotor Hamiltonian (3.53) can be written in a convenient form for setting up the Hamiltonian in a symmetric rotor basis:

$$\begin{aligned} H_r = {} & \tfrac{1}{2}(a + b)\mathbf{J}^2 + [c - \tfrac{1}{2}(a + b)]J_c^2 - \Delta_J(\mathbf{J}^2)^2 - \Delta_{JK}\mathbf{J}^2 J_c^2 - \Delta_K J_c^4 \\ & + (J_a^2 - J_b^2)[\tfrac{1}{4}(a - b) - \delta_J \mathbf{J}^2 - \delta_K J_c^2] \\ & + [\tfrac{1}{4}(a - b) - \delta_J \mathbf{J}^2 - \delta_K J_c^2](J_a^2 - J_b^2) \end{aligned} \qquad (6.105)$$

If we write $H_0 = aJ_a^2 + bJ_b^2 + cJ_c^2$ then (6.105) can be written as[315]

$$\begin{aligned} H = {} & H_0 - d_J(\mathbf{J}^2)^2 - d_{JK}\mathbf{J}^2 J_c^2 - d_K J_c^4 - d_{EJ} H_0 \mathbf{J}^2 \\ & - \tfrac{1}{2} d_{EK}(H_0 J_c^2 + J_c^2 H_0) \end{aligned} \qquad (6.106)$$

where the coefficients, d, are related to the coefficients $a \ldots \Delta \ldots \delta \ldots$.[315] We can evaluate this expression in the $|J\,\tau\rangle$ basis in which H_0 is diagonal, thus if $E_0 = \langle J\,\tau|H_0|J\,\tau\rangle$ we obtain

$$\begin{aligned} E = {} & E_0 - d_J J^2(J + 1)^2 - d_{JK} J(J + 1)\langle J_c^2\rangle - d_K\langle J_c^4\rangle \\ & - d_{EJ} E_0 J(J + 1) - d_{EK} E_0 \langle J_c^2\rangle \end{aligned} \qquad (6.107)$$

This form is convenient for treating the centrifugal distortion effects as first-order correction terms to the unperturbed rigid-rotor energy E_0. Watson[316] and Kirchhoff[160] have discussed the application of these results to experimentally observed data in detail. The relations between the various Δ, δ and d coefficients are given by Watson. Dowling[71] has discussed the planarity conditions and Pierce, DiCianni and Jackson have obtained the force field of F_2O purely from the rotational spectrum.[238] The derivation of force-field information from centrifugal distortion data has been discussed by several authors.[332,100] Sextic coefficients have also been discussed.[317,42]

6.6 The Interpretation of Vibration–Rotation α Constants

In the asymmetric rotor where there are no degenerate vibrational modes, a calculation using perturbation theory in essentially the same way as in

Sections 6.2a and 6.4 yields the following expression for α_r^b the coefficient of $(v_r + \frac{1}{2})J(J + 1)$ to $O(\kappa^4)$

$$\alpha_r^b = -\frac{2b^2}{\omega_r}\left[\sum_\alpha \frac{3(a_r^{b\alpha})^2}{4I_\alpha} + \sum_{r'} (\zeta_{rr'}^{(b)})^2 \frac{3\omega_r^2 + \omega_{r'}^2}{\omega_r^2 - \omega_{r'}^2} + \frac{1}{2}\sum_{r'} \varphi_{rrr'} a_{r'}^{(bb)}\left(\frac{\omega_r}{\omega_{r'}^{3/2}}\right)\right] \quad (6.108)$$

This α term can be considered as the sum of three terms as implied by the order in (6.108) as

$$\alpha_r = \alpha_r^{(\text{har})} + \alpha_r^{(\text{cor})} + \alpha_r^{(\text{anh})} \quad (6.109)$$

Although there are problems (as discussed below) associated with how strictly we interpret these superscripts they are nevertheless a useful aid in discussing the contributions to α.

$\alpha^{(\text{har})}$ corresponds to the first term in (6.108) and is the 'so-called' direct harmonic contribution; it originates from the first-order perturbation correction involving the term $\frac{3}{8}u^2\boldsymbol{J}^2\boldsymbol{q}^2$ of $O(\kappa^4)$ in Table 6.4. It is evaluated in essentially the same way as the result (6.34b) for the diatomic molecule. By taking advantage of the general sum rule[227]

$$\sum_r (a_r^{(\alpha\alpha)})^2 = 4I_{\alpha\alpha}$$

which in this simple case is easily evaluated, we can in fact obtain (6.36) directly.

$\alpha^{(\text{cor})}$ corresponds to the second-order coriolis contribution and originates from a second-order perturbation treatment of the term $-\mu_e \boldsymbol{J}\boldsymbol{\pi}$ in Table 6.4. The perturbation energy correction is of $O(\kappa^2 \times \kappa^2) = O(\kappa^4)$. As there are no degenerate vibrations the vibrational angular momentum is given by (6.64) which only in cases of accidental degeneracy when $\omega_r \sim \omega_{r'}$ reduces to (6.65). In these cases we get coriolis resonances and a direct diagonalization must be carried out and the contribution included in $E^{(*)}$.

$\alpha^{(\text{anh})}$ corresponds to a contribution from the cubic anharmonicity of the vibrational motion. This term is essentially identical with that derived for the diatomic molecule except that we must now sum over all the cubic constants including those of the type $\varphi_{rrr'}$.

The corresponding α expressions for a symmetric rotor are essentially of a similar form[197] though rather more involved due to the complications caused by degenerate vibrational modes.

The α relations for linear molecules, particularly triatomic ones, are worthy of some comment because these systems have been the prototypes for most vibration–rotation calculations. The problem was first discussed by Adel and Dennison[1] and reviewed by Dennison.[64] Further treatments have been given by Nielsen,[214] Dorman and Lin[69] and Nakagawa and Morino.[211] All these studies used a *basis* vibrational Hamiltonian in which the displacements were assumed to be rectilinear. In the case of the bending frequency of an unsymmetric triatomic molecule α_2 is given by

$$\alpha_2 = \frac{B_e^2}{\omega_2}\left[1 + \frac{4\zeta_{21}^2\omega_2^2}{\omega_1^2 - \omega_2^2} + \frac{4\zeta_{23}^2\omega_2^2}{\omega_3^2 - \omega_2^2}\right] - (2B_e)^{3/2}\left[\frac{\zeta_{23}\varphi_{122}}{2\omega_1^{3/2}} - \frac{\zeta_{21}\varphi_{322}}{2\omega_3^{3/2}}\right] \tag{6.110}$$

Plíva[242] noted that if the rectilinear displacement basis Hamiltonian was used then $\alpha^{(\mathrm{anh})}$ really contains direct 'harmonic' contributions. As a simple example of the reasoning, consider a symmetric triatomic molecule such as CO_2. In this case $\zeta_{21} \to 0$, $\zeta_{23} \to 1$ and $\varphi_{322} \to 0$, so simplifying (6.110). The motion associated with the harmonic bending vibration corresponds to essentially the $\widehat{\mathrm{OCO}}$ angle bend with a fairly constant C—O bond length.[290,291] During this vibration we see that the projection of the atomic coordinates on the principal axis will be such that the projected bond length will shorten with increasing amplitude. We would thus expect that B_{0v0} would *increase* with v even if there were zero 'true' anharmonicity. The rectilinear basis model however will yield no direct harmonic contribution and in this treatment the variation of B_{0v0} with v filters in a rather unnatural way through the cubic anharmonic constant φ_{122}. This problem has been discussed further by Lide and Matsumura[180] who have reinterpreted α_2 in (6.110) in the light of this reasoning. The problem has been further discussed by Hougen, Bunker and Johns[135] who discuss the problem in a way that allows highly anharmonic potentials to be catered for.

6.7 The Inertial Defect

Consider the principal moments of inertia of a molecule; with the aid of the definitions (2.26) we see that

$$I_{\gamma\gamma} - I_{\alpha\alpha} - I_{\beta\beta} = -2\sum_n m_n r_{n\gamma}^2 \tag{6.111}$$

where $r_{n\gamma}$ is the distance of the nth particle (which may include electrons) from the $\alpha\beta$ plane. For a rigid molecule this equation holds exactly and in the case of a rigid planar molecule vanishes if we neglect electrons. When vibrational motion is taken into account however (6.111) no longer holds exactly and does not quite vanish for a planar molecule. For planar systems we define the inertial defect Δ^0 as

$$\Delta^0 = I_C^0 - I_A^0 - I_B^0 \tag{6.112}$$

for the zeroth vibrational level. $\Delta^{[v]}$ is usually a strong function of vibrational quantum number. For the ground vibrational state Δ^0 is usually a small positive number of the order of 0·05–0·5 amu Å^2 for 'well-behaved' planar molecules. If the molecule is non-planar then as implied by (6.111) Δ^0 will be negative.

Mecke[196] first noted that (6.112) did not vanish for the case of H_2O. Darling and Dennison[62] showed that Δ was independent of the cubic anharmonic force constants and thus the vibrational contribution to Δ could be calculated directly from the harmonic force field. Oka and Morino[227] have also considered the centrifugal distortion and electronic contributions to Δ. It is convenient to split the function up as

$$\Delta^0 = \Delta_v^0 + \Delta_{el}^0 + \Delta_{cd}^0 \tag{6.113}$$

where Δ_v^0, Δ_{el}^0 and Δ_{cd}^0 are the vibrational, electronic and centrifugal distortion contributions respectively. In a fairly typical molecule such as F_2CO, Oka and Morino[228,229] show that

$$\Delta_{\text{calc}}^0 = 0{\cdot}1506 = (0{\cdot}1554 + (-0{\cdot}0052) + 0{\cdot}0004)$$

where the individual contributions are in the same order as in (6.113) and the units are amu Å^2. Note that the out-of-plane electronic density gives rise to a significant negative contribution and also that the vibrational contribution is an order of magnitude larger than the rest.

We can calculate the general form of Δ_v^0 starting with the relation

$$I_\alpha^0 = I_\alpha^e + \varepsilon_\alpha \tag{6.114}$$

where application of the binomial theorem to (6.72) indicates that to a good degree of approximation

$$\varepsilon_b = \frac{1}{4b^2}\sum_r \alpha_r^{(b)} \tag{6.114a}$$

in the case of an asymmetric molecule (the degeneracy factor $d_r = 1$). If we now substitute (6.114) into (6.112) and note that Δ_v^e vanishes by definition for a planar molecule, we see that

$$\Delta_v^0 = \varepsilon_C - \varepsilon_A - \varepsilon_B \tag{6.115}$$

If we use the zeta sum rule

$$\tfrac{1}{4}a_r^{(\alpha\gamma)}(I^{e-1})_{\gamma\delta}\, a_{r'}^{(\delta\beta)} = A_{rr'}^{(\alpha\beta)} - \sum_{r''} \zeta_{rr''}^{(\alpha)}\zeta_{r'r''}^{(\beta)} \tag{6.116}$$

which implicitly relates (6.40) with (6.40a), it allows us to rephrase the expression (6.108) for α as

$$\alpha_r = -\frac{2b^2}{\omega_r}\left[3A_{rr}^{(bb)} + \sum_{r'} (\zeta_{rr'}^{(b)})^2 \frac{4\omega_{r'}^2}{\omega_r^2 - \omega_{r'}^2} + \tfrac{1}{2}\sum_{r'} \varphi_{rrr'}a_{r'}^{(bb)}\left(\frac{\omega_r}{\omega_{r'}^{3/2}}\right)\right] \tag{6.117}$$

If we now substitute (6.117) into (6.114a) we obtain the result

$$\varepsilon_b = \sum_r \frac{-2}{\omega_r}\left[\tfrac{3}{4}A_{rr}^{(bb)} + \sum_{r'} (\zeta_{rr'}^{(b)})^2 \frac{\omega_{r'}^2}{\omega_r^2 - \omega_{r'}^2} + \tfrac{1}{8}\sum_{r'} \varphi_{rrr'}a_r^{(bb)}\left(\frac{\omega_r}{\omega_{r'}^{3/2}}\right)\right] \tag{6.118}$$

According to (6.115) and (6.114a) we will have to sum the three individual terms in (6.117) over all vibrations, index r, and it is convenient to consider them individually. Oka and Morino[227] have shown that the A constants obey the following simple relations: $A_{ss}^{(CC)} = A_{ss}^{(AA)} + A_{ss}^{(BB)} = 1$ where s represents in-plane vibrations only and $A_{tt}^{(AA)} = A_{tt}^{(BB)} = 1$ and $A_{tt}^{(CC)} = 0$ where t represents the out-of-plane vibrations. As a result of these relations we see immediately that contributions to Δ_v^0 (6.115) involving A_{ss} constants vanish whereas those involving A_{tt} constants sum to -2, i.e.

$$A_{tt}^{(CC)} - A_{tt}^{(AA)} - A_{tt}^{(BB)} = -2 \qquad (6.119)$$

As we noted previously the cubic anharmonic contribution vanishes. We can see this by considering the first derivative of (6.111) with respect to Q_r (about the equilibrium configuration). This yields, using (2.23),

$$a_r^{(\gamma\gamma)} - a_r^{(\alpha\alpha)} - a_r^{(\beta\beta)} = -4\sum_n m_n r_{n\gamma}^e \left(\frac{\partial r_{n\gamma}}{\partial Q_r}\right)_e \qquad (6.120)$$

where $r_{n\gamma}^e$ is the out-of-plane equilibrium coordinate, which is zero for a planar molecule; this function thus vanishes. As a result we obtain the following expression for Δ_v^0‡

$$\Delta_v^0 = \sum_t \frac{3}{\omega_t} - \sum_r \sum_{r'} \frac{2\omega_{r'}^2}{\omega_r(\omega_r^2 - \omega_{r'}^2)}[(\zeta_{rr'}^{(C)})^2 - (\zeta_{rr'}^{(B)})^2 - (\zeta_{rr'}^{(A)})^2] \qquad (6.121)$$

Oka and Morino[227] have shown that

$$\Delta_{el}^0 = -\frac{m}{M}[I_{cc}g_{cc} - I_{aa}g_{aa} - I_{bb}g_{bb}] \qquad (6.122)$$

where m is the mass of the electron and M the mass of a proton. The gs are the components of the g-tensor for the rotational magnetic moment.

The centrifugal distortion contribution, Δ_{cd}^0, can be understood by considering the results of Section 6.5. In this calculation we found that some small terms in τJ^2 filtered down into the rigid-rotor constants as in (6.98). In a similar way small τ correction terms occur in the asymmetric rotor case (6.103). In the planar asymmetric rotor, distortions due to vibrational motion cannot tilt the plane of the molecule due to the Eckart conditions; any such motion must be a rotation and thus $u_r^{(ac)} = u_r^{(bc)} = 0$. On the other hand an in-plane tilt may occur and thus $u_r^{(ab)} \neq 0$. As a result, $\tau_{bcbc} = \tau_{acac} = 0$ and the μ' relations in the set (6.103) become

$$a' = a - \tfrac{1}{2}\tau_{abab}, \qquad b' = b - \tfrac{1}{2}\tau_{abab}, \qquad c' = c + \tfrac{3}{4}\tau_{abab} \qquad (6.123)$$

‡ If we include the conversion factor $\hbar^2/hc\,(\mathrm{cm}^{-1})$ this result is the same as that derived by Oka and Morino.[227]

As a result we see that

$$\Delta^0_{cd} = -\tau_{abab}\left[\frac{3}{4}\frac{I_C}{C} + \frac{1}{2}\frac{I_B}{B} + \frac{1}{2}\frac{I_A}{A}\right] \tag{6.124}$$

The final result is that Δ^0 is given by (6.113), where the various terms are (6.121), (6.122) and (6.124) respectively. As the ωs and ζ constants depend only on the harmonic force field it is possible to calculate the first and last terms in (6.113) directly from the results of a force-field calculation to a good degree of accuracy[228,229] for a well behaved molecule.

Certain molecules such as formamide[55,161] can present a problem as this type of treatment does not apply. In these cases the out of plane vibration may be highly anharmonic and in some cases there may be a double minimum. For molecules such as H_2CO and F_2CO the inertial defect can be calculated using these relations[228] and only when $\Delta_{calc} \sim \Delta_{obs}$ can a positive inertial defect be taken as a confirmation of planarity.

Herschbach and Laurie[121] have also discussed the problem as well as the general consequences in non-planar molecules.

6.8 *l*-type Doubling

In the case of linear and symmetric top molecules there is a vibration–rotation interaction term which becomes particularly important because of the presence of degenerate vibrations. It is convenient to treat this term which causes *l-type doubling* under the $E^{(*)}$ heading. Consider the term

$$H' = -2B_e(J_a\pi_a + J_b\pi_b) \tag{6.125}$$

which occurs for a linear molecule and is of $O(\kappa^2)$ in Table 6.3. The vibrational angular momentum components are made up of sums of terms of the form

$$\pi_a = \sum_{st} \zeta^{(a)}_{st_1}[(\omega_t/\omega_s)^{1/2}q_s p_{t_1} - (\omega_s/\omega_t)^{1/2}q_{t_1}p_s] \tag{6.126a}$$

$$\pi_b = \sum_{st} \zeta^{(b)}_{st_2}[(\omega_t/\omega_s)^{1/2}q_s p_{t_2} - (\omega_s/\omega_t)^{1/2}q_{t_2}p_s] \tag{6.126b}$$

In a linear triatomic molecule $s = 1$ and 3 and $t = 2$. In general one can determine which constants are non-zero by group theory[115] in the way discussed in Section 6.5. In this case $\zeta^{(a)}_{st_1} = -\zeta^{(b)}_{st_2}$ and thus (6.126a) and (6.126b) can be substituted into (6.125) and rephrased to yield ($\zeta^{(a)}_{st_1} \to \zeta_{st}$):

$$H' = -\sum_{st} B_e\zeta_{st}[(\omega_t/\omega_s)^{1/2}q_s(J^+p^+ + J^-p^-) - (\omega_s/\omega_t)^{1/2}p_s(J^+q^+ + J^-q^-)] \tag{6.127}$$

Now if we consider the matrix elements of this operator we see that they connect states such as $|J\ K\ v_t\ l_t\ v_s\rangle$ with $|J\ K \pm 1\ v_t \pm 1\ l_t \pm 1\ v_s \pm 1\rangle$. In the case of a linear molecule $K = \Sigma l_t$. The complete set of interaction terms can be accounted for by second-order perturbation theory as shown by Oka[224] (see also Grenier–Besson[102,103] and Nielsen[216]). The result is that when $K = l = \pm 1$, the l-doubling effect can be accounted for by a matrix element which splits the degeneracy of the two states $(1/\sqrt{2})[|\pm 1 \pm 1\rangle \pm |\mp 1 \mp 1\rangle]$, where the quantum numbers are in the order $|K\ l\rangle$. The matrix element is

$$\begin{aligned}\langle J\ K\ v_t\ l_t|\tilde{H}|J\ K-2\ v_t\ l_t-2\rangle &= \langle J\ K-2\ v_t\ l_t-2|\tilde{H}|J\ K\ v_t\ l_t\rangle \\ &= \tfrac{1}{4}q_t\{[(v_t+1)^2-(l_t-1)^2] \\ &\quad \times [J(J+1)-K(K-1)] \\ &\quad \times [J(J+1)-(K-1)(K-2)]\}^{1/2}\end{aligned} \tag{6.128}$$

In the case of a linear triatomic molecule $\frac{1}{2}q_t$ is given by the first bracketed relation in (6.110).[180] In the special case $l = K = \pm 1$ for linear and symmetric top molecules we obtain from (6.128) the result that an additional $E^{(*)}$ type contribution must be introduced into (6.74) and (6.73) of the form

$$E^{(*)} = \pm\tfrac{1}{4}q_t(v_t+1)J(J+1) \tag{6.128a}$$

The result is that a term of $O(\kappa^4) \sim O(\alpha)$ will split the l degeneracy and should be much larger than the centrifugal distortion terms which are of $O(\kappa^6)$. The effect bears a close similarity to the first-order asymmetry splitting of $|K| = 1$ asymmetric rotor states with which it is tenuously related. The experimental manifestation of this effect in both linear and symmetric top molecules is shown in Section 6.10.

From the results of Section 6.4c and relations (6.73) and (6.128a) we see that we can write the overall energy of a molecule in an excited degenerate vibrational state as

$$\begin{aligned}E(vJKl) = E(vl) &+ B_vJ(J+1) + (A_v - B_v)K^2 - 2A_v\zeta_vKl \\ &- D_JJ^2(J+1)^2 - D_{JK}J(J+1)K^2 - D_KK^4 \\ &+ \eta_{tJ}J(J+1)Kl + \eta_{tK}K^3l \pm \delta_{Kl1}\tfrac{1}{4}q_tJ(J+1)(v+1) + \cdots\end{aligned} \tag{6.129}$$

Note that when $Kl = 1$ then $\delta_{Kl1} = 1$ and zero otherwise. The resulting energy-level pattern is shown in Figure 6.5. In this diagram the pure vibrational energy, $E(vl)$, and the term $(A_v - B_v)K^2$ have both been dropped. They do not contribute to the rotational spectrum because these transitions occur within the v, l and K manifolds, i.e. Δv, Δl and ΔK are all zero.

The major splitting shown in Figure 6.5 is due to the first-order coriolis interaction discussed in Section 6.4c. The states for which the product Kl is positive are split apart from those for which this product is negative. Which way they go in energy depends on the sign of ζ. The l-doubling interaction causes a further splitting of those states for which $K = l = \pm 1$. Interactions

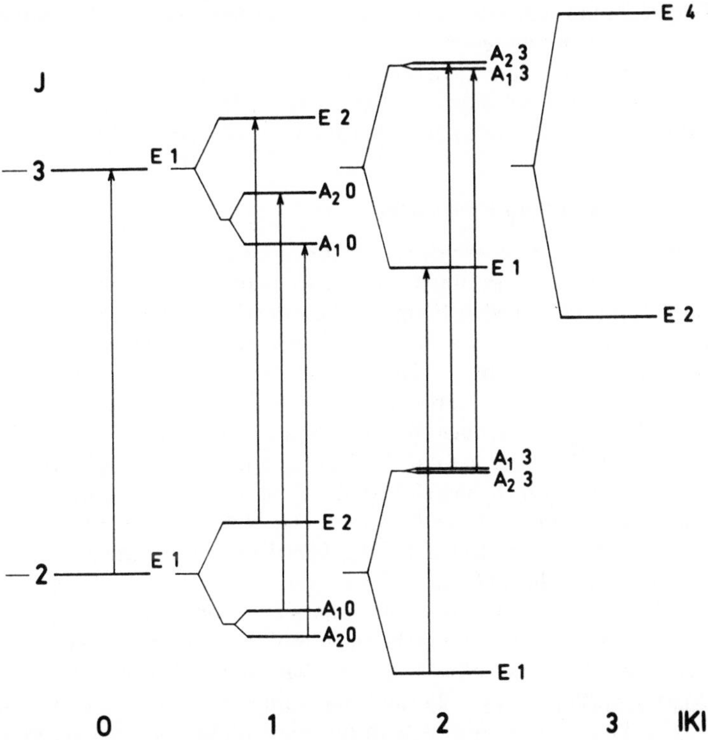

Figure 6.5 Schematic representation of the $J = 2$ and 3 levels in a singly excited degenerate vibrational state of a C_{3v} symmetric top. The energy-level contributions are drawn to varying scales in order to give a meaningful representation (see text). The pure vibrational energy and the term $(A - B)K^2$ have been dropped. The major splitting is due to the first-order coriolis interaction. The splitting between the A_1 and A_2 states when $K = l = \pm 1$ is due to the l-doubling interaction. Other A_1, A_2 pairs can be split but the interaction is of higher order, is very small, and even though it is shown in the diagram it can usually be neglected. The allowed transitions are indicated and the spectrum which results is discussed in Section 6.10b.

of order κ^8 can split A_1 and A_2 states in other cases but this effect is very small indeed. This splitting has been included in Figure 6.5. Figure 6.5 has not been drawn to scale and it is important to get a feel for the magnitudes of the various energy contributions. In the case of SiH_3CN,[33] $A \cong 83$ GHz, $B_0 = 4{\cdot}973$ GHz and for the $v_8 = 1$ vibrational level $q_t = 7{\cdot}343$ MHz. The levels have been labeled by their rotational–vibrational symmetry which is determined by the pseudo-quantum number $|\bar{K}| = |K - l|$ which can replace $|K|$ in the symmetry discussion given in Section 3.11c. The value of $|\bar{K}|$ is also given in Figure 6.5.

The allowed transitions have been indicated in Figure 6.5 and the resulting rotational spectrum is discussed in Section 6.10b. Direct transitions between the l-doublets can also occur as is also discussed in Section 6.10b.

6.9 Fermi and Coriolis Resonances

The rotational energy levels can sometimes be affected by terms in $E([\tilde{v}])$ which do not appear to exhibit any explicit $\boldsymbol{J}$ character. Consider the classic case of CO_2 first explained by Fermi,[86] where the vibrational levels associated with ν_1 and $2\nu_2$ should on a simple harmonic basis be almost exactly degenerate (within $\sim 3\ \mathrm{cm}^{-1}$). The frequencies given by Gordon and McCubbin[96,97] are $\omega_1 = 1336{\cdot}97$ and $\omega_2 = 667{\cdot}19$, cm^{-1}, and thus $2\omega_2 = 1334{\cdot}38$. The observed frequencies are $\nu_1 = 1388{\cdot}17$ and $2\nu_2^0 = 1285{\cdot}40$[96,97] the mean of which is $\sim 1336{\cdot}8\ \mathrm{cm}^{-1}$. The two states have been pressed apart by an interaction which has shifted them by $\pm 50\ \mathrm{cm}^{-1}$. Suzuki[290] has considered the complete force field for CO_2. Note that there is also a Δ state, $2\nu_2^2$, which is unaffected by the Fermi resonance interaction. The superscript is the value of l.

For CO_2 the only non-zero cubic anharmonic constants are $\varphi_{111}, \varphi_{122}$ and φ_{133}. All other cubic terms in the Hamiltonian are non-totally symmetric. If we consider the effect of φ_{122} we see that matrix elements of the type $\langle v_1\ v_2|\frac{1}{6}\varphi_{122}q_1q_2^2|v_1 \pm 1\ v_2 \mp 2\rangle$ link the vibrational states $|1\ 0\ 0\rangle$ and $|0\ 2^0\ 0\rangle$. Taking account of this term directly with the aid of (1.40) we see that the vibrational states have energies shifted to

$$E = \tfrac{1}{2}(E^0_{020} + E^0_{100}) \pm \tfrac{1}{2}[(E^0_{020} - E^0_{100})^2 + (\tfrac{1}{6}\varphi_{122})^2|\langle 0\ 2^0\ 0|q_1q_2^2|1\ 0\ 0\rangle|^2]^{1/2} \tag{6.130}$$

If we set $a = \cos\theta$ and $b = \sin\theta$ in the resulting eigenvector relation we see that the observed B values ($\tilde{B}$) must be corrected[303] according to

$$\tilde{B}_{100} = a^2B^0_{100} + b^2B^0_{020} \tag{6.131a}$$

$$\tilde{B}_{020} = b^2B^0_{100} + a^2B^0_{020} \tag{6.131b}$$

where $a^2 + b^2 = 1$ and $\tilde{B}_{100} + \tilde{B}_{020} = B^0_{100} + B^0_{020}$. This type of correction has been applied in several cases. Morino and Saito[208] discuss this effect in asymmetric rotors in their analysis of the F_2O spectrum. The result is that the vibrational energy levels are separated whereas the effective rotational constants are mixed and closer in value than otherwise (6.131).

In the case of coriolis resonance two states which are close in energy may be connected by a matrix element which corresponds to the term $\frac{1}{2}\mu\zeta\boldsymbol{Jpq}$ in Table 6.4. In such a case the second-order perturbation result is no longer valid. Thus the appropriate rr' term in $\alpha_r^{(cor)}$, (6.109) and (6.108), must be dropped and the interaction taken into account by a direct diagonalization procedure. As the off-diagonal term is essentially of the form $-B\pi_\alpha J_\alpha$ it connects only states with $\Delta K = \pm 1$ or 0. This type of resonance occurs between ν_1 and ν_3 in ozone.[295]

6.10 Spectra of Semi-rigid Molecules

In this section we consider the ways in which some of the interactions we have just discussed manifest themselves in the observed spectra.

6.10a Centrifugal Distortion Effects in Symmetric Top Molecules

In Section 6.4 we developed the energy-level expression (6.97) for the ground vibrational state of a symmetric top which takes centrifugal distortion into account. Application of the selection rule $\Delta J = +1$, $\Delta K = 0$ yields the frequency relation

$$\Delta E(JK) = 2B_0(J+1) - 2D_{JK}(J+1)K^2 - 4D_J(J+1)^3 \qquad (6.132)$$

The centrifugal distortion terms are of $O(\kappa^6)$ and thus are small compared with the value of B. Under high resolution we would expect each J 'line' of the spectrum of CH_3D shown in Figure 5.3 to consist of $(J+1)|K|$ components. In Figure 6.6 part of this K-type substructure is shown for the $J = 16 \leftarrow 15$ transition of the ground vibrational state of trimethyl silyl isocyanate.[32] The displacements from the $K = 0$ line should be proportional to K^2 and thus the separation of adjacent lines should follow the sequence ratio: 1, 3, 5, 7, ... We can obtain the value of D_{JK} from these spacings and the value of D_J if other $J + 1 \leftarrow J$ transitions are observed. D_K (and A also for

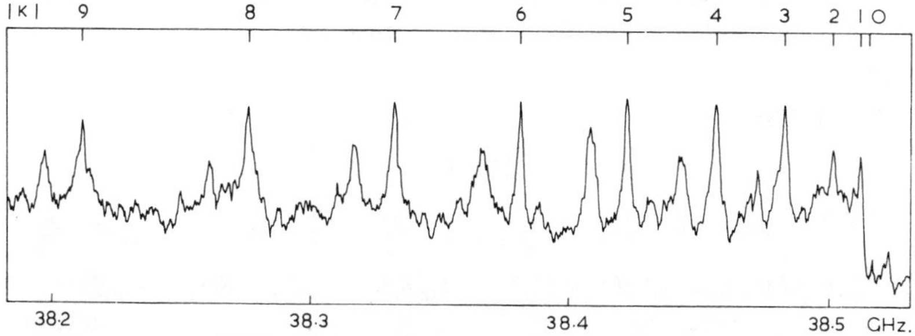

Figure 6.6 The microwave spectrum of the $J = 16 \leftarrow 15$ transition of trimethyl silyl isocyanate observed by Careless, Green and Kroto.[32] The K sub-band structure is well resolved as indicated by the leading lines. Satellites belonging to one of the torsionally excited states lie ~20 MHz to lower frequency of each line. The relative spacing between the lines should follow the sequence 1, 3, 5, 7 etc. (See also Figure 9.12)

that matter) which can only be determined if $\Delta K \neq 0$ transitions occur is sometimes obtained from the analysis of high resolution vibrational spectra in the infra-red. A similar spectrum is shown in Figure 6.9.

Watson[320] has shown that in spherical top molecules, such as CH_4, centrifugal deformation of the structure can give rise to a non-negligible dipole moment. This centrifugal-induced dipole moment allows pure rotational transitions to take place. The spectrum of CH_4 has been observed by Rosenberg, Ozier and Kudian[262,232] in the far infra-red and by Curl and Oka[60] using double resonance techniques. The spectra of SiH_4 and GeH_4 have also been observed.[211,260] It should also be possible to observe such spectra using molecular-beam techniques.

Chu and Oka have observed $\Delta K = \pm 3$ transitions in PH_3 ($\Delta|K| = 1$) using a standard microwave spectrometer.‡ These transitions become weakly allowed due to centrifugal distortion because of mixing by $\Delta K = \pm 3$ matrix elements (see Section 6.5a and particularly the footnote to Table 6.8). Such transitions allow the rotational constant about the symmetry axis to be determined accurately.

Kirchhoff[60] has reviewed the problem of analysing the rotational spectra of asymmetric rotors with the inclusion of centrifugal distortion corrections. In general at low J values the energy levels are often adequately described using a rigid-rotor Hamiltonian and only at high J are centrifugal effects significant.[161]

It is sometimes necessary to take vibration–rotation effects into account in Raman spectroscopy. When centrifugal distortion effects are important we need to apply the selection rule $\Delta J = \pm 2$ to the more flexible energy-level expression (6.35) and thus obtain

$$|\Delta E| = (4B_0 - 6D)(J + 3/2) - 8D(J + 3/2)^3 \qquad (6.133)$$

One can determine the centrifugal distortion constant D by similar methods to those discussed in Section 6.10a. In symmetric top spectra the appropriate shift expressions are

R and *P* branches

$$|\Delta E| = 2B_0(J + 1) - 2D_{JK}K^2(J + 1) - 4D_J(J + 1)^3 \qquad (6.134a)$$

S and *O* branches

$$|\Delta E| = (4B_0 - 6D_J)(J + 3/2) - 4D_{JK}K^2(J + 3/2) - 8D_J(J + 3/2)^3 \qquad (6.134b)$$

Expression (6.134a) applies for levels with $|K| > 0$ and (6.134b) for all $|K|$ values. As a consequence the resulting spectra consist of sets of overlapping

‡ These can mix states with $K = \pm 2$ with $K = \mp 1$ etc.[47a]

unresolved lines.[123,285] Because the R and P branches have spacings of $2B$, and the O and S branches spacing of $4B$ the composite spectrum often shows a complicated intensity alternation.[123 (p. 36)] In general the centrifugal distortion constant D_{JK} is too small to allow the K substructure to be resolved by present techniques though it is often possible to determine D_J.[285]

6.10b *Vibrational Satellites, l-doubling*

As a result of the vibrational dependence of the rotational constants (6.73a) we see that we should in general observe a set of vibrational satellite lines if the associated vibrational frequencies are low enough for reasonable population factors to occur. In the spectrum of NCN_3 shown in Figure 5.6 we see that the ground-state set of lines for the $J = 6 \leftarrow 5$ is accompanied by almost identical sets displaced by intervals of 100 to 150 MHz. Each successive set is less intense than the previous one because of the Boltzmann factor. This relatively simple satellite pattern belongs to the $v = 1, 2, 3, \ldots$ vibrational levels of the low frequency ($\sim 180 \pm 30\,\text{cm}^{-1}$) $\widehat{\text{CNN}}$ angle bending vibration. All other frequencies are much higher[10] and the associated satellites are consequently not so prominent.

In the case of linear molecules similar relations apply as indicated in Section 6.4. However when degenerate vibrational levels are excited the most striking features are the result of l-type doubling. If we combine the splitting term (6.128) together with (6.74) and use the selection rules $\Delta J = 1$ and $\Delta l = 0$ (essentially the symmetric top selection rules) we obtain, for the $v_t = 1$ state, the following transition frequency expression

$$\Delta E(J\ v_t = 1\ |l| = 1) = 2B_{[v]}(J + 1) \pm q_t(J + 1) - 4D(J + 1)^3 \quad (6.135)$$

where the q_t term is of $O(\kappa^4)$ which is the same as that of the α terms. This indicates that the rotational levels of the $\Pi^{\pm}$ states for $v = 1$ $(1/\sqrt{2})[|+1\rangle \pm |-1\rangle]$; the specification here is l) are split into two stacks with somewhat different effective rotational constants. The resulting effect on the spectrum is that the $v_t = 1$ satellite is a doublet the mean of which is centred roughly where we would expect this satellite if l-doubling is absent (Figure 6.7). The $v_t = 2$ state levels (Figure 6.3) do not contain $|l| = 1$ levels and thus the effects are second order and the set of lines are close to the expected position. The separation of the l-doublets in Figure 6.7 gives $2q_t(J + 1)$ directly which in this case gives $q_t = 6{\cdot}54$ MHz. A very accurate value was determined by analysis of direct transitions between these two levels observed by Lafferty[168] and discussed below.

The effects of l-doubling tend to dominate the vibrational satellite patterns of symmetric top molecules also. The interaction and transition energy level diagram is shown in Figure 6.5 for the $J = 3 \leftarrow 2$ transitions and is discussed in Section 6.8. The transitions which can occur are governed by the selection

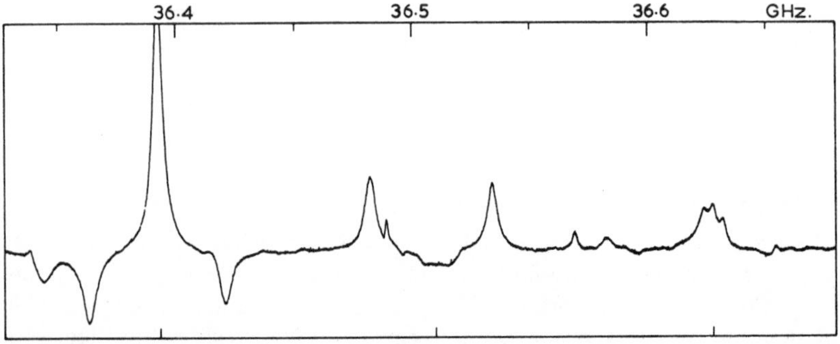

Figure 6.7. The microwave spectrum of the $J = 4 \leftarrow 3$ transition of H—C≡C—C≡N. The frequency of the ground state line is 36.393 GHz. The *mean* of the $v_4 = 1$ l-doublets lies essentially half-way between the ground state line and the $v_4 = 2$ complex

rules $\Delta J = 1, \Delta K = 0$ and $\Delta l = 0$ and the transition energy expression when $v_t = 1$ and $|l| = 1$ is

$$\Delta E(JK) = 2(B_v \pm \delta_{Kl1}\tfrac{1}{2}q_t - D_{JK}K^2 + \eta_{tJ}Kl)(J + 1) + \cdots \quad (6.136)$$

Terms in $(J + 1)^3$, etc. have been dropped. The most important term in (6.136) apart from B_v is the term in q_t associated with the l-doubling interaction, and occurs only when $K = l = \pm 1$ as is indicated by the Kronecker δ. In Figure 6.8 the spectrum of the $J = 3 \leftarrow 2$ transition of SiH_3CN is depicted where the vibrational satellites associated with a low frequency ($\sim 235\ cm^{-1}$) bending vibration are quite strong. In general bending vibrations tend to shorten the molecule and the effective B value increases causing the resulting satellites to lie to the high-frequency side of the ground state lines. With the aid of Figure 6.5 and the first two terms of (6.136) we can interpret the gross features depicted in the spectrum, Figure 6.8. The $v_8 = 1$ satellite pattern which is labeled 1_8^1, consists of a bunch of lines displaced by approximately $2\alpha(J + 1)$ from the ground state K-complex and flanked symmetrically to either side by the two l-doubling lines. The strong l-doubling interaction occurs when the contribution to the component of overall angular momentum along the axis is due only to the vibrational motion and $|l| = 1$.

If we refer to Figure 6.4 we see that when $v = 2$ there are two resulting states of differing symmetry. One is an A_1 state with $l = 0$ and the other is an E state with $|l| = 2$. The A_1 state has no vibrational angular momentum and behaves like the ground state as long as small interaction terms (with $\Delta l = \pm 2$, etc.) with the E state are neglected. The E state is subject to more complex interactions associated with the coupling of vibrational and rotational angular momenta. The strong l-doubling interaction does not occur because the two components of the E state differ by 4 in l. They can interact by a second-order term via the A_1 state which has the form $\langle E|H|A_1\rangle\langle A_1|H|E\rangle/[E(E) - E(A_1)]$

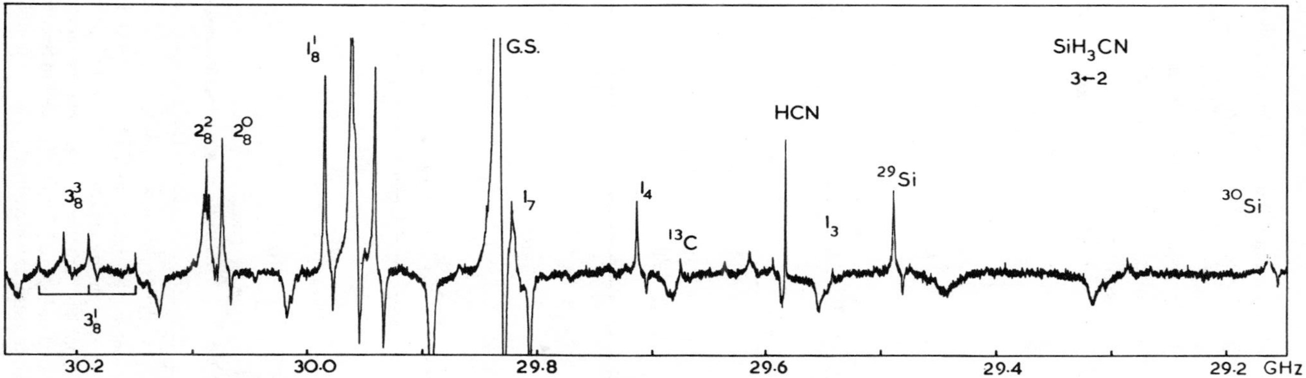

Figure 6.8 The microwave spectrum of SiH_3CN observed by Careless and Kroto.[33] The $J = 3 \leftarrow 2$ transition of the vibrational ground state is shown together with the associated vibrationally excited and isotopic satellites. The lines have been assigned using the shorthand notation v_n^l. Thus 3_8^3 means that $v_8 = 3$ and $l = 3$. Species with ^{28}Si(92·3%), ^{29}Si(4·7%), ^{30}Si(3·1%), ^{13}C(1.1%) are observed in natural abundance. The l-doubling transition of HCN with $J = 12$ occurs in this spectrum as an impurity at 29·585 GHz

and involves $\Delta l = \pm 2$ matrix elements. The pattern which results is quite complicated. When $v = 3$, the vibrational states which result are an E state with $|l| = 1$ and an $A_1 + A_2$ state with $|l| = 3$, as can be seen from Figure 6.4. Only the E state is subject to a strong l-doubling interaction and, as indicated by (6.128a), the resulting splitting is twice that of the $v = 1$ state. All these points are more clearly shown in Figure 6.9 where the spectra associated with the $J = 9 \leftarrow 8$ transition of methyl diacetylene are depicted.[112a,166a]

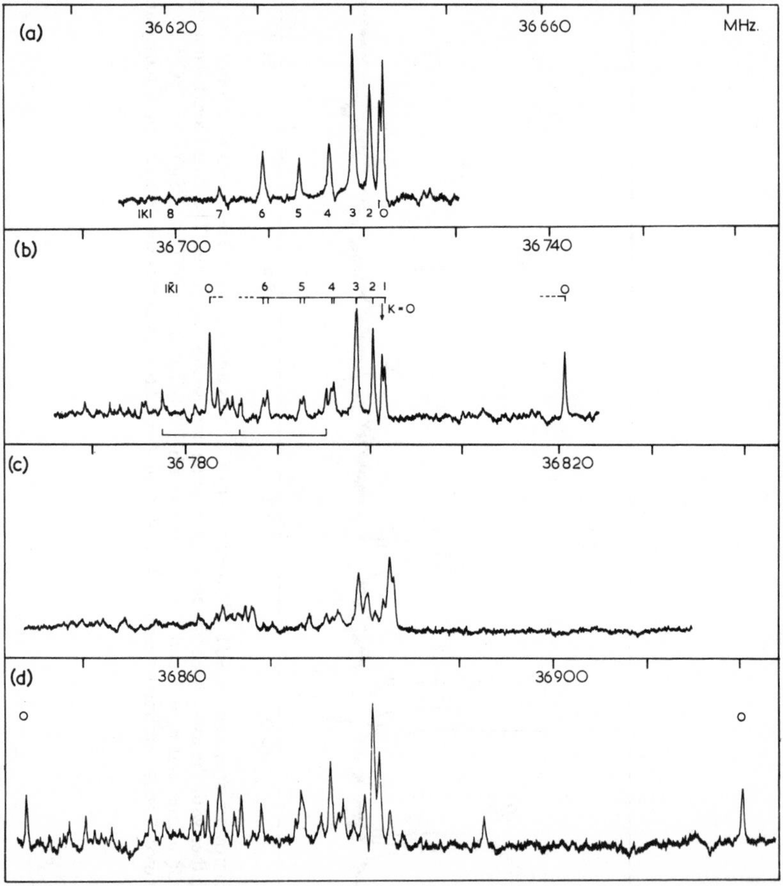

Figure 6.9 The spectrum of the $J = 9 \leftarrow 8$ transition of methyl diacetylene, $CH_3C{\equiv}C{-}C{\equiv}C{-}H$, as observed by Kroto and Maier[166a] (a) The ground state spectrum showing K-substructure split by centrifugal distortion. (b) The $v = 1$ vibrational satellite pattern. The l-doublets flank the main lines at the center of the spectrum. The assignment is also indicated in Figure 7.7 and discussed in Sections 6.10b and 7.4. (c) The $v = 2$ vibrational satellite patterns which do not exhibit l-doubling. (d) The $v = 3$ vibrational satellite pattern. The l-doublets are split by twice the separation of the l-doublets in the $v = 1$ state

Some of the spectral features can be assigned with the aid of the Stark effect as is discussed in Chapter 7. The lines observed in the $v_t = 1$ state are shown in Figures 6.9(b) and 7.7, and in order to understand the observed pattern it is useful to re-phrase (6.136) by introducing the pseudo quantum number $\bar{K} = K - l$. This quantum number can be considered as an effective quantum number associated with the rotation of the molecular frame only. Thus for the l-doublet split states, where $K = l = \pm 1$, we see that $\bar{K} = 0$ and there is essentially no contribution to J_A from the rotation of the nuclear frame. To make the resulting expressions more general we can also include a higher-order perturbation term which replaces the l-doubling term when $Kl \neq l$.[104,57] This term can be written when $l = \pm 1$ as

$$\varepsilon\{[(J+1)^2/\bar{K}l] - \bar{K}l\}(J+1)$$

where $\varepsilon = -q_t^2/4[A(1-\zeta_t) - B]$. The form of ε demonstrates its second-order origins. As a result we obtain

$$\begin{aligned}\Delta E(J\bar{K}) = 2[(B_v + \eta_{tJ} - D_{JK}) - D_{JK}\bar{K}^2 + (\eta_{tJ} - 2D_{JK} + \tfrac{1}{2}\varepsilon)\bar{K}l \\ \pm \delta_{Kl1}(\tfrac{1}{2}q_t) - (1 - \delta_{Kl1})\tfrac{1}{2}\varepsilon(J+1)^2/\bar{K}l](J+1)\end{aligned} \tag{6.136a}$$

If we use this relation in conjunction with Figures 6.9(b) and 7.7 we see that for $\bar{K} \neq 0$ the lines behave like those of the ground state in that the pattern is governed by D_{JK}; the role of K has however now been taken over by $\bar{K}$. If we can neglect the term in $(\bar{K}l)^{-1}$ we see that as $\bar{K}$ increases a splitting of the $\bar{K}$ substructure occurs, caused by the term in $2(\eta_{tJ} - 2D_{JK} + \frac{1}{2}\varepsilon)\bar{K}l$. This term is positive or negative depending on the sign of $\bar{K}l$. The splitting that results is clearly seen in Figures 6.9(b) and 7.7. At low $\bar{K}$ the term in $(\bar{K}l)^{-1}$ may become important and this will depend on the size of ε which in turn is strongly dependent on how close ζ_t is to unity; in this case ε is quite small. Note that when $\bar{K} = \pm 1$ there are two different types of state. One has $l = \mp 1$ and $K = 0$ in which the vibrational and frame angular momenta cancel. In the other $K = \pm 2$ and $l = \pm 1$. The $K = 0$ state is easily identified by the Stark effect as shown in Figure 7.7.

As in the case of asymmetric tops where transitions between K-doublets can often be observed, so also can transitions between l-doublets occur. Shulman and Townes[276,327,301,303] first observed such transitions for HCN and in Table 6.6 these transitions and some of the transitions observed by Lafferty[168] for HC≡C—C≡N are listed. This table also includes transitions in the $v_t = 3$ state with $l_t = \pm 1$ for HC≡C—C≡N. These transitions give very accurate values for q_t because of the large J weighting factor. The $J = 11$ direct l-doubling transition of HCN lies at 29 585·1 MHz and appears as an impurity line in the spectrum of SiH_3CN shown in Figure 6.8.

Mizushima and Venkateswarlu[200,198] showed that spherical top molecules in degenerate vibrational states may possess a non-zero dipole moment. Curl

Table 6.9 The Frequencies (MHz) of Direct *l*-Doubling Transitions for $HCN^{327,301}$ and $HCCCN^{168}$

J	$HCN(v_2 = 1)$	J	$HCCCN(v_7 = 1)$	J	$HCCCN(v_7 = 3)$
6	9 423·32	35	8 212·93	25	8 539·04
7	12 562·32	36	8 680·72	26	9 220·14
8	16 148·55	37	9 161·28	27	9 927·14
9	20 181·4	38	9 654·62	28	10 659·58
10	24 660·4	39	10 160·80	29	11 417·71
11	29 585·1	40	10 679·69	30	12 201·26
12	34 953·5	⋮		⋮	
		58	22 186·79	40	21 412·36

and Oka have used an infra-red laser-microwave double resonance technique to detect the resulting rotational transitions between rotational levels of the excited state.[60]

6.11 Structure Determination

6.11a The Equilibrium Structure

Perhaps the most important data that can be obtained from rotational spectra is molecular structural information. Rotational constants can often be obtained with high accuracy; for example in the case of AlF, B_0 was determined in Section 6.2 to about 6 kHz or about 1 part in 10^6. Molecular-beam measurements can give even higher accuracy. The derivation of well defined structural parameters from the rotational constants of polyatomic molecules to an accuracy better than 1 part in 100 however turns out to be a problem. In Section 6.2 we saw that the term *bond length* in a dynamic system can be interpreted in many different ways and the result depends on the way in which a measurement is carried out. We shall thus examine the concept of *structure* and in particular note the general problems which rotational spectroscopists face in analysing their data to obtain such information. We should also examine the relationship between this structural data and that obtained by other techniques such as electron diffraction. The raw structural parameters are not necessarily the same and it is by an analysis of the underlying reasons for this that we gain a true understanding of the term *structure*.

In the case of the diatomic molecule we saw that the term r_e was defined as the bond length at the bottom of the internuclear potential which governs the vibrational motion in the Born–Oppenheimer approximation. It will be the same—or almost the same[294]—for all isotopic species whereas the value of r_0 will not. Thus r_e is, almost obviously, the structural parameter which one should aim to determine. It therefore seems reasonable that we should aim at the geometrical parameters which characterize the equilibrium configuration of a polyatomic molecule. However, apart from a few very simple cases this has not so far been possible and in general we are forced to compromise.

We can focus on the central problem with the aid of the diatomic molecule results. For a diatomic molecule we can use (6.114) to show that $I^0 \simeq I^e(1 + \alpha/2B)$ and as a result we see that we can write

$$r_e = r_0(1 + \alpha/2B)^{-1/2} = r_0(1 - \alpha/4B) \tag{6.137}$$

where r_0 is the apparent bond length we obtain from I^0 or B_0. In the particular case of AlF, $B_0 = 16\,488{\cdot}325$ MHz, $B_e = 16\,562{\cdot}49$ MHz and $\mu = 11{\cdot}1485$ amu. From this information we see that $r_0 = 1{\cdot}65810$ Å and $r_e = 1{\cdot}65436$ Å.[344] Using (6.137) we see that $\delta r \sim r_0\alpha/4B = 0{\cdot}00372$ Å whereas the accurately determined value is $0{\cdot}00374\,(\pm 0{\cdot}00002)$ Å. We can thus determine a guide to the discrepancy between bond lengths determined directly from B_0 (i.e. r_0) and the equilibrium value, r_e. Using (6.36b) and the empirical observation that a_1 usually lies in the range -2 to -4 (say ~ -3) we see that $\alpha \sim 12B^2/\omega$ and thus

$$(r_0 - r_e)/r_0 \sim \alpha/4B \sim 3B/\omega \tag{6.138}$$

This discussion has clear general implications as far as molecular structure determinations obtained from rotational data are concerned. In the case of a polyatomic molecule the problem is compounded because we have contributions to the observed rotational constants from all the vibrations.

We can thus define an equilibrium or r_e structure as the structure of the equilibrium reference configuration which lies at the minimum of the multi-dimensional vibrational potential surface. If we neglect all higher terms such as those in γ in (6.72) which appears to be a good assumption[205,265] we see that this structure is the one we obtain using I^e given by

$$I_\alpha^e = I_\alpha^0 - \varepsilon_\alpha \tag{6.139}$$

according to (6.114). Thus if the I^0s and if *all* the αs are known for a sufficient number of isotopic species we can obtain an estimate of the r_e structure. This estimate is certainly close to the r_e structure.[321]

One of the main limitations to the use of this method is that we cannot often detect all the vibrational satellites because the Boltzmann factor may be small when the vibrational frequency is high. In a large molecule assignment also becomes a problem. Another factor which can cause problems is that we cannot neglect resonances which must be allowed for in general. Complete r_e determinations have been carried out for a small number of very small molecules.[203]

Now that we have defined the ideal procedure for structure determination and are aware of the difficulty, and in general the impossibility, of reaching this goal it is perhaps worthwhile considering what criteria any alternative procedures should fulfil. Perhaps the main requirements are that the procedure be well defined, fairly general in applicability and the resulting structure bear a well defined relationship with the equilibrium structure. This last point implies that we know roughly how far we are from the equilibrium structure.

No general structure determination is really complete without some consideration of the dynamic aspects of molecular configurations. As an extreme example the structure of NH_3 cannot be discussed in isolation from the fact that it appears to be inverting through a planar configuration at $\sim$24 GHz. Thus in the more flexible systems with low-frequency motions, potential surface data should be used to complement our understanding of the problem.

There is one final inherent problem, from which rotational structure determinations suffer, that cannot be overlooked. When an atom lies close to a principal axis the associated moment of inertia is rather insensitive to the atom's position. As a result, any structural data concerning this atom which depends on this constant may be subject to a larger error than usual.

6.11b The r_s Substitution, and r_m Methods

The most acceptable alternative procedure to a complete r_e determination has in general been to make use of a set of structure relations developed in general form by Kraitchman.[165] The method makes use of the fact that when an isotopic substitution has been made on an atom, Kraitchman's equations yield the coordinates of that atom with respect to the principal axes of the original molecule directly. These relations, which hold only for rigid structures, are valid exactly only for the equilibrium structure. Costain[52] showed that for a diatomic molecule the resulting so-called substitution or r_s bond length was the mean of r_e and r_0 to a good degree of approximation, i.e.

$$r_s \simeq \tfrac{1}{2}(r_0 + r_e) \tag{6.140}$$

Costain also showed that in some small polyatomic molecules the use of Kraitchman's equations yielded more consistent sets of structural data (Table 6.10) than r_0 determinations[289] and that a similar relation to (6.140) appeared to hold for some small systems. Watson[321] has shown that in fact the relation

$$I^s \simeq \tfrac{1}{2}(I^e + I^0) \tag{6.141}$$

which is closely related to (6.140) holds in general.

The essence of the r_s method can be seen in a simple analysis of the effects of substitution in a diatomic molecule, Figure 6.10. The total mass is

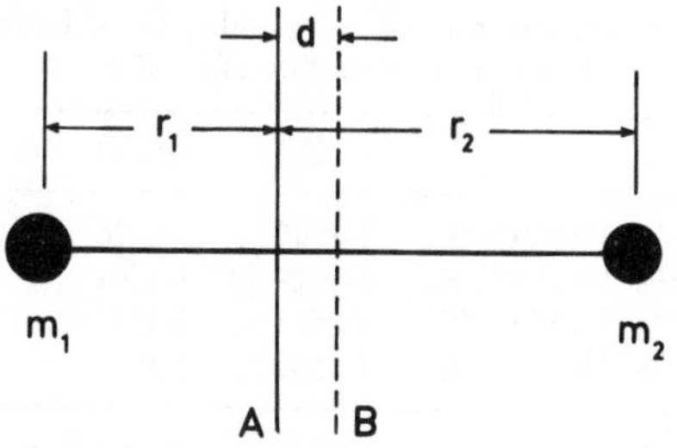

Figure 6.10 The effect of substitution on the moment of inertia and centre of mass in a diatomic molecule

$m_1 + m_2 = M$ and the moment of inertia about the axis A through the centre of gravity is I. If we add an extra mass Δm to m_2 the centre of gravity will shift so that the new axis is B and the moment of inertia about B is I'. We can apply the parallel axes theorem[322] then we find that the moment of inertia of the new system (with the extra mass) about the old axis, A, is

$$I'' = I' + (M + \Delta m)d^2 \tag{6.142}$$

We know that $I'' = I + \Delta m r_2^2$ and thus combining this with (6.142) and substituting for d using the first moment equation, we obtain the result

$$I' - I = \mu r_2^2 \tag{6.143}$$

where $\mu = M\Delta m/(M + \Delta m)$, *the reduced mass for substitution.* Such a relation clearly applies to substitution in any system where the substitution does not tilt the inertial axes. Kraitchman has developed the relations which apply in the general cases and these are given in Section A6.

The improvement, which Costain noted,[52] in the consistency of r_s structural data over r_0 data for small molecules is exemplified by Table 6.10. In this case the variations of the CO and CS bond lengths of OCS was reduced from ~0·01 Å to 0·001 Å—about an order of magnitude improvement. In this particular case the complete r_s structure appears to differ from the r_e structure by ~0·001 Å.[321] This reliability of the structural data is obtainable when an atom is of the order of 1 Å away from the centre of mass. There is always a problem with atoms close to the centre, and in the celebrated case of $^{15}N^{14}N^{16}O$ one obtains an imaginary solution for the position of the central N atom.[52] For coordinates greater than 0·02 Å from an axis Costain estimates that the reliability $\delta r \sim 0{\cdot}0012/r$ for the coordinate with respect to that axis.[53]

The meaning of the substitution moments of inertia I^s has been clarified by Watson.[321] If we substitute (6.114) into (A6.2) taking note of (6.114a) we

Table 6.10 Comparison of r_0, r_s, r_e and r_m Bond Length Determinations of the Structure of OCS (in Å)

	r_0(CO)	r_0(CS)	Reference
$^{16}O^{12}C^{32}S$, $^{16}O^{12}C^{34}S$	1·1647	1·5576	52, 289
$^{16}O^{12}C^{32}S$, $^{16}O^{13}C^{32}S$	1·1629	1·5591	
$^{16}O^{12}C^{34}S$, $^{16}O^{13}C^{32}S$	1·1625	1·5594	
$^{16}O^{12}C^{32}S$, $^{18}O^{12}C^{32}S$	1·1552	1·5653	
Mean	1·1613	1·5604	
Range	0·0095	0·0077	
Basis molecule	r_s(CO)	r_s(CS)	Reference
$^{16}O^{12}C^{32}S$	1·16012	1·56020	52, 289
$^{18}O^{12}C^{32}S$	1·15979	1·56063	
$^{16}O^{13}C^{32}S$	1·16017	1·56008	
$^{16}O^{12}C^{34}S$	1·16075	1·55963	
Mean	1·16021	1·56014	
Range	0·00096	0·0010	
	r(CO)	r(CS)	Reference
r_m	1·1588	1·5592	321
r_e	1·1594	1·5588	29
r_e	1·1545	1·5630	207

can generate the relation

$$[r_\alpha^s(i)]^2 = [r_\alpha^e(i)]^2 + \frac{\partial \varepsilon}{\partial m_i} + \left[\frac{1}{M}\frac{\partial \varepsilon}{\partial m_i} + \frac{1}{2}\frac{\partial^2 \varepsilon}{\partial m_i^2}\right]\Delta m_i + \cdots \qquad (6.144)$$

where $\Delta\varepsilon$ has been expanded as a Taylor series in Δm_i. If we retain only terms independent of Δm_i and note that $[r_\alpha^e(i)]^2 = \partial I^e/\partial m_i$ then we obtain the interesting result

$$[r_\alpha^s(i)]^2 = [r_\alpha^e(i)]^2 + \frac{\partial \varepsilon}{\partial m_i} = \frac{\partial I^0}{\partial m_i} \qquad (6.145)$$

By taking advantage of the homogeneous functional dependence of ε on the masses, Watson showed that (6.141) is true in general.[321] Using this general expression, a mass dependence estimate of I^e called I^m can be defined

$$2I^s - I^0 = I^m \sim I^e \qquad (6.146)$$

This implies that the resulting structural data (r_m) is a good estimate of the r_e structure and can be obtained if enough substitution moments of inertia are known. The method has been tested successfully for a few small molecules. Both the r_s and the r_m methods suffer from rather large errors in the estimate of the positions of H atoms.

Other procedures including multiple substitution methods are reviewed by Wollrab[343] and Gordy and Cook.[100]

6.11c General Problems in Structure Determinations

With the aid of a few simple but typical examples, we shall discuss some practical problems which frequently face the spectroscopist when r_e, r_m or complete r_s structures cannot be obtained. Consider the molecule F_2CS;[34] a complete r_s structure is not possible because there is only one stable fluorine isotope and a complete determination using vibrational satellites is also not feasible because the spectrum is too weak. We must therefore compromise.

The inertial defect is small and positive (+0·203 amu Å^2) and close to that expected for such a molecule[230] if planar. The ^{34}S isotope has a natural abundance of 4 % and as a result A_0, B_0 and C_0 for two isotopic species could be determined.[34] This allows us to determine the r^s coordinate of the S atom. We can however determine this coordinate in two ways using either I_B^0 or I_C^0 together with (A6.3) and they yield different values as shown in Table 6.11. Without the data for the ^{13}C species, which was too weak to detect, it is still possible to make an estimate of the F and C atom positions using the *first moment equations* as follows. The value of I_A^0 determines the r^0 distance of the fluorines from the symmetry axis (A-axis). There are two values for I_A^0 but they are very close in this case and the ambiguity is well within the accuracy of the determination.

Table 6.11 Structural Data[a] for F_2CS [34]

Route[b]	r_B^0(F)	r_A^s(S)	r_A^0(C)	r_A^0(F)	r(CF)	r(CS)	$\widehat{FCF}$
1 + 2a + 3a	1·0575	1·3317	0·2616	1·0379	1·3118	1·5933	107·44
1 + 2b + 3a	1·0575	1·3307	0·2535	1·0396	1·3177	1·5852	106·75
1 + 2a + 3b	1·0575	1·3317	0·2508	1·0413	1·3203	1·5815	106·44
1 + 2b + 3b	1·0575	1·3307	0·2428	1·0430	1·3261	1·5735	105·77

[a] Bond lengths in Å, angles in degrees.
[b] 1 Use I_A^0 to determine r_B^0(F).
2a Use ΔI_B^0 to determine r_A^s(S).
2b Use ΔI_C^0 to determine r_A^s(S).
3a Use I_B^0 + first moment equation to determine r_A^0(C) and r_A^0(F).
3b Use $I_C^0 - I_A^0$ together with first moment equation to determine r_A^0(C) and r_A^0(F).

We can now determine the coordinates of the C and F atoms along the symmetry axis from the centre of mass using the first moment equation (centre-of-mass condition) together with either I_B^0 or I_C^0—there is again a choice. The four resulting sets of structural parameters are given in Table 6.11. One might at first think that one could dispense with the first moment equation in this last step and use I_B and I_C as two separate pieces of data. This is not possible because I_A, I_B and I_C are not all independent because of the inertial defect relation (6.112) with $\Delta^0 \sim 0$.

The second example concerns certain aspects of the structure of trimethylsilylisocyanate[32] and focuses attention on the important differences between microwave and electron diffraction measurements. Previous microwave studies on RNCO molecules showed a wide variation in the angle at the N atom as a function of the group R. For instance the following estimates of this angle are: HNCO, 130°;[144] CH_3NCO, 150°;[175] and SiH_3NCO, 180°.[93] The spectra of these molecules are rather interesting in that this angle appears to be unusually flexible and the spectra more complex than usual. The microwave spectrum of the ground vibrational level of $Si(CH_3)_3NCO$ was shown in Figure 6.6. As discussed in Section 6.10c this structure is consistent with that of a symmetric top molecule. On a rigid-rotor picture, a departure of at least 5° from a linear Si—N=C=O chain is necessary for the first-order asymmetry splitting of the $|K| = 1$ lines to be observable. On the other hand electron diffraction measurements on this molecule are consistent with an angle at the N atom of ~150°.[157] At first sight there appears to be a contradiction which on closer inspection disappears.

A major feature of the microwave spectrum of $Si(CH_3)_3NCO$[32] is that many of the vibrational levels associated with the chain bending frequency are populated (Figure 9.11). The intensities of the satellites indicate that the vibrational separations are of the order of 60 cm^{-1}. This low frequency together with the $(v + 1)$ degeneracy factor causes many molecules to be in highly excited levels of the bending vibration at room temperature. Electron diffraction measurements yield r_g, the value of r averaged over all the vibrational states populated at the temperature of the measurement. The r_g bond length is defined by

$$r_g = \int rP(r)\,dr \Big/ \int P(r)\,dr \tag{6.147}$$

$$P(r) = \sum_v \psi_v^2\, e^{-(E_v - E_0)/kT} \tag{6.147a}$$

We can now reconcile the discrepancy between the two measurements and at the same time develop a deeper understanding of the meaning of structural data. In the classical picture the wave functions associated with the higher vibrational levels correspond to dynamic geometries which on average are close to the classical turning points.[122, Figure 42] Thus, as many of the high vibrational states are populated at room temperature the electron diffraction measurements yield a bent average structure. Microwave spectroscopy can in principle probe the individual vibrational states and indicates that in this case the ground state is close to linear. Electron diffraction data on SiH_3NCO[94] indicate there may be a small hump in the potential curve associated with this angle and this data should be obtainable from an analysis of the vibrational satellites in the microwave spectrum.

It is perhaps also worth noting that the harmonic contributions average out when we determine $\langle r \rangle$ as indicated in Section 6.2 and it is possible to relate electron diffraction data to microwave data in some cases.[100,229,206]

A second consequence of the type of motion which occurs in $Si(CH_3)_3NCO$ is that care must also be taken when we consider the significance of the bond lengths. The Si—N length appears to be rather long whereas the N—C bond appears to be rather short. This type of effect has been very clearly characterized by Winnewisser and Winnewisser in their studies on HCNO.[339] A careful analysis of the microwave and infra-red spectra of this molecule indicates that it is in fact quasi-linear with a small potential hump in the bending potential which is well below the ground vibrational level. The spectrum of the ground state is consistent with a linear HCNO chain with an unusually short H—C bond length ~1.027 Å. We can understand this if we realize that the motion associated with the $\widehat{HCN}$ bending vibration has a large amplitude. The resulting rotational constant corresponds to the averaging of the projection of the C—H bond length on the equilibrium principal *A*-axis. A bent equilibrium structure with a C—H bond length of 1·060 Å is consistent with the data.

Chapter 7

The Stark Effect

7.1 The Interaction Between an Electric Charge Distribution and an Electric Potential

In Chapter 4 an expression (4.1) was given for the energy of a system in the presence of an electric field. This expression is an expansion in powers of the components of a *uniform* field and neglects the variation of the field over the space of the molecule. In this chapter we consider the first-order energy of interaction of an electric charge distribution described by $\rho(r)$ and an electric potential whose spatial properties are described by $V(r)$. The Hamiltonian for the interaction if no polarization of ρ is assumed is then

$$H_E = \int \rho(r)V(r)\,\mathrm{d}\tau \tag{7.1}$$

On expansion of $V(r)$ we obtain (in space-fixed coordinates)

$$V(r) = V^{(0)} + \sum_i r_i\left[\frac{\partial V}{\partial r_i}\right]_0 + \frac{1}{2!}\sum_{ij} r_i r_j\left[\frac{\partial^2 V}{\partial r_i \partial r_j}\right]_0 + \cdots \tag{7.2}$$

If $V_i^{(1)} = (\partial V/\partial r_i)_0$ and $V_{ij}^{(2)} = (\partial^2 V/\partial r_i \partial r_j)_0$ etc. then we can re-express H_E as

$$H_E = V^{(0)}\int \rho(r)\,d\tau + \sum_i V_i^{(1)}\int \rho(r)r_i\,d\tau + \frac{1}{2!}\sum_{ij} V_{ij}^{(2)}\int \rho(r)r_i r_j\,d\tau + \cdots \tag{7.3}$$

We can deal with each term in (7.3) individually.

The first term, $V^{(0)}\int \rho(r)\,d\tau$, is the familiar energy of a charge in a uniform potential and can be rewritten as $V^{(0)}q$, where the charge density integral can be treated as if the charge is concentrated at a single point, i.e. $\int \rho(r)\,d\tau = q$.

The second term has been discussed in Chapter 4 and here we have a physical description. It is the scalar product of the potential gradient (field) $-V_i^{(1)}$ and the dipole moment of the charge distribution, $\mu_i = \int \rho(r)r_i\,d\tau$. This term represents the Stark energy.

The third term is a little trickier to visualize and deal with. It represents the interaction of the electric quadrupole moment of $\rho(r)$ with the field gradient. If we set the integral $\int \rho(r)r_i r_j\,d\tau = F_{ij}^{(2)}$ then the energy can be written as the scalar product of two second-rank tensors $V_{ij}^{(2)}F_{ij}^{(2)}$. Note that when we come to consider these quantities in more detail, only the orientation dependent parts $V_{ij}^{(2)}$ and $F_{ij}^{(2)}$ are used. Higher-order multipole interaction terms are clearly generated in a straightforward fashion and one can thus rewrite (7.3) as

$$H_E = V^{(0)}F^{(0)} + V_i^{(1)}F_i^{(1)} + \tfrac{1}{2}V_{ij}^{(2)}F_{ij}^{(2)} + \tfrac{1}{6}V_{ijk}^{(3)}F_{ijk}^{(3)} + \cdots \qquad (7.4)$$

where $F^{(0)} = q$, $F_i^{(1)} = \mu_i, \cdots$, etc.

7.2 The Stark Interaction

We consider a rigid rotating molecule whose energy of interaction with a static field is given by the second term in (7.4). This is not too convenient a form as on rotation the space-fixed vector components $\mu_i(\equiv F_i^{(1)})$ are time dependent. We can, however, use the direction cosine matrix **S** to transform to molecule-fixed components, μ_α, which are molecular constants and setting $V_i^{(1)} \equiv -E_i$, the electric field, we obtain

$$H_E = -E_i S_{\alpha i} \mu_\alpha \qquad (7.5)$$

μ_α are the components of the *molecular* dipole moment defined by

$$\mu_\alpha = \int \rho(r) r_\alpha \, d\tau \qquad (7.6)$$

We can use (7.5) to evaluate the contribution to the molecular energy due to the Stark effect by perturbation theory.

It is perhaps worthwhile to list the reasons for the importance of the Stark effect as far as rotational spectroscopy and microwave spectroscopy in particular are concerned:

(i) The Stark pattern, if well resolved, can be used to identify the J value of a line.

(ii) Q branch ($\Delta J = 0$) and R or P branch ($\Delta J = \pm 1$) lines can be distinguished by the intensity distribution in the Stark pattern.

(iii) Quantitative analysis of the Stark shift yields the dipole moment of the molecule in a particular vibrational state.

(iv) Symmetric top levels with $K \neq 0$ exhibit first-order Stark effects (the energy varies linearly with field) whereas in linear and asymmetric molecules it is second order (quadratic in field). The qualitative patterns are characteristically different. Care must be taken when levels lie close together, as in the case when a molecule is only very slightly asymmetric and the asymmetry K splitting is small. This happens in F_2CS,[34] as is discussed later. The $v = 1$, $l = \pm 1$ levels of the bending vibration of a linear molecule can show first-order effects for the same reason.

(v) Experimentally, the Stark effect is generally used as a means of modulating the signal such that phase-sensitive detection techniques can be used.

We have thus to solve for the eigenvalues of the field-dependent Hamiltonian

$$H = H_r + H_E \tag{7.7}$$

where H_r is the field-free Hamiltonian solved in Chapter 3. The effects of H_E will be determined by perturbation theory as far as second order. In fact this problem serves as an excellent illustration of the perturbation method.

7.3 First-order Stark Effect

If we use the perturbation procedure summarized in Chapter 1 we see that the first-order correction to the energy due to the field is given by taking matrix elements of H_E over the *unperturbed* wave functions. As all types of rotor can be described by summations over the symmetric rotor wavefunction $|J\,K\,M\rangle$ we see from (7.5) that we need to evaluate the term

$$-E_i\mu_\alpha\langle J\,K\,M|S_{\alpha i}|J\,K\,M\rangle \tag{7.8}$$

The dipole moment of a symmetric top molecule lies along the axis of symmetry. Standard microwave spectroscopy measurements are usually made using the experimental arrangement where the field is linearly polarized and thus only one field component is non-zero. Thus we need determine only one tensor component, say S_{cz}, if we follow convention and assign E_z and μ_c as the non-zero parameters. Evaluating $\langle J\,K\,M|S_{cz}|J\,K\,M\rangle$ we find that

$$E_E^{(1)} = -E_z\mu_c\frac{KM}{J(J+1)} \tag{7.9}$$

Note the K dependence of this term which indicates that the $K = 0$ stack of symmetric rotor eigenstates does not show a first-order Stark effect, nor in fact do linear molecules (in their ground vibrational states) which also have $K = 0$. In the Hughes–Wilson Stark cell arrangement[138,275] $\Delta M = 0$ and thus for a $J+1 \leftarrow J$ transition ($\Delta K = 0$ as usual) we find that the frequency of a particular transition in the presence of an electric field is shifted by an amount given by

$$\Delta E_E^{(1)} = E_z\mu_c\frac{2KM}{J(J+1)(J+2)} \tag{7.10}$$

For E_z in volts/cm, μ_c in Debyes and ΔE in MHz the appropriate expression is

$$\Delta E_E^{(1)} = (0{\cdot}50344)E_z\mu_c\frac{2KM}{J(J+1)(J+2)} \tag{7.11}$$

The resulting first-order Stark pattern is shown schematically in Figure 7.1 for a $3 \leftarrow 2$ transition of a symmetric top. There are a couple of points worthy of mention:

(a) The K structure of a symmetric top transition is not always well resolved, especially at low J where the centrifugal distortion may not be significant. The resulting Stark pattern will be a composite sum over the various substructures. A typical example is shown in Figure 7.1 for the $3 \leftarrow 2$ transition of SiH_3Cl if we neglect the effects of quadrupole interactions.

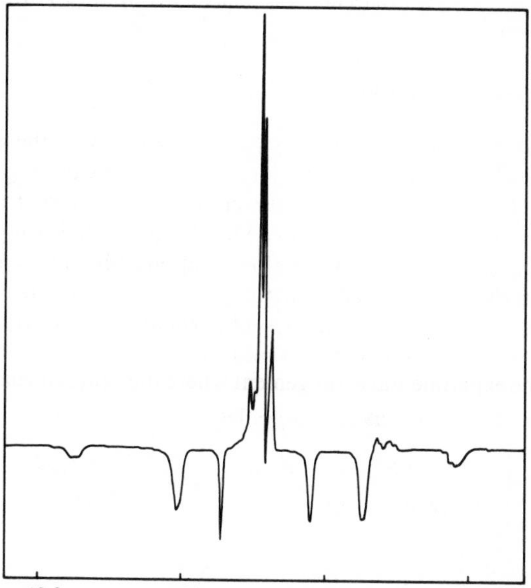

Figure 7.1 Stark effect pattern for the $J = 3 \leftarrow 2$ line of SiH_3Cl. To a first approximation we can neglect the quadrupole effect. The overall symmetric top Stark pattern which is observed is composed of the patterns belonging to the $|K| = 1$ and 2 lines as seen from (7.11). The outermost lobes have $K = M = \pm 2$. Those with $K = \pm 1$ and $M = \pm 2$ overlap those for which $K = \pm 2$ and $M = \pm 1$ and are displaced by $\frac{1}{2}$ the distance of the outer lobes. The lobes with $K = \pm 1$ and $M = \pm 1$ lie at $\frac{1}{4}$ this distance and the $K = 0$ and $M = 0$ lobes are only weakly modulated

(b) It is also worth noting that in this case there are nine stark components (in Figure 7.1) for which K or $M = 0$ and are subject only to 'slow' second-order shifts. These components in this fairly typical case account for more than half the overall intensity of the line. Such a transition will therefore not be anything like fully modulated at low voltages.

(c) When the Stark pattern is not resolved with a Stark spectrometer two other methods may be used to determine μ. One is the method of Lindfors and Cornwell which utilizes the rate of growth of the Stark pattern with applied field.[181] The other is a double resonance technique applied by Martins and Wilson to $XeOF_4$.[193] The accuracy of these techniques appears to be of the order of 10–20%.

(d) Note that the symmetry of this pattern will be lost at higher fields due to the fact that second-order effects which we are about to discuss will start to become important.

7.4 Second-order Stark Effect

The second-order corrections may be carried out with the aid of (1.46c) using the symmetric top wave functions. The results can be generalized to linear molecules by putting $K = 0$ as well as the symmetric top levels with $K = 0$. As asymmetric top properties may be calculated using symmetric-rotor eigenfunctions the results are directly applicable to this problem also.

We can use the interaction diagram, Figure 7.2, as a simple book-keeping device to aid the calculation and also get a feel for this type of problem and the jargon which is used in these types of problem such as 'mixing' and 'pushing'.

Expression (1.46c) for this case becomes

$$E_E^{(2)} = \mu_c^2 E_z^2 \left[\frac{|\langle J\ K\ M|S_{cz}|J+1\ K\ M\rangle|^2}{E(J) - E(J+1)} + \frac{|\langle J\ K\ M|S_{cz}|J-1\ K\ M\rangle|^2}{E(J) - E(J-1)} \right] \tag{7.12}$$

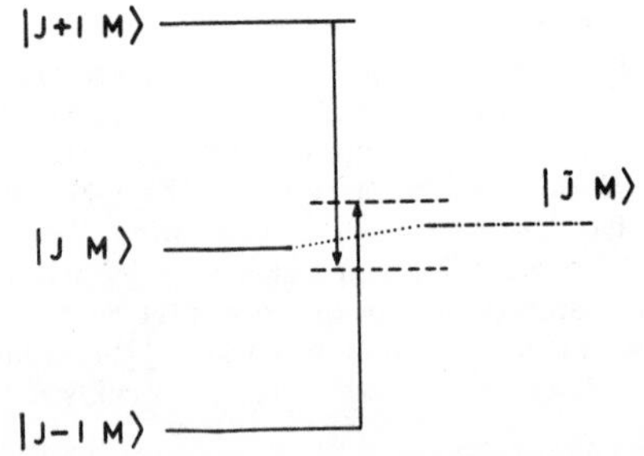

Figure 7.2 Perturbation pushing diagram. Second-order perturbation interactions from above and below combine to determine the position of the perturbed energy level, $|\tilde{J}M\rangle$

which with the aid of Table 4.1 becomes

$$E_E^{(2)} = \frac{\mu_c^2 E_z^2}{2B}\left[-\frac{[(J+1)^2 - K^2][(J+1)^2 - M^2]}{(J+1)^3(2J+1)(2J+3)} + \frac{[J^2 - K^2][J^2 - M^2]}{J^3(2J+1)(2J-1)}\right] \tag{7.13}$$

If we evaluate (7.13) for $K = 0$ we obtain the Stark energy shift for a linear molecule and also that of the $K = 0$ stack of a symmetric top as:

$$E_E^{(2)} = \frac{\mu_c^2 E_z^2}{2B}\left[\frac{J(J+1) - 3M^2}{J(J+1)(2J-1)(2J+3)}\right] \tag{7.14}$$

For the special case of the $J = 0$ level only the first term inside the bracket in (7.13) occurs and for this case we obtain

$$E_E^{(2)} = \frac{\mu_c^2 E_z^2}{2B}\left[-\frac{1}{3}\right] \quad (J = 0) \tag{7.15}$$

In the interaction diagram, Figure 7.2, we can see the result of the Stark interaction qualitatively. Under the influence of the field the eigenstates are 'mixed'. The meaning of this expression is quantitatively implied by the non-diagonal matrix elements in the expression (7.12). Qualitatively any particular eigenstate is 'pushed' from above and below by the adjacent eigenstates, whether the energy goes up or down depends on the relative values of J and M as in (7.14). It is interesting to consider this effect in a semi-classical way with the aid of Figure 7.3 where the two limiting cases for a linear molecule are depicted. Initially the molecule is assumed to be in a stationary state. The motion of the molecule in such a state is considered to be constant and is said to be conserved. In Figure 7.3(a), the high M case, J, is pointing approximately along the field direction. In this state there is essentially no motion in the direction of the field and the same arguments apply as in a static model. We find that the field will exert a torque which will tend to tilt the molecule such that we obtain an average dipole moment component in the direction of the field. The energy of the state will thus be *lowered* as expected! In the low M case, Figure 7.3(b), the field will exert a torque which will tend to slow the dipole down as it moves away from the field direction and speed it up as it moves towards it. The net effect is that it will be moving more quickly during the half-period that it points in the field direction than the other half-period. Thus on *time average* one obtains a dipole moment opposing the field and a consequent *increase* in energy. For the special case where $J = 0$ one can consider that in the presence of the field the molecule oscillates about the field direction, never quite making it round—thus one gets a lowering of energy as in the static model case.

We can also note at this point that we have discussed an example which bears out a general property of quantum mechanical systems.[169] A system

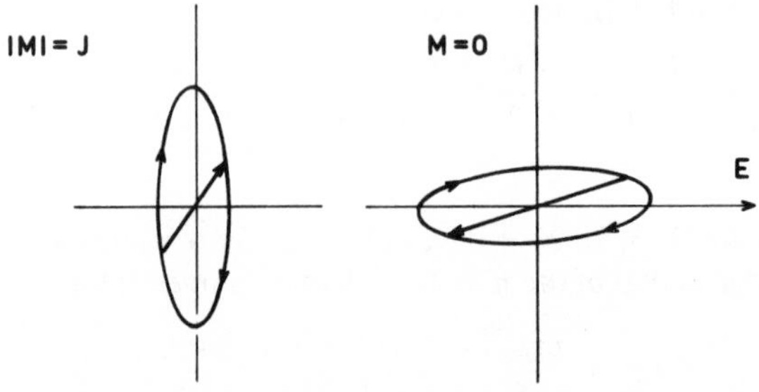

Figure 7.3 Classical model for the rotation of a dipole under the influence of a field. (a) The dipole, represented by the arrow, is rotating in a plane perpendicular to the field direction when $|M| = J$. The field will tend to tip the dipole towards the field direction thus inducing a dipole moment component in the field direction. (b) In the case where $M = 0$, $\boldsymbol{J}$ is perpendicular to the field, and as discussed in the text, in this case the field perturbs the motion to induce a dipole moment component which is *opposed* to the field direction

may only possess an average dipole moment if a degeneracy exists. This can be seen to be the case here. In free space a linear molecule rotates uniformly such that on time average the space-fixed dipole moment is zero. A symmetric top on the other hand has a K degeneracy and can rotate such that its symmetry axis (in which μ lies) precesses about a particular direction and can thus have a dipole moment which will not average to zero. This is why symmetric tops show first-order effects. In the linear molecule on the other hand the field must first interfere with the molecular motion (i.e. a sort of polarization effect occurs) such that a dipole moment is induced and then interaction can take place. This is the significance of the second-order contributions which come when matrix elements are taken over perturbed eigenfunctions. The $J = 0$ level must go down in energy as there is no level below it to counteract the pushing effect of the level above it associated with the negative term inside the brackets in (7.13).

In Figure 7.4 an energy-level diagram is drawn for the first few levels of a linear molecule.[137,288] The diagram shows the rapid decrease of the Stark effect with J as is implied by (7.14). Using (7.14) we find that the Stark energy shift can be written in a convenient form as:

$$\Delta E(J, M) = 0{\cdot}25345\frac{\mu_c^2 E_z^2}{B} f(JM) \tag{7.16}$$

where E_z is in MHz, μ_c in Debyes, E_z in volts/cm and

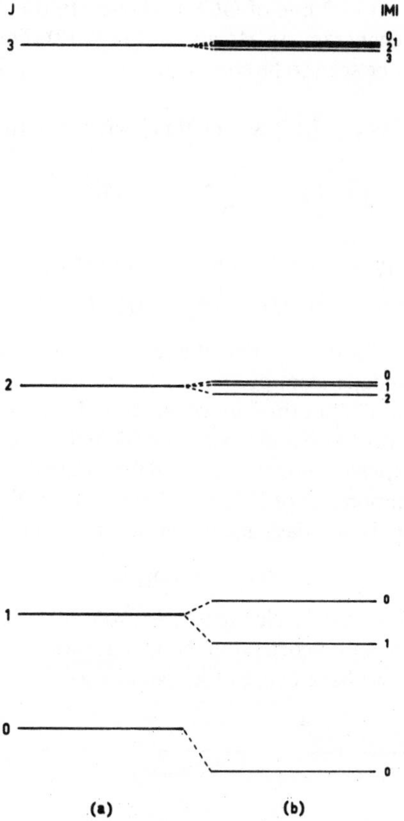

Figure 7.4 The energy levels of a linear molecule. (a) The field-free energy levels. (b) The energy levels in the presence of a field. The splittings of the $|M|$ components have been exaggerated by a factor of ~1000 compared to the field-free spacings (for a field of average strength ~100–1000 volts/cm)

$$f(JM) = \frac{3M^2(8J^2 + 16J + 5) - 4J(J + 1)^2(J + 2)}{J(J + 1)(J + 2)(2J - 1)(2J + 1)(2J + 3)(2J + 5)} \tag{7.17}$$

For the first few levels $f(JM)$ is

$$f(00) = -0{\cdot}26667, \quad f(10) = -0{\cdot}07619, \quad f(11) = 0{\cdot}06190$$

$$f(20) = -0{\cdot}01270, \quad f(21) = -0{\cdot}003571, \quad f(22) = 0{\cdot}02381$$

In Figure 7.5 the 3 ← 2 line of OCS is shown. In this case the field does not lift the $\pm M$ degeneracy as M^2 appears in (7.17). The displacements of the Stark lobes can be seen to be consistent with the ratios of the coefficients listed above.

The energy expression (7.14) is sometimes written in the form

$$E_E^{(2)} = \frac{\mu_c^2 E_z^2}{B}(A_J + B_J M^2) \tag{7.18}$$

and thus we can express $f(JM)$ in the transition energy shift (7.16) as

$$f(JM) = \Delta A_J + \Delta B_J M^2 \tag{7.18a}$$

Note has been taken of the fact that the second-order Stark shift is the sum of two terms, one independent of M the other proportional to M^2.

The evaluation of (7.5) in the case of linear and symmetric top molecules is fairly straightforward as symmetry reduces (7.5) to but one term, $-E_z S_{cz} \mu_c$. Also closed form expressions can be derived for ΔE. In the case of asymmetric rotors all three components of (7.5) may have to be evaluated if there is no symmetry and as H_E is not diagonal in J the matrix can be quite complicated.

$$H_E = -E_z S_{\alpha z} \mu_\alpha \tag{7.19}$$

We thus need the matrix elements $\langle J\ \tau\ M|S_{\alpha z}|J'\ \tau'\ M\rangle$ to evaluate the second-order perturbation correction. Note that these elements are diagonal in M (Table 4.1) as we have chosen the spatial z-azis.

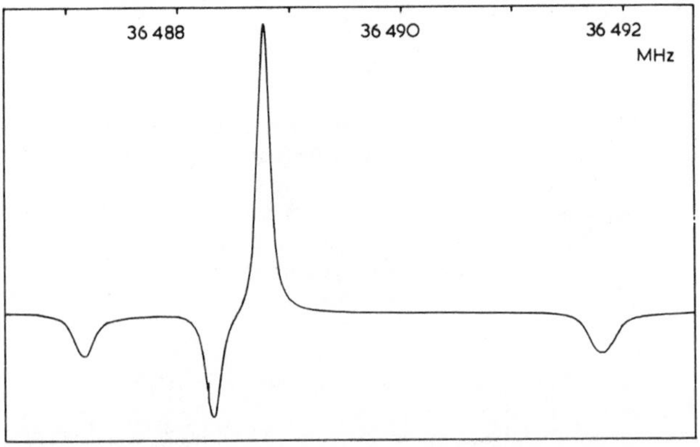

Figure 7.5 Stark effect observed for the $J = 3 \leftarrow 2$ line of OCS. The field-free line is displayed positively and the three field-split components (lobes) with $|M| = 0$, 1 and 2 are displayed negatively. The separations between the $M = 0$ lobe (at 36 487 MHz) and the others should be proportional to M^2 (7·18). The relative intensities of the lobes should, according to (7.24), be $9 : 8 \times 2 : 5 \times 2$ for $|M| = 0 : 1 : 2$ respectively

$$E_E^{(2)} = E_z^2 \sum_\alpha \mu_\alpha^2 \sum_{J'\tau'}{}' \frac{|\langle J\,\tau\,M|S_{\alpha z}|J'\,\tau'\,M\rangle|^2}{E_{J\tau} - E_{J'\tau'}} \tag{7.20}$$

The summation is restricted to *exclude* diagonal terms with $J\tau M = J'\tau'M$. There are also symmetry restrictions on the matrix elements of (7.19) which may have non-zero values. Thus the product

$$\Gamma(\psi_{J\tau}) \times \Gamma(S_{\alpha z}) \times \Gamma(\psi_{J'\tau'}) \tag{7.21}$$

must transform as A species under the operations which characterize D_2 (Table 3.7). We can thus structure the Hamiltonian matrix according to this recipe.[95]

We can use the expression (4.61) to factorize out the M dependence in (7.20) and so obtain

$$E_z^2 \sum_\alpha \mu_\alpha^2 \left[\frac{J^2 - M^2}{J^2(4J^2-1)} \sum_{\tau'} \frac{|\langle J\,\tau|S_{\alpha z}|J-1\,\tau'\rangle|^2}{E_{J\tau}^0 - E_{J-1\tau'}^0} + \frac{M^2}{J^2(J+1)^2} \sum_{\tau'}{}' \frac{|\langle J\,\tau|S_{\alpha z}|J\,\tau'\rangle|^2}{E_{J\tau}^0 - E_{J\tau'}^0} \right.$$
$$\left. + \frac{(J+1)^2 - M^2}{(J+1)^2(2J+1)(2J+3)} \sum_{\tau'} \frac{|\langle J\,\tau|S_{\alpha z}|J+1\,\tau'\rangle|^2}{E_{J\tau}^0 - E_{J+1\tau'}^0} \right] \tag{7.22}$$

This expression is essentially the same as that given by Golden and Wilson[95] with the phase and notation changes indicated in the footnote of Table 4.1. Golden and Wilson have also given the resulting shift relation as

$$E_E^{(2)} = \frac{E_z^2}{A + C} \sum_\alpha \mu_\alpha^2 [A_{J\tau} + M^2 B_{J\tau}] \tag{7.23}$$

in which the M dependent and independent parts have been conveniently separated as in (7.18). One can evaluate (7.22) in most cases by using the line strengths λ defined in (4.84) which have been tabulated as a function of κ.[59,346] Using relations (4.85) the tabulated numbers can be substituted directly into (7.22), as discussed in Section 7.6.

7.5 Stark Effect Patterns and Line Assignments

In the case of a first-order Stark effect the pattern of the Stark components will be symmetric about the field-free line as shown in Figure 7.1. On the other hand the second-order effects will not show this symmetry. Examination of (7.17), (7.18) and (7.23), which govern the second-order behaviour indicates that the various $|M|$ components will form a characteristic pattern in which their separations from the position of the $M = 0$ component will be proportional to M^2. In the spectrum of OCS in Figure 7.5 these ratios are 1:4 and in Figure 6.7 the 4 ← 3 transition of HC≡C—C≡N is shown where the spacings are 1:4:9, as expected.

The intensity distribution among the Stark components is characteristic of the transition branch type. In Section 4.4c line intensities were discussed and in particular we see that according to (4.84) and (4.83)

$$I\,(R \text{ branch}) \propto [(J+1)^2 - M^2] \tag{7.24a}$$

$$I\,(Q \text{ branch}) \propto M^2 \tag{7.24b}$$

$$I\,(P \text{ branch}) \propto [J^2 - M^2] \tag{7.24c}$$

Often the highest M value components are subject to the strongest Stark shifts and will thus be the first components to be modulated. Thus for R and P branch transitions, for which $\Delta J = \pm 1$, we see that the highest M and most strongly displaced components will in general be weaker than the lower M components. Note that each $|M|$ component is doubly degenerate except for $M = 0$. The components in Figure 7.5 appear to have approximately the correct intensity ratios ($|M| = 0$, 1 and 2 should be 9:16:10 respectively). For Q branch lines we see that according to (7.24b) the most strongly displaced lines will usually be the strongest. An excellent example of this type of pattern is exhibited by the $17_{3\,14} \leftarrow 17_{3\,15}$ transition of H_2CO shown in

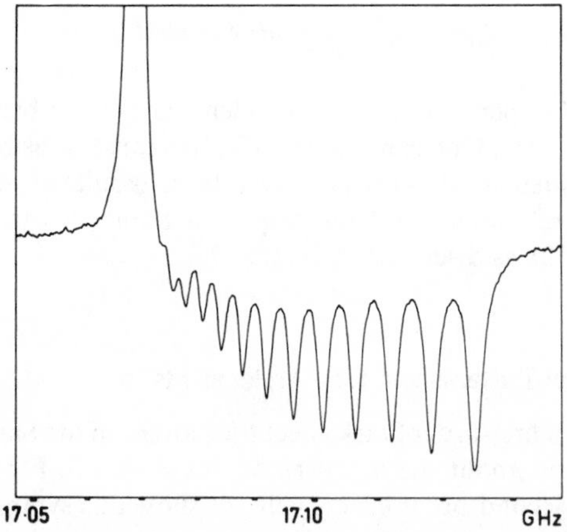

Figure 7.6 The Q-branch type Stark effect as observed on the $17_{3\,14} \leftarrow 17_{3\,15}$ line of H_2CO. The intensity is seen to be consistent with (7.24b). The line has not been fully modulated even at a field of 2000 Volts/cm

Figure 7.6. Even if the components are not resolved the contours of the Stark pattern can serve as an indication of the branch type.

If all the components are resolved then the J value of the line is immediately determined. When $\Delta J = +1$ there are $J + 1$ components, when $\Delta J = -1$ there are J components and when $\Delta J = 0$ there are J components. Transitions for which $\Delta J = 0$ and $M = 0$ are forbidden. If, as frequently occurs, the Stark pattern is incompletely resolved then the number of components determines a lower limit to the J value of the line. Thus for the line shown in Figure 7.6 we see that there are thirteen resolved components and there may be other components which have still not been pulled out by the field. As it is a Q branch line the minimum value of J is thus 13.

This discussion applies only to microwave spectra for which $\Delta M = 0$. In some spectrometers the field and the electric vector of the radiation are perpendicular. In these cases $\Delta M = \pm 1$ transitions occur and the Stark patterns are more complex.[100]

Use can often be made of the fact that on a Stark spectrometer the larger the *effective* dipole moment associated with a particular line, the lower will be the strength of the electric field required for it to be modulated. As long as the Stark shift is smaller than the line width, the line shape will be distorted, and if it is very small the line may not be observed at all. In Figure 7.7 is

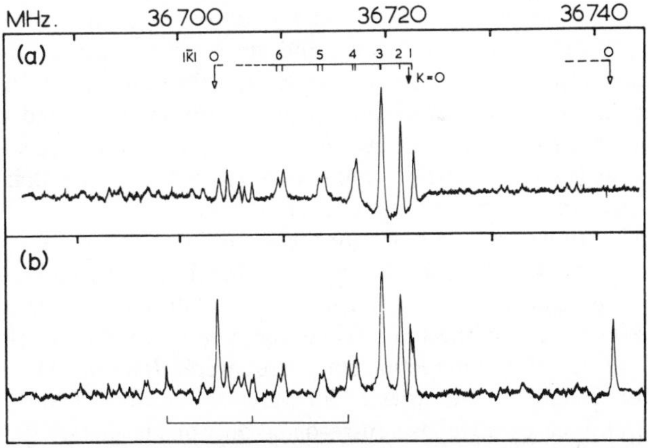

Figure 7.7 (a) The spectrum of the $v = 1$ vibrational satellite of the methyldiacetylene $J = 9 \leftarrow 8$ transition observed at 20 volts per cm on a Stark spectrometer.[166a] The $K = 0$ line for which $\bar{K} = \pm 1$ and $l = \pm 1$, and the $K = \pm 1$ lines for which $\bar{K} = 0$ and $l = \pm 1$ do not show first order Stark effects and are not modulated at this low field strength. (b) The same spectrum observed at 1600 volts/cm. The lines which show a slow Stark effect appear. The $v = 1$ satellite of a second vibration appears weakly and can be identified by the Stark behaviour. The pattern is indicated in (b)

shown the $v = 1$ vibrational satellite of the $J = 9 \leftarrow 8$ transition of methyldiacetylene at 20 volts per cm in (a) and at 1600 volts per cm in (b). At the lower voltage the $K = 0$ line is not modulated as the associated states can exhibit only second-order Stark effects. The $K = l = \pm 1$ states are split by the l-doubling interaction as discussed in Chapter 6 and depicted in Figure 6.5. This interaction destroys the degeneracy which gives rise to the first-order Stark effect. As a consequence these lines also are only slowly modulated and do not appear on the low-voltage spectrum, Figure 7.7(a).

7.6 Dipole Moment Determination

We can take a simple example to highlight many of the features of dipole moment determinations and the Stark effect itself. Consider the molecule F_2CS which has a very small dipole moment (0·08 Debye). There are two main consequences of this small dipole moment. One is that absorption lines are very weak $\sim$150 times weaker than if the dipole moment was $\sim$1 Debye. A second and quite serious problem on a Stark spectrometer is that the lines are not well modulated. To be well modulated the Stark shift should be of the order of the line width or greater. On such a spectrometer, the line and Stark components are displayed with opposite phase and thus will tend to cancel. Some energy levels may however be nearly coincident and linked by non-zero transition moments. As a consequence the second-order expression (7.22) might break down and we must diagonalize the 2×2 Stark interaction matrix for the two levels exactly using (1.40): Physically what happens is that the Stark interaction energy increases and becomes large compared to the unperturbed level separation. These levels then are essentially degenerate relative to the perturbation and the Stark effect becomes first order as in the symmetric rotor situation.

Such a situation occurs in F_2CS for the $K_A = 3$ doublet of the $J = 4 \leftarrow 3$ transition where the lower levels are only 23·1 MHz apart and the upper levels 159·8 MHz apart. We can restrict our calculation to the evaluation of two 2×2 matrices for the interactions represented by $\langle 4_{32}|H_E|4_{31}\rangle$ and $\langle 3_{31}|H_E|3_{30}\rangle$. All other interacting states are several thousand MHz distant and introduce shifts of the order of 0·1 MHz or less at $\sim$1000 V/cm field strength. In this case only μ_A is non-zero and only one term in (7.19) need be considered, $H_E = -E_z S_{Az}\mu_A$. For these levels we set up two matrices of the form (1.39) with the following data $\kappa = -0{\cdot}6264$. Thus extrapolation of the λs given in tables[346] yields:

$$\lambda^A(4_{31} - 4_{32}) = 4{\cdot}005, \qquad \lambda^A(3_{30} - 3_{31}) = 5{\cdot}232$$

$$\Delta E = 159{\cdot}8 \text{ MHz}, \qquad \Delta E = 23{\cdot}1 \text{ MHz}$$

Using Table 4.1 together with the $\Delta J = 0$ expression (4.85b) we see that

$$\alpha(\text{MHz}) = \langle J\,\tau\,M|H_E|J\,\tau'\,M\rangle = -(0{\cdot}50344)E_z\mu_A M\left[\frac{\lambda^A(J\tau, J\tau')}{J(J+1)(2J+1)}\right]^{1/2} \tag{7.25}$$

The conversion factor is defined in (7.11). The dipole moment was determined by diagonalizing the two interaction matrices for various values of the dipole moment and comparing the resulting Stark behaviour with experiment. In Figure 7.8 is shown the observed and calculated behaviours and in Figure 5.11 these particular lines are shown; note the mirror-image Stark pattern which for $\delta \to 0$ approaches that of a symmetric top. The result of this is, that $\mu_A = 0{\cdot}08$ Debye. Using this value for instance we see that

$$\alpha(3_{30} - 3_{31}) = -1{\cdot}0053 \times 10^{-2}\,EM \quad (\text{MHz}) \tag{7.26a}$$

$$\alpha(4_{31} - 4_{32}) = -0{\cdot}6009 \times 10^{-2}\,EM \quad (\text{MHz}) \tag{7.26b}$$

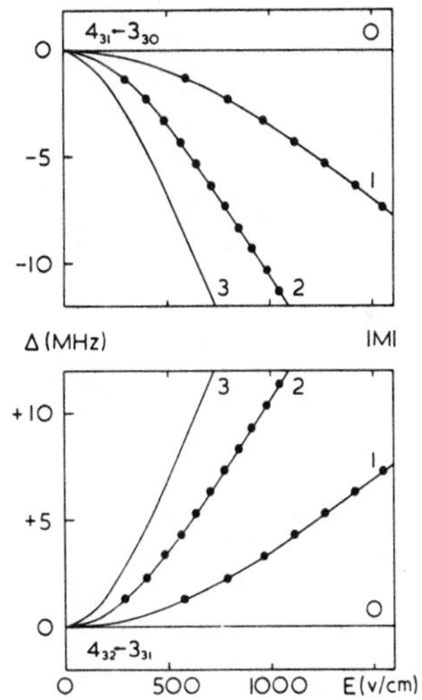

Figure 7.8 Plot of the observed and calculated frequency shifts (Δ) for the various Stark components ($|M|$) of the $4_{31} \leftarrow 3_{30}$ and $4_{32} \leftarrow 3_{31}$ transitions of F_2CS as a function of electric field (E). The solid lines represent the calculated behaviour for a dipole moment of 0·080 Debye and the points represent the experimental measurements[34]

It is worthwhile examining the behaviour of the individual levels as the field increases. Using these values of α the plot of shift versus field shown in Figure 7.9 was obtained. We note immediately that the $J = 3$ levels show the stronger field dependence. There are two reasons for this, not only are these levels closer together (more nearly degenerate, Figure 3.5) but also the Stark effect decreases with increasing J and in this case we see that it is roughly proportional to $1/J^3$.

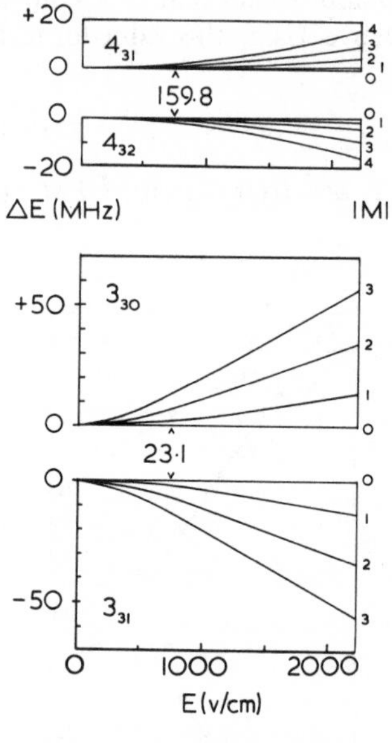

Figure 7.9 Plot of the calculated behaviours of the energy levels 4_{31}, 4_{32}, 3_{30} and 3_{31} of F_2CS as a function of electric field E

Finally one can check that the formula is right by comparing the result one obtains using $\lambda(\kappa \to -1{\cdot}0)$ with the first-order result (7.10) for a true symmetric top. When $\kappa = -1$ the tabulated values for the line strengths are $(4_{32} - 4_{31}) = 4{\cdot}05 = 81/20$ and $\lambda(3_{31} - 3_{30}) = 5{\cdot}25 = 21/4$. These values can be substituted into (7.25) and if we note that in the symmetric top limit the levels are degenerate ($\delta = 0$ in (1.40)) then we see that the shift is simply $\pm\alpha$. This result is identical with that given by (7.9) using the appropriate values of J and K. In the high field limit $\Delta E \sim \pm\alpha$ as can be seen by comparing (7.26a) with Figure 7.9 for $J = 3$.

Chapter 8

Nuclear Quadrupole Hyperfine Interactions

8.1 Electric Interactions Involving the Nucleus

Under the high resolution afforded by microwave techniques, effects of interactions involving the atomic nuclei can in certain cases be detected. Some nuclei possess spin angular momentum which causes the charged particles (protons) in the nucleus to sweep out a charge distribution which possesses a time average cylindrical symmetry about the spin axis. The distribution can be considered continuous because the rotation is fast compared with many other molecular processes. It is also a good approximation to assume that the relative positions of atoms in the molecule are fixed and that the electronic charge is well described by a static charge distribution relative to the molecular axes. As we are going to do our calculations in a space-fixed coordinate system, we must take into account the fact that the potential $V(\mathbf{r})$ witnessed by any particular nucleus is due to such a charge distribution averaged over the rotational motion of the molecule.

Under these conditions the interaction may be described by an expression of the form (7.4) involving integrals over the nuclear volume. A monopole term of the type $V^{(0)}F^{(0)}$ is independent of orientation and can be neglected. Terms of the type $V_i^{(1)}F_i^{(1)}$ can also be neglected as they involve a nuclear electric dipole moment. As nuclear eigenstates are non-degenerate they possess a definite parity[253,169] and thus integrals of the type

$$\int \psi_{\text{nuc}}\, \mu_i \psi_{\text{nuc}}\, d\tau \tag{8.1}$$

vanish because μ_i is odd. This problem is related to the case of the average value of $\boldsymbol{\mu}$ discussed in Section 7.4 on the Stark effect. The rule is rather more general than this, as all *odd* electric multipole moments are zero for a nucleus in a spin eigenstate.[253]

In this section we are going to deal with contributions to the molecular energy which originate in the third term of (7.4) namely $V_{ij}^{(2)}F_{ij}^{(2)}$. Before going further it is worth introducing a few useful techniques developed by Racah and Wigner, for handling tensor quantities. Monopoles and in general all scalar quantities are defined as zero-rank tensors. Similarly, vectors (and therefore dipole moments) are first-rank tensors and have a single index, i.e. μ_i, r_i, etc. Quantities with double indices, such as the moment of inertia tensor, $I_{\alpha\beta}$, as well as $V_{ij}^{(2)}$ and $F_{ij}^{(2)}$ are second-rank cartesian tensors. Complete discussions of tensor properties have been given by Racah,[247,248,249] Wigner,[330] Tinkham[298] and Rose.[259]

8.2 Spherical Tensors

A general cartesian second-rank tensor T_{ij} is reducible in the sense that it can be decomposed according to the expression

$$T_{ij} = \tfrac{1}{3}\delta_{ij}T^t + T_k^a + T_{ij}^s \tag{8.2}$$

involving tensors of the same or lower rank. In (8.2) T^t is the trace of $T_{ij}(\equiv T_{ii})$ and is independent of orientation. T_k^a is defined by

$$T_k^a = \tfrac{1}{2}(T_{ij} - T_{ji}) \tag{8.3}$$

and is an antisymmetric second-rank tensor with zero trace. Note that T_k^a vanishes if T_{ij} is symmetric, as is often the case. The tensor (2.21) is of this type. T_{ij}^s is a symmetric tensor also with zero trace:

$$T_{ij}^s = \tfrac{1}{2}(T_{ij} + T_{ji}) - \tfrac{1}{3}\delta_{ij}T^t \tag{8.4}$$

Note that there are one, three and five independent components associated with T^t, T_k^a and T_{ij}^s respectively, making nine in all as in the most general cartesian tensor T_{ij}. The usefulness of this decomposition lies in the different transformation properties (symmetries) possessed by the reduced tensors. As T^t, T_k^a and T_{ij}^s transform in the same way that the spherical harmonics with Y_m^l with $l = 0$, 1, 2 respectively do, these tensors are often called spherical tensors or irreducible tensors.

Many calculations involving tensors are facilitated by transforming from the cartesian basis to a spherical basis. As an introduction, consider the set of operators defined by

$$T_1^{(1)} = -\frac{1}{\sqrt{2}}T_+, \qquad T_0^{(1)} = T_z, \qquad T_{-1}^{(1)} = \frac{1}{\sqrt{2}}T_- \tag{8.5}$$

This set of tensor components transforms in the same way as the spherical harmonics Y_m^l with $l = 1$ and $m = +1$, 0 and -1 respectively. We have already used some of the useful properties of operators in this form, i.e. $J_\pm$. As an example it is often convenient in atomic problems to take the set of p orbital electronic eigenfunctions: $(p_x + ip_y)$, p_z and $p_x - ip_y$ as this set are all eigenfunctions of J_z and are associated with $m = +1$, 0 and -1 respectively.

A general spherical tensor component will be specified by the symbol $T_m^{(l)}$ where l denotes the rank and m the component. m takes on all values from $l, l-1, \ldots, -l$ and the set (8.5) are the complete set of irreducible spherical components for $l = 1$. Racah[247] uses the following commutation relations to *define* the $T_m^{(l)}$

$$[J_\pm, T_m^{(l)}] = \sqrt{(l \mp m)(l \pm m + 1)}\, T_{m\pm 1}^{(l)} \tag{8.6a}$$

$$[J_z, T_m^{(l)}] = mT_m^{(l)} \tag{8.6b}$$

Using these relations we see that the set (8.5) are generated by taking $T_0^{(1)} = T_z$ and the T-class operator definition (4.32a). Tensors of other rank can be generated from the set $T_m^{(1)}$ using the tensor product multiplying rule.

$$T_M^{(L)}(A_1A_2) = \sum_{m_1m_2} a_{Mm_1m_2}^{Ll_1l_2} T_{m_1}^{(l_1)}(A_1)T_{m_2}^{(l_2)}(A_2) \tag{8.7}$$

where the a coefficient is the Clebsch–Gordon (CG) or Wigner coefficient. There are unfortunately as many different notations for these coefficients as authors. The main reason for this is that there are redundancies in the indices on the CG coefficient as written in (8.7). These redundancies arise from the vector coupling relations which hold among L, l_1 and l_2 which must sum according to the triangle rule $\Delta(Ll_1l_2)$. This means that the three numbers must be able to form the three sides of a triangle.[259] This is the requirement that follows from the associated vector relation $\boldsymbol{L} = \boldsymbol{l}_1 + \boldsymbol{l}_2$.

In particular we note that because $M = m_1 + m_2$ the CG coefficients vanish for $m_2 \neq M - m_1$, by which m_2 can thus be replaced in (8.7). A tabulation of these coefficients is given by Condon and Shortley[49] where $a_{Mm_1m_2}^{Ll_1l_2} \equiv (l_1l_2m_1m_2|l_1l_2LM)$ in their notation. Tinkham[298] and Rose[259] discuss these coefficients and their properties in detail.

The CG coefficients for the scalar product $T_0^{(0)}(A_1A_2)$ generated by (8.7) require $l_1 = l_2(\equiv l)$ and $m_1 = -m_2$ and are given by the formula $(-1)^{l-m}(2l+1)^{-1/2}$ and thus

$$T_0^{(0)}(AB) = \frac{(-1)^l}{\sqrt{2l+1}} \sum_m (-1)^m T_m^{(l)}(A)T_{-m}^{(l)}(B) \tag{8.8}$$

Now $\boldsymbol{T}^{(l)}(A) \cdot \boldsymbol{T}^{(l)}(B) \equiv (-1)^l\sqrt{2l+1}\, T_0^{(0)}(AB)$ and for $l = 1$ the scalar product or invariant is just $\equiv -\sqrt{3}\, T_0^{(0)}$. As a result we can write

$$\mathscr{I}(AB) = \sum_{-m}^{m} (-1)^m T_m^{(l)}(A)T_{-m}^{(l)}(B) \tag{8.9}$$

Using (8.7) and Table 2[3] of Condon and Shortley[49] yields the following forms for the second-rank tensors.

$$T^{(2)}_{\pm 2}(AB) = A_{\pm 1}B_{\pm 1} = \tfrac{1}{2}A_{\pm}B_{\pm} \tag{8.10a}$$

$$T^{(2)}_{\pm 1}(AB) = \frac{1}{\sqrt{2}}(A_{\pm 1}B_0 + A_0B_{\pm 1}) = \mp\tfrac{1}{2}(A_{\pm}B_z + A_zB_{\pm}) \tag{8.10b}$$

$$T^{(2)}_0(AB) = \frac{1}{\sqrt{6}}(2A_0B_0 + A_1B_{-1} + A_{-1}B_1) = \frac{1}{\sqrt{6}}(2A_zB_z - A_xB_x - A_yB_y) \tag{8.10c}$$

In this case the CG coefficients are those for which $L = 2$ and $l_1 = l_2 = 1$. Thus $a^{2\,1\,1}_{M\,m\,M-m}$ is:

$$\begin{array}{ll} 1 & (M = 2,\quad m = \pm 1) \\ 1/\sqrt{2} & (M = 1,\quad m = \pm 1) \\ 2/\sqrt{6} & (M = 0,\quad m = 0) \\ 1/\sqrt{6} & (M = 0,\quad m = \pm 1) \end{array}$$

For instance, using (8.9) the electric dipole–field interaction Hamiltonian ($\equiv -\boldsymbol{E}\cdot\boldsymbol{\mu}$) may be expressed as

$$H_E = -[E^{(1)}_0\mu^{(1)}_0 - E^{(1)}_{-1}\mu^{(1)}_1 - E^{(1)}_1\mu^{(1)}_{-1}] \tag{8.11}$$

8.3 The Quadrupole Hamiltonian and Energy-level Expressions

Using the results of Section 8.2 we see that both $V^{(2)}_{ij}$ and $F^{(2)}_{ij}$ can be decomposed according to the expression (8.2) into zero, first- and second-rank tensors. We are interested in the interaction energy represented by the scalar product of $V^{(2)}_{ij}$ and $F^{(2)}_{ij}$ and thus from (8.8) we see that such a zero-rank tensor is formed only by combining tensors of equal rank. The interaction energy due to the combination of the traces may be neglected in our case for two reasons: (1) only S electrons may possess a non-negligible charge distribution within the nucleus and as this distribution is spherically symmetric, no orientation dependent interaction will occur, (2) for a potential due to a charge outside the nucleus the Laplace equation, $\nabla^2 V = V^t = 0$, applies. The antisymmetric tensor components of $F^{(2)}_{ij}$ and $V^{(2)}_{ij}$ both vanish because $[r_i, r_j] = 0$ and $(\partial^2 V/\partial r_i\partial r_j - \partial^2 V/\partial r_j\partial r_i) = 0$. We are thus left with an orientation dependent interaction energy which can be written as:

$$H_Q = \tfrac{1}{2}\sum_{ij} V^s_{ij}F^s_{ij} \tag{8.12}$$

where

$$F^s_{ij} = \int \rho(r)[\tfrac{1}{2}(r_ir_j + r_jr_i) - \tfrac{1}{3}r^2\delta_{ij}]\,d\tau \tag{8.13}$$

It is usual to set $3F^s_{ij} = Q_{ij}$ and call Q_{ij} the *quadrupole moment tensor* of the nucleus. As $V^s_{ij} = V^{(2)}_{ij}$ because the Laplace equation holds, and as the components of $\boldsymbol{r}$ commute we can write the Hamiltonian for the quadrupole interaction as

$$H_Q = \tfrac{1}{6}\sum_{ij} V^{(2)}_{ij} Q^{(2)}_{ij} \tag{8.14}$$

where

$$Q^{(2)}_{ij} = \int \rho(r)[3r_i r_j - r^2 \delta_{ij}]\, d\tau \tag{8.15}$$

Though $Q^{(2)}_{ij}$ has been simplified by application of the commutation properties of r_i we shall need to re-invoke the more correct form implied by (8.13) at a later stage. $V^{(2)}_{ij}$, the second derivative of the potential, is equal to $-\partial E_i/\partial r_j$, the negative of the first derivative of the field.

According to (8.9) we can express H_Q in spherical tensor form as

$$H_Q = \tfrac{1}{2}\sum_{-m}^{m} (-1)^m V^{(2)}_m F^{(2)}_{-m} \tag{8.16}$$

where the appropriate tensor components are listed in Table 8.1.

Table 8.1 Spherical Tensor Relations[a]

$F^{(2)}_{\pm 2} = \frac{1}{2}\int \rho(r)(r_x \pm ir_y)^2\, d\tau$	$V^{(2)}_{\pm 2} = \frac{1}{2}(V_{xx} - V_{yy} \pm 2iV_{xy})$
$F^{(2)}_{\pm 1} = \mp \int \rho(r) r_z(r_x \pm ir_y)\, d\tau$	$V^{(2)}_{\pm 1} = \mp(V_{xz} \pm iV_{yz})$
$F^{(2)}_0 = \frac{1}{\sqrt{6}}\int \rho(r)(3r_z^2 - r^2)\, d\tau$	$V^{(2)}_0 = \frac{1}{\sqrt{6}}(3V_{zz} - \sum_i V_{ii})$

[a] These tensor components are normalized such that they are consistent with the standard tensor multiplying relation (8.8). In some texts H_Q is written in the form

$$\sum_m (-1)^m V'_m Q'_{-m}$$

where $Q'_m = \sqrt{6/4}F^{(2)}_m$ and $V'_m = \sqrt{1/6}V^{(2)}_m$. Note that when the Laplace relation applies $\sum_i V_{ii} = 0$ in $V^{(2)}_0$.

We have now obtained the quadrupole energy expression in terms of operators which fairly readily yield matrix elements. We now recall some of the results of Chapter 4 where we discussed the matrix elements of T class operators and drew attention to the general property that the dependence of the matrix elements on the projection quantum numbers was the same for all operators in the class. This property is essentially the point made by the

Wigner–Eckart theorem[259] and we can use it to express the quadrupole Hamiltonian directly in terms of the angular momentum operators $\boldsymbol{I}$ and $\boldsymbol{J}$.

The classical expression for the quadrupole moment, Q, of a cylindrically symmetric nuclear charge distribution is

$$Q = \frac{1}{e}\int_{\text{nuc}} \rho(r)(3r_z^2 - r^2)\,d\tau \tag{8.17}$$

where e is the charge of a proton.

As implied by the Wigner–Eckart theorem we can write

$$\langle I\ m_I|eQ|I\ m_I\rangle = C_I\langle I\ m_I|3I_z^2 - \boldsymbol{I}^2|I\ m_I\rangle \tag{8.18}$$

The left-hand side of (8.18) would be equal to eQ if $\boldsymbol{I}$ and I_z were coincident. Quantum mechanical considerations indicate that this is never quite so, though it will approach this value at high values of I when $m_I = I$. It is standard convention to define this limiting case as 'the quadrupole moment of the nucleus'

$$eQ = \langle I\ m_I = I|\rho(r)(3r_z^2 - r^2)|I\ m_I = I\rangle \tag{8.19}$$

Evaluating the right-hand side of (8.18) and combining the result with this definition we see that

$$C_I = eQ/I(2I - 1) \tag{8.20}$$

An essentially identical argument allows us to express $V_0^{(2)}$ in terms of $\boldsymbol{J}$, the angular momentum of the molecular frame exclusive of $\boldsymbol{I}$. The *total* angular momentum is usually given the symbol $\boldsymbol{F}$ (with quantum number F) and is defined by

$$\boldsymbol{F} = \boldsymbol{I} + \boldsymbol{J} \tag{8.21}$$

We define the quantity q_J which is the value of the molecular field gradient in the z direction averaged over the state $|J\ m_J = J\rangle$ and given by

$$q_J = \langle J\ m_J = J|V_{zz}|J\ m_J = J\rangle \tag{8.22}$$

Using this relation the coefficient C_J in

$$\langle J\ m_J = J|\sqrt{6}V_0|J\ m_J = J\rangle = C_J\langle J\ m_J = J|3J_z^2 - \boldsymbol{J}^2|J\ m_J = J\rangle \tag{8.23}$$

becomes

$$C_J = q_J/J(2J - 1) \tag{8.24}$$

Using these results we can cast the Hamiltonian into the following cartesian representation as

$$H_Q = \tfrac{1}{6}C_I C_J[\tfrac{3}{2}(I_iI_j + I_jI_i) - \boldsymbol{I}^2\delta_{ij}][\tfrac{3}{2}(J_iJ_j + J_jJ_i) - \boldsymbol{J}^2\delta_{ij}] \tag{8.25}$$

As we are using a zeroth-order basis set in which we have assumed $[I_i, J_j] = 0$, on expanding (8.25) terms of the type $I_iI_jJ_iJ_j$ rearrange to $I_iJ_iI_jJ_j \equiv (\boldsymbol{I}\cdot\boldsymbol{J})^2$. If we use the commutation relations (3.4) we can reduce the most complicated terms as follows

$$I_iI_jJ_jJ_i + I_jI_iJ_iJ_j = 2I_iJ_iI_jJ_j - ie_{ijk}J_k[I_i, I_j] \tag{8.26}$$

As indicated in (A1.16), $e_{ijk}e_{ijl} = 2\delta_{kl}$ and thus (8.26) is equivalent to

$$2[(\boldsymbol{I}\cdot\boldsymbol{J})^2 + \boldsymbol{I}\cdot\boldsymbol{J}] \tag{8.26a}$$

We thus obtain the result

$$H_Q = \frac{1}{2}\frac{eQq_J}{I(2I-1)J(2J-1)}[3(\boldsymbol{I}\cdot\boldsymbol{J})^2 + \tfrac{3}{2}(\boldsymbol{I}\cdot\boldsymbol{J}) - \boldsymbol{I}^2\boldsymbol{J}^2] \tag{8.27}$$

We can easily evaluate the diagonal matrix elements of (8.27) because the components of $\boldsymbol{I}$ commute with those of $\boldsymbol{J}$. Thus from (8.21) we see that

$$\boldsymbol{I}\cdot\boldsymbol{J} = \tfrac{1}{2}[\boldsymbol{F}^2 - \boldsymbol{I}^2 - \boldsymbol{J}^2] \tag{8.28}$$

whose matrix elements in the $|J\ I\ F\rangle$ scheme are

$$\langle J\ I\ F|\boldsymbol{I}\cdot\boldsymbol{J}|J\ I\ F\rangle = \tfrac{1}{2}[F(F+1) - I(I+1) - J(J+1)] = \tfrac{1}{2}C \tag{8.29}$$

Using this we see that we can write

$$\langle F\ I\ J|H_Q|F\ I\ J\rangle = \tfrac{1}{2}\,eQq_J\left[\frac{\tfrac{3}{4}C(C+1) - I(I+1)J(J+1)}{I(2I-1)J(2J-1)}\right] \tag{8.30}$$

We now want to relate q_J which is the expectation value of the space-fixed component V_{zz} averaged over the rotational motion (for the state with $m_J = J$) to the molecular parameters $V_{\alpha\beta}$. They are related by the expression

$$V_{zz} = S_{\alpha z}S_{\beta z}V_{\alpha\beta} \tag{8.31}$$

where $S_{\alpha z}$ etc., are the direction cosines. It is again useful to call upon a spherical tensor form of V_{zz}. Thus we can use the relation (8.9) remembering however that we are now permuting molecule-fixed axes. Written out explicitly we obtain

$$\begin{aligned} V_{zz} = {} & \tfrac{1}{4}S_z^{+\,2}(V_{aa} - V_{bb} - 2iV_{ab}) + S_{cz}S_z^{+}(V_{ac} - iV_{bc}) + \tfrac{1}{6}(3S_{cz}^2 - \mathbf{S}^2)(3V_{cc}) \\ & + S_{cz}S_z^{-}(V_{ac} + iV_{bc}) + \tfrac{1}{4}S_z^{-\,2}(V_{aa} - V_{bb} + 2iV_{ab}) \end{aligned} \tag{8.32}$$

where $S_z^{\pm} = S_{az} \pm iS_{bz}$. In this expression a, b and c are the principal axes of the molecule's inertia tensor. *Only if the molecule possesses symmetry are they necessarily aligned with any of the principal axes of the field-gradient tensor.* Essentially the same point holds for the more familiar dipole moment of a molecule. The matrix elements of V_{zz} given by (8.32) can be evaluated in the asymmetric top representation using the eigenvectors (3.86) in which

K is either all even, or all odd for any particular eigenstate. Using Table 4.1 we see that the second and fourth terms in (8.32) connect functions with $\Delta K = \pm 1$ and their first-order contributions vanish in this representation. We can also drop the imaginary terms which yield a contribution of the type $\frac{1}{2}iV_{ab}(S_z^{+2} - S_z^{-2})$ which vanishes when we average over the basis functions. We are thus left with the simple relation

$$V_{zz} = q[\tfrac{1}{2}(3S_{cz}^2 - \mathbf{S}^2) + \tfrac{1}{4}\eta(S_z^{+2} + S_z^{-2})] \tag{8.33}$$

where $q = V_{cc}$ is called the *field gradient* and $\eta = (V_{aa} - V_{bb})/V_{cc}$ is called the *asymmetry parameter*. These are the molecular constants which characterize those electrical properties of a molecule which can be obtained by analysis of the quadrupole hyperfine structure. We can note immediately that $\eta \to 0$ if the molecule is a symmetric top and the quadrupolar nucleus is on the symmetry axis. This always applies to rigid linear molecules.

If we recall our definition of q_J we see that we need the matrix elements of (8.32) in the $|J\ K\ M = J\rangle$ basis. We are only going to calculate the first-order contribution and thus we can limit ourselves to those matrix elements diagonal in J. Using Table 4.1 we see that

$$\langle J\ K\ M = J|S_{cz}^2|J\ K\ M = J\rangle = \frac{2K^2}{(J+1)(2J+3)} + \frac{1}{2J+3} \tag{8.34}$$

If we also remember that $\langle \mathbf{S}^2 \rangle = 1$ (Chapter 5) we see that the diagonal elements are given by

$$\begin{aligned}\langle J\ K\ M = J|V_{zz}|J\ K\ M = J\rangle &= q\left[\frac{3K^2 - J(J+1)}{(2J+3)(J+1)}\right] \\ &= q\frac{J}{2J+3}\left[\frac{3K^2}{J(J+1)} - 1\right]\end{aligned} \tag{8.35}$$

There are off-diagonal elements in K which come from the term in η. These can be written as

$$\langle J\ K\ M = J|V_{zz}|J\ K+2\ M = J\rangle = \tfrac{1}{4}\eta\frac{f(K\ K+2)}{(2J+3)(J+1)} \tag{8.36}$$

where

$$f(K\ K+2) = \sqrt{(J-K)(J+K+1)(J-K-1)(J+K+2)} \tag{8.37}$$

Having evaluated these matrix elements we are now in a position to obtain the energy-level expressions.

In the case of a *diatomic molecule* $\eta \to 0$ and also $K = 0$. As a result (8.35) becomes simply

$$q_J = -q[J/(2J+3)] \tag{8.38}$$

On substitution of this result into (8.30) we obtain the expression which is conventionally written as

$$E_Q(JIF) = -eQqY(JIF) \tag{8.39}$$

where $Y(JIF)$ is known as Casimir's function and is given by

$$Y(JIF) = \frac{\frac{3}{4}C(C+1) - I(I+1)J(J+1)}{2I(2I-1)(2J-1)(2J+3)} \tag{8.40}$$

Tables of this function are to be found all over the place.[303,100]

For *symmetric top molecules* in which the quadrupolar nucleus is on the symmetry axis, the resulting energy expression is

$$E_Q(JKIF) = eQq\left[\frac{3K^2}{J(J+1)} - 1\right]Y(JIF) \tag{8.41}$$

If the coefficients of the *asymmetric rotor* eigenfunctions in the symmetric top representation are known, say by diagonalization of H_r (Chapter 3), then

$$q_J = \frac{q}{(J+1)(2J+3)}\sum_K \{a_{JK}^2[3K^2 - J(J+1)] + \tfrac{1}{2}\eta a_{JK}a_{JK+2}f(K\ K+2)\} \tag{8.42}$$

and this expression can be substituted into (8.30) to yield the hyperfine pattern. The asymmetric-rotor problem has been discussed by Bragg.[26]

Numerous other ways of expressing the quadrupole energy for an asymmetric rotor are on the market. Perhaps the most useful is one which utilizes the average values of the squared angular momentum components discussed in Chapter 3. We can expand (8.31) directly in cartesian form noting that the $V_{\alpha\beta}$ are constants and take average values over the asymmetric top basis. We can use group theory to show that the terms in the Hamiltonian with $\alpha \neq \beta$ vanish as they do not possess a_1 symmetry. In the last column of the D_2 group character table are given the species of the products of the cartesian vectors. As **S** transforms in the same way as $\boldsymbol{r}$ as far as the operations of the D_2 group are concerned, we see that only those coefficients (i.e. the field gradient components $V_{\alpha\beta}$) of $S_{\alpha z}S_{\beta z}$ with $\alpha = \beta$ may be non-zero. We can thus write

$$\langle V_{zz}\rangle = \langle S_{az}^2\rangle V_{aa} + \langle S_{bz}^2\rangle V_{bb} + \langle S_{cz}^2\rangle V_{cc} \tag{8.43}$$

We can apply the Wigner–Eckart theorem (p. 75) at this point, using a generalization of the result implied in (8.34). Thus

$$\langle J\,\tau\,M = J|S_{\alpha z}^2|J\,\tau\,M = J\rangle = \frac{2}{(J+1)(2J+3)}\langle J\,\tau|J_\alpha^2|J\,\tau\rangle + \frac{1}{2J+3} \tag{8.44}$$

Using this result together with the Laplace relation we see that

$$q_J = \frac{2}{(J+1)(2J+3)}[V_{aa}\langle J_a^2\rangle + V_{bb}\langle J_b^2\rangle + V_{cc}\langle J_c^2\rangle] \tag{8.45}$$

We thus obtain the following commonly used version of the quadrupole energy by substituting the relations evaluated in Section 3.10d for the average values $\langle J_\alpha^2\rangle$.

$$E_Q = \frac{Y(J\,I\,F)}{J(J+1)}\left[\chi_{aa}\left(J(J+1) + E(\kappa) - (\kappa+1)\frac{\partial E(\kappa)}{\partial\kappa}\right) + 2\chi_{bb}\frac{\partial E(\kappa)}{\partial\kappa} + \chi_{cc}\left(J(J+1) - E(\kappa) + (\kappa-1)\frac{\partial E(\kappa)}{\partial\kappa}\right)\right] \tag{8.46}$$

where $\chi_{\alpha\alpha} = eQq_{\alpha\alpha}$.

Another useful form of E_Q, especially when the molecule is a near symmetric top, can be obtained from (3.83) together with (3.80)

$$E_Q = \frac{Y(J\,I\,F)}{J(J+1)}\chi[3K^2 - J(J+1) - 3C_2b^2 + 6C_3b^3 + \cdots + \eta(C_1 + 2C_2b + 3C_3b^2 + 4C_4b^3 + \cdots)] \tag{8.47}$$

where $\chi = eQq_{cc}$.

The most satisfactory way of carrying out a quadrupole interaction analysis, when the first-order treatment is no longer adequate, is to diagonalize the complete Hamiltonian H_Q using a computer.[18]

8.4 Spectra with Quadrupole Hyperfine Structure

As an example of the type of complications that quadrupole interactions can cause in rotational spectroscopy we shall first consider the $J = 4 \leftarrow 3$ transition of SiH_3Br. In Figure 8.1 the transitions in the region 34·2 to

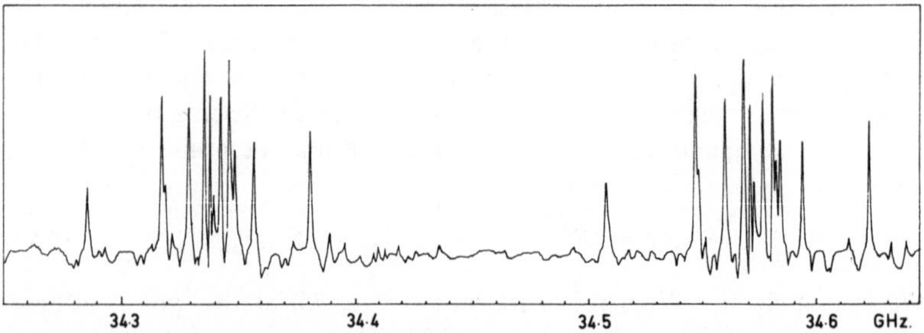

Figure 8.1 The microwave spectrum of SiH_3Br in the region 34·25 to 34·65 GHz. The two sets of lines shown are the $J = 4 \leftarrow 3$ transitions belonging to $SiH_3^{79}Br$, at high frequency, and $SiH_3^{81}Br$ at low frequency. The $K = 0$ lines are not fully modulated (Chapter 7) and are weaker and narrower than otherwise

34·7 GHz are shown. There are two sets of transitions belonging to the two isotopic species containing ^{79}Br (50·5 %) and ^{81}Br (49·5 %). In Figure 8.2 is shown a set of computer simulated patterns for various $J + 1 \leftarrow J$ transitions of a symmetric top molecule with a spin 3/2 nucleus on the symmetry axis. These simulations have been developed from the energy expression (8.41) together with the selection rules

$$\Delta J = +1, \qquad \Delta K = 0, \qquad \Delta I = 0, \qquad \Delta F = 1, 0 \text{ and } -1 \qquad (8.48)$$

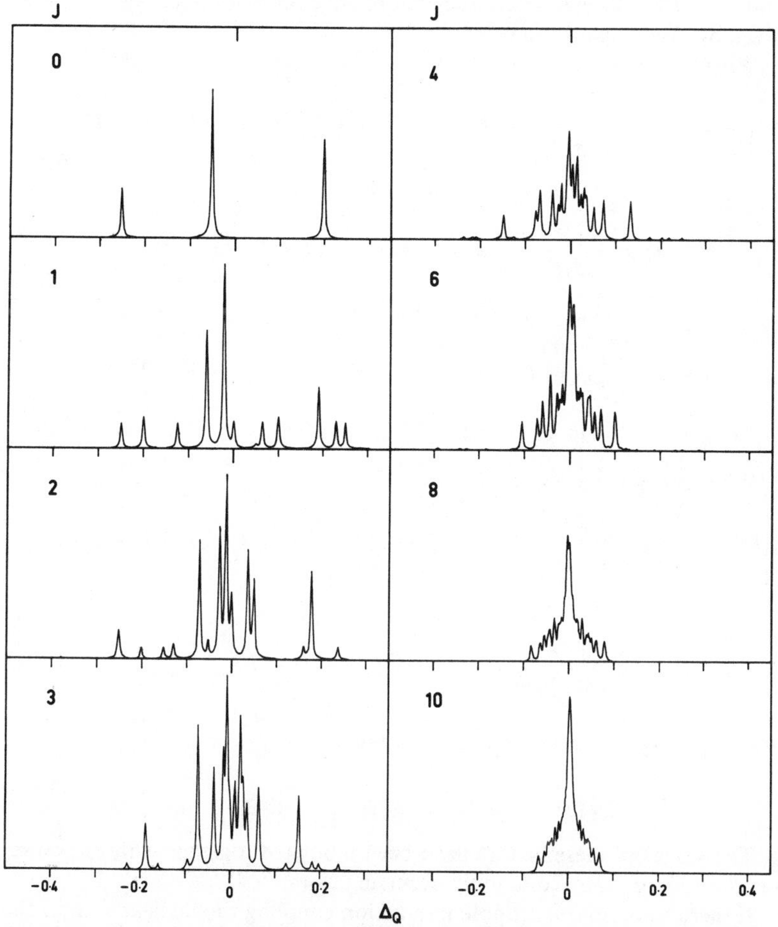

Figure 8.2 A set of basic computer simulated patterns for the quadrupole structure associated with various symmetric top transitions with $I = \frac{3}{2}$. The scale factor Δ_Q defined in (8.49) is plotted on the abscissa. Centrifugal distortion has been neglected and the relative intensities are governed by (8.50) only. Note how the strongest lines tend to lie below the origin at low J, and at high J pile up at the origin

The transition-energy shift relation which results can be written as

$$\Delta E_Q = eQq\Delta_Q \tag{8.49}$$

where

$$\Delta_Q = \left[\frac{3K^2}{(J+1)(J+2)} - 1\right] Y(J+1\,I\,F') - \left[\frac{3K^2}{J(J+1)} - 1\right] Y(J\,I\,F'') \tag{8.49a}$$

and $F' - F'' = 0, \pm 1$. The associated relative intensity relations have been given by Townes and Schawlow.[303]

When $\Delta J = +1$, i.e. $J + 1 \leftarrow J$

$$I(F+1 \leftarrow F) \propto \frac{1}{F+1}(J+F+I+3)(J+F+I+2)(J+F-I+2) \times (J+F-I+1) \tag{8.50a}$$

$$I(F \leftarrow F) \propto -\frac{(2F+1)}{F(F+1)}(J+F+I+2)(J+F-I+1) \times (J-F+I+1)(J-F-I) \tag{8.50b}$$

$$I(F-1 \leftarrow F) \propto \frac{1}{F}(J-F+I+2)(J-F+I+1)(J-F-I+1) \times (J-F-I) \tag{8.50c}$$

When $\Delta J = 0$, i.e. $J \leftarrow J$

$$I(F+1 \leftarrow F) \propto \frac{-1}{F+1}(J+F+I+2)(J+F-I+1)(J-F+I) \times (J-F-I-1) \tag{8.51a}$$

$$I(F \leftarrow F) \propto \frac{2F+1}{F(F+1)}[J(J+1)+F(F+1)-I(I+1)]^2 \tag{8.51b}$$

$$I(F-1 \leftarrow F) \propto \frac{-1}{F}(J+F+I+1)(J+F-I)(J-F+I+1) \times (J-F-I) \tag{8.51c}$$

The values of these factors have been tabulated together with the values of $Y(J\,I\,F)$ for many commonly occurring cases.[303,100]

If there were no quadrupole interaction coupling the nuclear spin to the rotational angular momentum the nuclear spin component would be unaffected by transitions among the J levels. When quadrupole interactions are not negligible all the transitions allowed by the general selection rule $\Delta F = \pm 1, 0$ are observed. In general when $J > I$ those for which $\Delta F \neq \Delta J$

are much weaker. This can be seen in a rough way by studying the intensity relations (8.50) and (8.51). In general I is not large and $J \sim F \gg I$ and by substituting J for F and neglecting I we see that the following *relative* intensity relations hold (very approximately):

$$(\Delta J = +1) \quad F+1 \leftarrow F : F \leftarrow F : F-1 \leftarrow F \quad \text{as} \quad J^3 : J : J^{-1} \tag{8.52a}$$

$$(\Delta J = 0) \quad F+1 \leftarrow F : F \leftarrow F : F-1 \leftarrow F \quad \text{as} \quad J : J^3 : J \tag{8.52b}$$

This rapid loss of intensity, for $\Delta J \neq \Delta F$ transitions, with increasing J can be understood by realizing that as the molecule rotates more and more rapidly the nuclear spin is not coupled strongly enough to follow the rotational vaguaries associated with transitions. This behaviour is illustrated clearly in the series of simulations shown in Figure 8.2. As J increases the intensity slowly funnels into the unshifted position. This can be seen approximately by noting that for high J, $Y(J+1\ I\ F+1) \sim Y(J\ I\ F)$ and as $J \sim F$ for high J, we see that $\Delta_Q \sim -6K^2/J^3$. Note also that the intensity drops away rapidly with increasing K in a symmetric rotor sub-branch as indicated in relation 5, Table 4.4 and illustrated in Figure 6.9.

In Figure 8.3 the way in which the various K substates with $J = 3$ and 4 for SiH_3Br are split by quadrupole interaction is indicated quantitatively. A comparison of Figures 8.2 and 8.3 with Figure 8.1 indicates that the two most widely separated strong lines are those for which $|K| = 3, F = 5/2 \leftarrow 3/2$ and $|K| = 3, F = 9/2 \leftarrow 7/2$. The separation of these lines gives a good estimate of eQq directly using Figure 8.2 and (8.49).

$$\Delta_Q(9/2 \leftarrow 7/2) - \Delta_Q(5/2 \leftarrow 3/2) = 0.34 \tag{8.53a}$$

from Figure 8.2, and for $SiH_3^{79}Br$

$$\Delta E_Q(9/2 \leftarrow 7/2) - \Delta E_Q(5/2 \leftarrow 3/2) = 114\ \text{MHz} \tag{8.53b}$$

from Figure 8.1, we see that (8.49) yields the result that $eQq \cong 335$ MHz.

A more careful calculation using (8.49) and either (8.40) or tables[303,100] to evaluate $Y(J\ I\ F)$ indicates that

$$\Delta_Q(9/2 \leftarrow 7/2) = (0{\cdot}35)(-0{\cdot}159091) - (1{\cdot}25)(-0{\cdot}166667) \tag{8.54a}$$

$$\Delta_Q(5/2 \leftarrow 3/2) = (0{\cdot}35)(0{\cdot}178571) - (1{\cdot}25)(0{\cdot}200000) \tag{8.54b}$$

which yields a more accurate value for relation (1.53a) of 0·340152. A more accurate measurement for the separation is 114·3 MHz which yields a value of $eQq = 336$ MHz. In Table 8.2 a complete set of shift terms and intensities calculated with this value of eQq is listed.

As a second example we will consider the $J = 3 \leftarrow 2$ transition of BrCN which is shown under medium resolution in Figure 8.4. This case is rather more complicated because both Br($I = 3/2$) and N($I = 1$) possess quadrupole moments. Townes, Holden and Merritt[302] have applied the theory developed

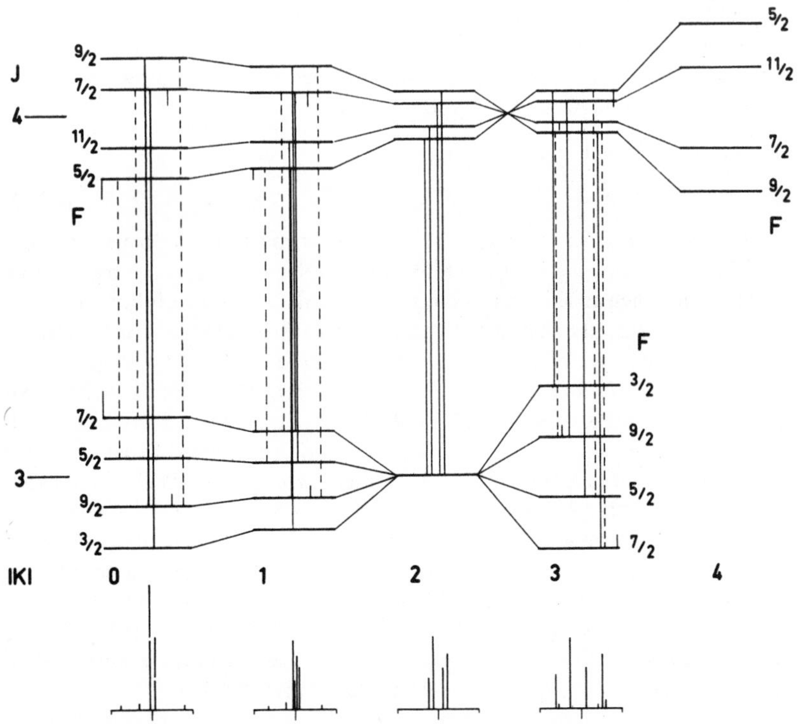

Figure 8.3 Energy-level and transition diagram for $SiH_3^{79}Br$ for which $I = \frac{3}{2}$ and eQq = 336 MHz. The quadrupole structure is shown for each individual $|K|$ set and $(A - B)K^2$ has been subtracted so that each $|K|$ set has the same horizontal origin. The data are given in Table 8.2. Transitions with: $\Delta F = 1$ —, $\Delta F = 0$ - - - and $\Delta F = -1$ are indicated by lead lines only. For $\Delta F = 1$ $K = 0$ two pairs of lines coincide

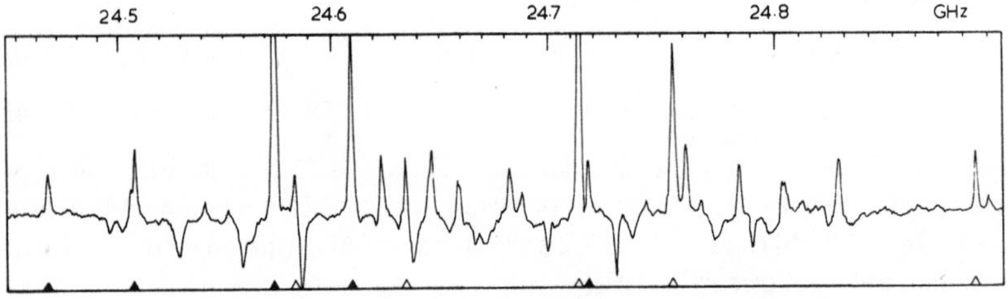

Figure 8.4 Medium resolution scan of $J = 3 \leftarrow 2$ transitions of $^{79}BrCN$ and $^{81}BrCN$. The frequency data is given in Table 8.3. The lines belonging to $^{79}BrCN$ in the vibrational ground state are identified by open triangles, those of $^{81}BrCN$ are identified by black triangles

Table 8.2 First-Order Quadrupole Shifts for the 4 ← 3 Transition of $SiH_3^{79}Br$ Calculated using (8.41) and (8.49) with $eQq = 336{\cdot}0$[a]

$K = 0$				$\|K\| = 1$			
F'	F''	ΔE	Intensity	F'	F''	ΔE	Intensity
5/2 ←	7/2	−116·0	0·09	5/2 ←	7/2	−93·0	0·09
5/2 ←	5/2	−76·8	2·29	5/2 ←	5/2	−63·6	2·29
7/2 ←	7/2	−32·0	3·02	7/2 ←	7/2	−21·6	3·02
9/2 ←	7/2	−2·5	25·46	11/2 ←	9/2	−5·0	33·33
11/2 ←	9/2	−2·5	33·33	5/2 ←	3/2	−0·6	14·29
5/2 ←	3/2	7·2	13·42	9/2 ←	7/2	3·5	25·46
7/2 ←	5/2	7·2	17·97	7/2 ←	5/2	7·8	19·13
7/2 ←	9/2	52·0	0·07	7/2 ←	9/2	41·4	0·07
9/2 ←	9/2	81·5	2·32	9/2 ←	9/2	66·4	2·32
$\|K\| = 2$				$\|K\| = 3$			
5/2 ←	3/2	−24·0	14·29	5/2 ←	3/2	−63·0	14·29
5/2 ←	5/2	−24·0	2·30	9/2 ←	9/2	−53·7	2·32
5/2 ←	7/2	−24·0	0·09	7/2 ←	9/2	−43·4	0·07
11/2 ←	9/2	−12·2	33·33	11/2 ←	9/2	−24·3	33·33
7/2 ←	7/2	9·6	3·02	7/2 ←	5/2	12·6	19·13
7/2 ←	9/2	9·6	0·07	5/2 ←	5/2	42·0	2·30
7/2 ←	5/2	9·6	19·13	9/2 ←	7/2	51·3	25·46
9/2 ←	7/2	21·4	25·46	7/2 ←	7/2	61·6	3·02
9/2 ←	9/2	21·4	2·32	5/2 ←	7/2	91·0	0·09

[a] Centrifugal distortion has been neglected in this calculation. The intensities have been calculated using (8.50) only.

by Bardeen and Townes[13] to handle molecules with two quadrupolar nuclei. In this case the quadrupole interaction is much larger in Br than in N and the spectrum can be discussed semi-quantitatively. In this case $eQq = 686{\cdot}5$, $573{\cdot}5$ and $-3{\cdot}83$ for ^{79}Br, ^{81}Br and ^{14}N respectively (MHz).[302] As a result, under medium resolution (Figure 8.4) only the Br quadrupole structure is obvious and is fairly easily assigned. In Table 8.3 the frequencies of the various components, for the ground vibrational state, are listed.[302] Under higher resolution each line shows fine structure due to the ^{14}N quadrupole moment. As an example in Figure 8.5 the structure of the strong line at 24755·2 MHz is shown. Neglecting the ^{14}N effect we can label the *unresolved* line with the associated values of F_1 which is the quantum number associated with $\boldsymbol{J} + \boldsymbol{I}_1$, where $\boldsymbol{I}_1$ is the spin of Br. Thus as seen from Table 8.3 it corresponds to the two quadrupole components. $J = 3 \leftarrow 2$, $F_1 = 3/2 \leftarrow 1/2$ and $J = 3 \leftarrow 2$, $F_1 = 5/2 \leftarrow 3/2$ for ^{79}BrCN. If we now go on to consider the further ^{14}N coupling we see that we can qualitatively vector couple the ^{14}N spin, $\boldsymbol{I}_2$ $(=1)$ with $\boldsymbol{F}_1$ which will split each level into three (as long as $F_1 > I_2$), i.e. the three values $F_1 + 1$, F_1 and $F_1 - 1$. Townes, Holden and

Table 8.3 The Frequencies (MHz) of the $J = 3 \leftarrow 2$ Transition of the Vibrational Ground State of BrCN

$F_1' \leftarrow F_1''$ [a]	^{79}BrCN	^{81}BrCN
3/2 ← 3/2	24 583·00	24 465·33
5/2 ← 5/2	24 633·71	24 507·38
7/2 ← 5/2, 9/2 ← 7/2	24 713·05	24 573·86
3/2 ← 1/2, 5/2 ← 3/2	24 755·22	24 608·92
7/2 ← 7/2	24 884·57	24 717·19

[a] F_1 is the quantum number associated with $J + I_1$ where I_1 is the spin of the Br nucleus. The structure due to the ^{14}N quadrupole moment is not resolved under medium resolution and the above frequencies represent the centers of the unresolved complexes.[302]

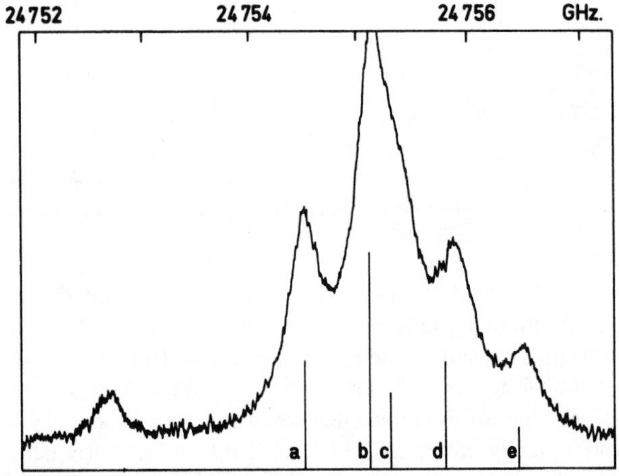

Figure 8.5 The fine structure associated with the line of ^{79}BrCN at 24 755·2 MHz. This complex belongs to the lines with $J = 3 \leftarrow 2$, $F_1 = 3/2 \leftarrow 1/2$ *and* $5/2 \leftarrow 3/2$. These lines show fine structure due to the nitrogen quadrupole moment. The assignment given by Townes, Merritt and Holden[302] is:

(a) 5/2 5/2 ← 3/2 3/2, 5/2 5/2 ← 3/2 5/2
(b) 3/2 3/2 ← 1/2 1/2, 3/2 3/2 ← 1/2 3/2, 5/2 7/2 ← 3/2 5/2
(c) 5/2 3/2 ← 3/2 1/2, 5/2 3/2 ← 3/2 3/2, 5/2 3/2 ← 3/2 3/2
(d) 3/2 5/2 ← 1/2 3/2
(e) 3/2 1/2 ← 3/2 1/2, 3/2 1/2 ← 1/2 1/2

The notation used is $F_1' F' \leftarrow F_1'' F''$

Merritt have discussed the structure of this line in detail.[302] The associated hyperfine levels which can occur are thus:

$$J = 3, \quad F_1 = 3/2, \quad F = 5/2, 3/2, 1/2$$

$$J = 2, \quad F_1 = 1/2, \quad F = 3/2, 1/2$$

$$J = 3, \quad F_1 = 5/2, \quad F = 7/2, 5/2, 3/2$$

$$J = 2, \quad F_1 = 3/2, \quad F = 5/2, 3/2, 1/2$$

where F is the quantum number associated with $\boldsymbol{J} + \boldsymbol{I}_1 + \boldsymbol{I}_2$. Numerous transitions can of course occur but in general the strongest components are those for which $\Delta F = +1$. The assignments of these lines given by Townes, Holden and Merritt[302] are also shown in Figure 8.5.

Chapter 9

Flexible Molecules

9.1 Definition of a Flexible Molecule

In this chapter we shall deal with the motions which occur in certain types of molecule for which the general procedure for handling semi-rigid systems (Chapter 6) is not adequate. We can trace the problem back as it is inherent in the assumptions considered in Section 2.2 when the transformation relation (2.9) was chosen to generate the separated form of the kinetic energy (2.29). The expediency of (2.9) lies to a great extent in the validity of the assumption that the vibrational displacement vectors d_i, which characterize the non-rigidity, are small. If ψ is the wavefunction which characterizes a particular type of motion and $d_i(\psi)$ are the associated displacements, then we can usefully consider the following cases:

$$\text{Semi-rigid molecule } d_i(\psi) \ll r_i^e \tag{9.1a}$$

$$\text{Flexible molecule } d_i(\psi) \sim r_i^e \tag{9.1b}$$

$$\text{Quasi molecule? } d_i(\psi) > r_i^e \tag{9.1c}$$

The relation (9.1b) essentially states that a flexible molecule is one for which the displacements which characterize the motion are of the same order of magnitude as the vectors, r_i^e, which specify the reference geometry. Fortunately, in many cases, the large amplitude motion can be separated allowing the rest of the system to be treated by the standard semi-rigid procedure. There are several different types of large amplitude motion and each type demands a theoretical approach which takes its particular properties carefully into account. In general a model for the motion is set up and the properties of the associated Hamiltonian are investigated and compared with experiment.

We shall discuss one particular type of motion—internal rotation—in more detail than the others as it is by far the most frequently observed case in rotational spectroscopy. A somewhat different type of flexibility is shown by molecules which can invert. The classic example is of course ammonia and in fact it was the transitions associated with inversion in this molecule which were observed in the first microwave spectroscopy experiments.[48] Apart from ammonia and related species a number of small strained-ring compounds such as trimethylene oxide[43] can invert through a planar configuration if the barrier is low enough. This type of motion is usually called ring puckering. A useful survey of the field of flexible molecules has been given by Laurie.[171]

Quasi-molecules will not be dealt with here. They may be thought of as collision complexes and perhaps are best handled with the aid of scattering techniques.[80]

9.2 Internal Rotation

In many molecules, such as ethane and acetone, there is a barrier which hinders the rotation of methyl groups relative to the rest of the molecule. Kemp and Pitzer[151] in 1936 concluded that such a barrier existed in ethane from a comparison of the experimentally observed entropy and the theoretically predicted value. In such molecules the observed values tended to lie between the theoretical values obtained for the two limiting cases; free rotor and rigid structure. For molecules such as dimethylacetylene, on the other hand, the observed value was found to be close to that calculated on the assumption that the methyl groups were free to rotate.

It is worthwhile to classify the cases that can occur. We will restrict our treatment to one top systems such as acetaldehyde. Two top systems such as dimethyl sulphide have been reviewed by Dreizler.[75] In Table 9.1 a representative table of the various subclassifications of the 'one top' cases is given. Up-to-date tables of barrier heights are available.[100] Note that the terms 'high or low barrier' need further qualification as discussed in Section 9.2c. The following treatment will only handle molecules of the types listed in Table 9.1 where the top at least is symmetrical. Systems in which both top and frame are asymmetric are much less easily handled.[246]

Table 9.1 Table of Barrier Heights[a] Determined for Various Different Types of Internal Rotation Systems[100]

	Low barrier	Intermediate barrier	High barrier
Symmetric top	$CH_3C{\equiv}CCH_3$ ($V_3 < 6$) $SiH_3C{\equiv}CCH_3$ ($V_3 < 3$)	CH_3SnH_3 ($V_3 = 650$) CH_3GeH_3 ($V_3 = 1239$)	CH_3CF_3 ($V_3 = 3480$) CH_3CH_3 ($V_3 = 2980$)
Symmetric rotor and Asymmetric frame	CH_3BF_2 ($V_6 = 14$) CH_3NCS ($V_3 = 304$)	CH_3CHO ($V_3 = 1168$) CH_3OH ($V_3 = 1070$)	CH_3NF_2 ($V_3 = 4170$) CF_3CHF_2 ($V_3 = 3510$)

[a] In cals/mole.

9.2a The Hamiltonian for Internal Rotation

Two approaches to the problem have been developed and each has its own advantages and disadvantages. They are somewhat complementary. Which approach is the more applicable depends on the barrier height, the geometry of the system and the mass distribution. One method called the Principal Axis Method (PAM) is rather more widely applicable and this is the one we shall discuss. The PAM has been developed by Kilb, Lin and Wilson[154] and Herschbach.[117] The second method (developed by Nielsen[215] and Burkhard and Dennison[30]) known as the Internal Axis Method (IAM) is applicable when the asymmetric group is light as in the case of methyl alcohol[30,163] where it is the more successful of the two methods.[117] The difference lies in the choice of the reference configuration. For a semi-rigid system the natural choice for the reference configuration is the equilibrium structure. One of the most important problems concerning flexible molecules is the choice of reference configuration and the main criterion is often the ease with which the resulting Hamiltonian is solved. In the IAM, one of the axes of the reference configuration is chosen to lie along the axis of the top. On the other hand in the PAM, the reference axes are the principal axes of the whole molecule. The IAM has the advantage over the PAM that the coupling terms between overall rotation and torsional motion are smaller. On the other hand it has the disadvantage that the axes do not coincide with the principal axes, thus introducing terms involving the products of inertia. Lin and Swalen[183] have given a review of the 'one top' problem.

The kinetic-energy expression (6.1) can be modified for a system of the acetaldehyde type as depicted in Figure 9.1. The system is considered to be rigid except for the freedom of the torsional angle τ. If we take the first term of

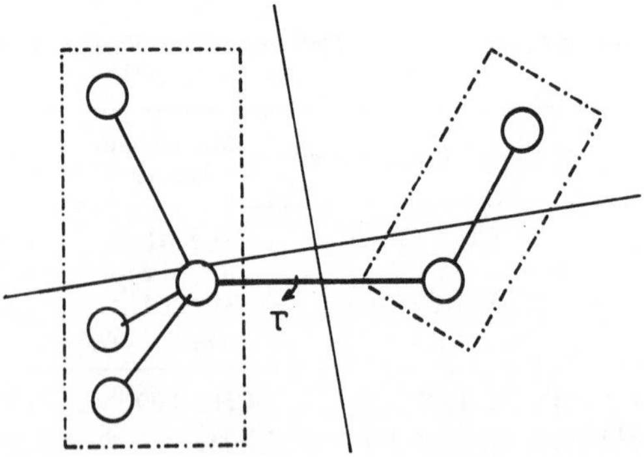

Figure 9.1 Schematic diagram of a typical model system for the PAM internal rotation calculation. Two of the principal axes are depicted

(6.1) it represents the total rotational energy as long as $\dot{\tau} = \omega_\tau = 0$. If the top is now allowed to rotate relative to the frame—the motion described by τ—part of the extra kinetic energy comes from the third term and as this is a function only of τ, by the methods of Chapter 2 it can be written as $T_\tau = \frac{1}{2}I_\tau\omega_\tau^2$. The only complicated term is the second, which represents a coupling between overall rotation and internal torsional motion. This term is essentially the scalar product between $\boldsymbol{\omega}$ and an 'internal' angular momentum defined by $\partial T_\tau/\partial\omega_\tau = I_\tau\omega_\tau$. In internal rotation theory the name internal angular momentum is reserved for $\partial T/\partial\omega_\tau$ as in (9.7) below. By introducing the direction cosines, λ_α, between the top axis and the frame the scalar product in (6.1) becomes

$$T_{r\tau} = I_\tau\omega_\tau \sum_\alpha \lambda_\alpha\omega_\alpha \tag{9.2}$$

We thus obtain the overall kinetic energy expression

$$T = \tfrac{1}{2}\sum_\alpha I_\alpha\omega_\alpha^2 + \tfrac{1}{2}I_\tau\omega_\tau^2 + I_\tau\omega_\tau\sum_\alpha \lambda_\alpha\omega_\alpha \tag{9.3}$$

which can be expressed in matrix notation as:

$$T = \tfrac{1}{2}\boldsymbol{\omega}^\dagger \mathbf{I}\boldsymbol{\omega} \tag{9.4}$$

where $\boldsymbol{\omega}$ and $\mathbf{I}$ are redefined as:

$$\boldsymbol{\omega} = \begin{bmatrix} \omega_a \\ \omega_b \\ \omega_c \\ \omega_\tau \end{bmatrix}, \qquad \mathbf{I} = \begin{bmatrix} I_a & 0 & 0 & \lambda_a I_\tau \\ 0 & I_b & 0 & \lambda_b I_\tau \\ 0 & 0 & I_c & \lambda_c I_\tau \\ \lambda_a I_\tau & \lambda_b I_\tau & \lambda_c I_\tau & I_\tau \end{bmatrix} \tag{9.5}$$

Using the energy expression (9.4) we obtain the following angular momentum definitions:

$$J_\alpha = \partial T/\partial\omega_\alpha = I_\alpha\omega_\alpha + \lambda_\alpha I_\tau\omega_\tau \tag{9.6}$$

$$p = \partial T/\partial\omega_\tau = I_\tau\omega_\tau + I_\tau\sum_\alpha \lambda_\alpha\omega_\alpha \tag{9.7}$$

Note that there is no implied summation in (9.6). Examination of (9.6) and (9.7) indicates that not only does the overall angular momentum $\boldsymbol{J}$ contain terms involving the motion of the top but also the momentum conjugate to τ includes contributions from the overall motion through $\boldsymbol{\omega}$. We shall call p the *internal angular momentum.*

If we introduce the function:

$$\pi = \sum_\alpha \rho_\alpha J_\alpha \tag{9.8}$$

where $\rho_\alpha = \lambda_\alpha I_\tau / I_\alpha$, we can combine J_α and p as given in (9.6) and (9.7) to form a 'relative' internal angular momentum function which depends only on ω_τ.

$$p - \pi = p - \sum_\alpha \rho_\alpha J_\alpha = rI_\tau \omega_\tau \tag{9.9}$$

where

$$r = 1 - I_\tau \sum_\alpha \lambda_\alpha^2 / I_\alpha = 1 - \sum_\alpha \lambda_\alpha \rho_\alpha \tag{9.10}$$

is a parameter which generates the 'reduced' moment of inertia for internal rotation, rI_τ. We can rephase the expression for T as

$$T = \tfrac{1}{2} \sum_\alpha \mu_{\alpha\alpha'} J_\alpha^2 + \Delta \tag{9.11}$$

where

$$\Delta = \tfrac{1}{2} rI_\tau \omega_\tau^2 = F(p - \pi)^2 \tag{9.11a}$$

where Δ can be thought of as a relative kinetic energy term which is defined such that $\partial\Delta/\partial\omega_\tau = p - \pi$. Thus we see that we can write the Hamiltonian for internal rotation as:

$$H = \tfrac{1}{2} \sum_\alpha \mu_{\alpha\alpha'} J_\alpha^2 + F(p - \pi)^2 + V(\tau) \tag{9.12}$$

where

$$F = \frac{1}{2rI_\tau} \tag{9.12a}$$

and $V(\tau)$ is the potential energy, as a function of τ, which we will discuss in the next section.

This Hamiltonian is now in a convenient form for separation into the following terms which serve as a convenient starting point for a perturbation calculation: Let

$$H = H_r' + H_\tau + H_{r\tau} \tag{9.13}$$

where

$$H_r' = H_r + F\pi^2 \tag{9.14a}$$

$$H_\tau = Fp^2 + V(\tau) \tag{9.14b}$$

$$H_{r\tau} = -2Fp\pi \tag{9.14c}$$

As H_r' only differs from H_r by terms which are quadratic in the components of $\boldsymbol{J}$ these terms essentially only alter the effective inertia tensor. H_r' is thus essentially still a rigid-rotor type Hamiltonian and can thus be handled in the standard way. H_τ is the torsional Hamiltonian and the way it is handled depends on the form of $V(\tau)$. It is dealt with in detail in the next section. Once H_r' and H_τ have been solved and the associated wave functions determined, the product $\Psi = \psi_r\psi_\tau$ serves as a basis for the determination of the effects of $H_{r\tau}$ by perturbation theory. This perturbation treatment indicates that characteristic spectral splittings and patterns will occur in the rotational spectra of asymmetric rotor molecules which can be analysed to yield information about $V(\tau)$.

9.2b *The Solutions of the Torsional Hamiltonian,* H_τ

In the sort of systems we are considering, such as acetaldehyde, $V(\tau)$ is non-zero and is periodic in τ such that $V(\tau) = V(\tau \pm 2\pi/n)$ for a C_n top. It is convenient to expand $V(\tau)$ in the following Fourier form

$$V(\tau) = \tfrac{1}{2} \sum_{m=1}^{\infty} V_{nm}(1 - \cos mn\tau) \tag{9.15}$$

The first two terms in the $n = 3$ case are thus:

$$V(\tau) = \tfrac{1}{2}V_3(1 - \cos 3\tau) + \tfrac{1}{2}V_6(1 - \cos 6\tau) + \cdots \tag{9.16}$$

This potential function has been found to be very rapidly convergent in practice, i.e., $|V_3| \gg |V_6| \gg |V_9|$, etc. (see Section 9.3). We will thus concentrate on the repercussions on the spectrum of the V_3 term, and set all higher terms to zero. The Hamiltonian (9.14b) thus can be written as

$$H_\tau = Fp^2 + \tfrac{1}{2}V_3(1 - \cos 3\tau) \tag{9.17}$$

This equation has usually been solved by writing the corresponding Schrödinger equation:

$$[Fp^2 + \tfrac{1}{2}V_3(1 - \cos 3\tau)]\psi(\tau) = E\psi(\tau) \tag{9.18}$$

and comparing this equation with the Mathieu equation:

$$\frac{d^2y}{dx^2} + (b - s\cos^2 x)y = 0 \tag{9.19}$$

for which solution procedures are known and tables have been published.[343,112,116,345] Comparison of (9.18) with (9.19) indicates that the following relations are necessary:

$$\begin{gathered} p = -i\frac{d}{d\tau}, \qquad \frac{d^2}{dx^2} = \frac{4}{n^2}\frac{d^2}{d\tau^2} \\ 2x = n\tau + \pi, \qquad V_n = \frac{n^2}{4}Fs, \qquad E = \frac{n^2}{4}Fb \end{gathered} \tag{9.20}$$

For a three-fold barrier we see that

$$V_3 = \tfrac{9}{4}Fs, \qquad E = \tfrac{9}{4}Fb \tag{9.21}$$

The solutions are usually given in terms of s, a barrier height parameter, and b, the eigenvalue parameter.

The solutions of the Mathieu equation depend on the boundary conditions. Rotation of a methyl group about the internal rotor axis gives the boundary condition upon τ. The Hamiltonian is invariant to the C_3 operations and the eigenfunctions, y, must be periodic in 2π in τ. They are thus conveniently written as the expansion

$$\psi(\tau) = \sum_l A_l e^{i(3l+\sigma)\tau} \tag{9.22}$$

where σ is an index which characterizes the symmetry of the function. For a threefold barrier, when $\sigma = 0$ the functions are non-degenerate (A species) and when $\sigma = \pm 1$ the functions are degenerate (E species). The A species wavefunctions are periodic in $2\pi/3$ in τ and according to the relations (9.20) can thus be seen to be periodic in π for x.

The general method of solving the Mathieu equation is to substitute (9.22) into (9.19) and by integrating with respect to τ from $0 \rightarrow 2\pi$ obtain a set of recurrence relations in terms of the coefficients A_l of the type:

$$A_{l+3} + MA_l + A_{l-3} = 0 \tag{9.23}$$

M is a function of b, s, l and $n(=3$ in this case).

This relation generates an infinite set of linear homogeneous simultaneous equations, whose determinant of coefficients must be zero for non-trivial solutions. From this the eigenfunction coefficients can be obtained.[345] The resulting energies of the lower eigenfunctions are plotted for the $n = 3$ case as a function of s ($=4V_3/9F$) in Figure 9.2. To put these results into clearer perspective we shall consider the two limiting cases which occur. The situation at the right-hand side of Figure 9.2 represents a torsional oscillator with $V_3 \rightarrow \infty$. On the other hand at the left-hand side the energy levels correspond to a free rotor as $V_3 \rightarrow 0$. Correlation of the levels from one case to the other facilitates an understanding of the intermediate situation.

In the *high-barrier limit* the torsional motion will be restricted to oscillations of small amplitude with $\tau \sim 0$. In these cases the series expansion for the cosine term

$$\cos(3\tau) = 1 - \frac{(3\tau)^2}{2!} + \frac{(3\tau)^4}{4!} - \cdots \tag{9.24}$$

can be truncated after the second term and the resulting simplified potential function on substitution into (9.17) yields:

$$H_\tau = Fp^2 + \tfrac{9}{4}V_3\tau^2 \tag{9.25}$$

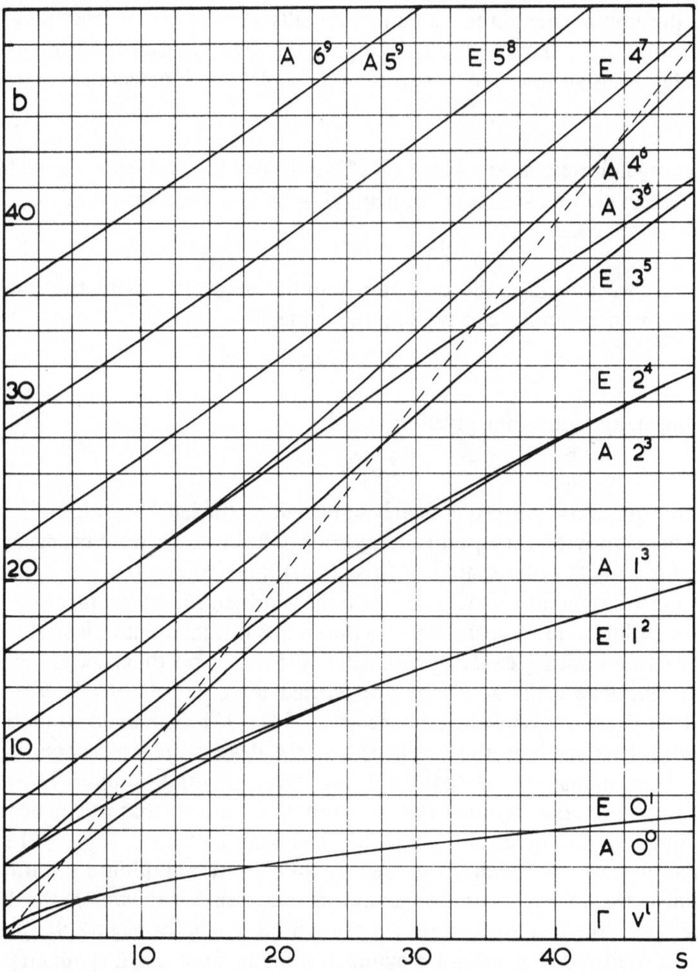

Figure 9.2 Quantitative correlation diagram of $b(v^l)$, the eigenvalues of the Mathieu equation (9.19), for a three-fold symmetric rotor as a function of the barrier height parameter s (9.21). The dashed line represents the barrier height

This Hamiltonian has the same form as the simple harmonic oscillator which was discussed in Chapter 1 and if we remember the associated characteristic frequency relation ($\alpha\beta = \frac{1}{4}\omega^2$ (1.9c)) we see that we obtain the following solutions:

$$E(v) = \omega_\tau(v + \tfrac{1}{2}) \tag{9.26}$$

$$\omega_\tau = 3\sqrt{FV_3} \tag{9.27}$$

In the low-barrier limit we can essentially neglect the barrier potential by setting $V(\tau) = 0$ in (9.14c). This implies that free-internal rotation occurs. The Hamiltonian thus becomes that of a simple one-dimensional rotator

$$H = Fp^2 \tag{9.28}$$

whose eigenfunctions are $\psi(\tau) = A\, e^{il\tau}$ (only one A_l coefficient in (9.22) is non-zero) which must satisfy the general boundary condition that

$$\psi(\tau) = \psi(\tau + 2\pi)$$

and thus we find that the quantum number associated with the internal angular momentum, p, can take on the values:

$$l = 0, \pm 1, \pm 2, \ldots \tag{9.29}$$

The eigenvalues are thus given by

$$E(l) = Fl^2 \tag{9.30}$$

and this quadratic pattern is clearly indicated by the left-hand side of Figure 9.2. There is now no zero-point energy and all the levels are doubly degenerate except for $l = 0$. Note that $l^2 = \frac{9}{4}b$ as (9.21) holds at $s = 0$.

In the intermediate-barrier situation the Mathieu equation must be solved and the solution must correlate with the simple results we have just obtained for the two limiting cases. A quantitative correlation diagram is given in Figure 9.2 where the solutions are plotted in terms of b $(=4E/9F)$ as a function the barrier height parameter s $(=4V_3/9F)$ as defined in (9.21). In the high-barrier limit the levels are triply degenerate corresponding to essentially uncoupled oscillations in the three potential minima. The usual quantum mechanical symmetry restrictions (C_3) indicate that we can describe the functions as corresponding to an A $(\sigma = 0)$ and an E $(\sigma = \pm 1)$ state. The levels can also be characterized by the vibrational quantum number v and also the internal rotation quantum number l with which the levels correlate in the free-rotor limit on the left-hand side of Figure 9.2. Levels with l a multiple of 3 possess A symmetry, the rest possess E symmetry.

As the situation changes and the value of s decreases the degeneracy is partially lifted, though the E states still remain degenerate. As the free-rotor limit $(s = 0)$ is approached, a particular A state becomes degenerate with an adjacent A state (a non-crossing rule applies) to become the two components of the free-rotor levels for which l is a multiple of three.

At this point the quantitative distinction between the high- and low-barrier limits is most easily defined. The rather loose terminology, 'high barrier' really means that we are considering a situation in a region where $b \ll s$ and represented in Figure 9.2 by the area in the bottom right-hand corner. A low-barrier situation is, on the other hand, defined by $b \gg s$ and this is represented in Figure 9.2 by the top left-hand corner. Here b represents the eigenvalues and s is a function which is essentially the ratio of V_n to F. This situation has its parallel in molecular vibration theory where the frequency of a system is a function of the ratio of k, the force constant, to μ, the reduced mass.

It is worthwhile for us to carry out the following 'back of an envelope' calculation which will hopefully clarify the meaning of the various terms that have so far been introduced. Some very nice results obtained by Weiss and Leroi,[324] who observed torsional transitions in the infra-red spectrum of ethane, show that even a very simple calculation is quite accurate in this case.

We need to calculate the reduced inertial constant, F. If we assume a tetrahedral $\widehat{HCC}$ angle and a C—H bond length of about 1·1 Å, in CH_3F it is 1·097 Å, then we see that the H atoms are close to 1 Å away from the A-axis. As

$$I_A = \sum_i m_i r_i^2 \sim 3 \text{ amu Å}^2‡$$

a good average value for $I_A(CH_3)$ is 3·168 amu Å^2 and in this case we see that $A = 5{\cdot}32 \text{ cm}^{-1}$.

F was defined in the last section in terms of the reduction parameter r. For molecules with coaxial symmetric rotors such as ethane and methyl silane, if I_1 is the moment of inertia of the top and I_2 that of the frame (symmetric in this case) then

$$r = 1 - \frac{I_1}{I_1 + I_2} = \frac{I_2}{I_1 + I_2} \tag{9.31}$$

$$rI_1 = \frac{I_1 I_2}{I_1 + I_2} \tag{9.32}$$

For molecules such as ethane and dimethylacetylene where $I_1 = I_2$, $r = 1/2$. One can get a feel for the meaning of rI_A in this simple case as it is the quantity one obtains for the reduced moment of inertia of a system consisting of two coaxial flywheels connected by a semi-rigid axle. The parallel with the reduced mass for the vibration of a diatomic molecule is worth noting (Chapter 1). In this case F is simply:

$$F = \frac{1}{2rI_1} = A_1 + A_2 \tag{9.33}$$

This yields a value of $10{\cdot}64 \text{ cm}^{-1}$ for F in ethane. The thermodynamically determined value of V_3 for ethane is ~3030 cal/mole which is equivalent to 1050 cm^{-1}. Substitution of these values into (9.27) implies that $\omega_\tau \sim 300 \text{ cm}^{-1}$. Weiss and Leroi[324] have observed not only the $v = 1 \leftarrow 0$

‡ Where r_i is the distance of the ith atom from the symmetry axis.

torsional transition at 289 cm^{-1} but also the $v = 2 \leftarrow 1$ transition to be split into an A component at 255 cm^{-1} and an E component at 258 cm^{-1}. These transitions are forbidden by the simple selection rules which appear to break down due to resonances. Substitution of the above value of F into (9.21) indicates that $b = E(\text{cm}^{-1})/23{\cdot}95$ and thus $b(v = 1) - b(v = 0) = 12{\cdot}07$. Such a separation can be seen in Figure 9.2 to correspond to a value of $s = 43{\cdot}5$. This very simple analysis thus yields a barrier height of 1042 cm^{-1}. Weiss and Leroi[324] use the extra information contained in the splitting of the $2 \leftarrow 1$ transition together with spectroscopically determined A rotational constant (which gives $F = 10{\cdot}705$ cm^{-1}) to derive a value of 1024 cm^{-1} or 2928 cals per mole for the barrier height.

In Figure 9.3 the barrier potential and energy levels are drawn to scale and show how the levels which are essentially three-fold degenerate in the depths of the well split into A and E torsional states in the vicinity of the barrier maximum. The splitting is usually described as due to tunneling through the barrier from one potential minimum to an adjacent one, and the frequency associated with splitting energy separation can be interpreted in terms of the frequency of the tunneling process in a similar way to that given in the case of ammonia inversion.[303]

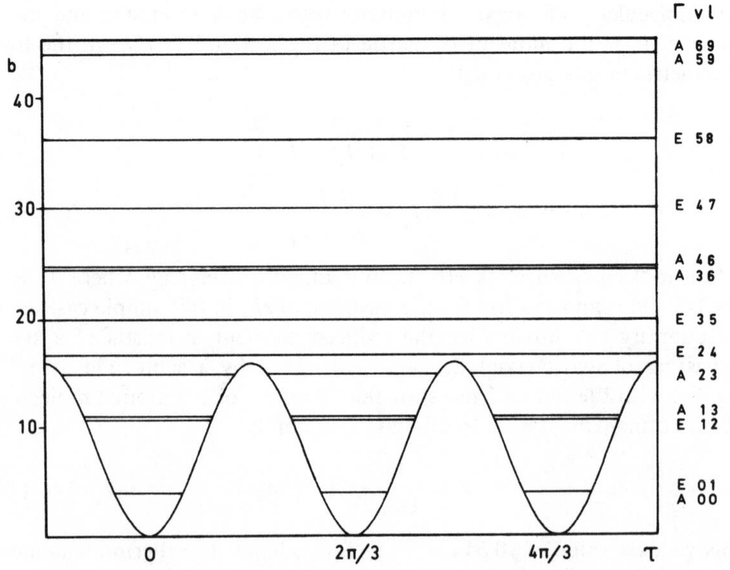

Figure 9.3 Simple cosine potential drawn to scale for a system with the barrier height parameter $s = 15$. In the ground state the internal rotation splitting is too small to be observable on this scale. Compare this diagram with Figure 9.2

9.2c *The Coupling of Internal Rotation with Overall Rotation*

Once the PAM Hamiltonian has been obtained in the form (9.14) it is a fairly straight-forward matter to generate an expanded form which readily lends itself to a perturbation treatment. If we take $H^{(0)}$ and $H^{(1)}$ as

$$H^0 = H_r + Fp^2 + V(\tau) \tag{9.34a}$$

$$H' = F\pi^2 - 2Fp\pi \tag{9.34b}$$

then the expanded form

$$[H_r] + [Fp^2 + V(\tau)] + [-2F\langle v\,\sigma|p|v\,\sigma\rangle\pi] + \left[F\left(1 + \sum_{v'}{}' \frac{4F|\langle v\,\sigma|p|v'\,\sigma\rangle|^2}{\Delta E}\right)\pi^2\right] + \cdots \tag{9.35}$$

can be developed as far as terms in π^2. Herschbach[117] writes the general result as

$$H_{v\sigma} = H_r + F\sum_{n=0}^{\infty} W_{v\sigma}^{(n)}\pi^n \tag{9.36}$$

where

$$W_{v\sigma}^{(0)} = \langle v\,\sigma|H_\tau|v\,\sigma\rangle = \frac{n^2}{4}b_{v\sigma} \tag{9.37a}$$

$$W_{v\alpha}^{(1)} = -2\langle v\,\sigma|p|v\,\sigma\rangle \tag{9.37b}$$

$$W_{v\sigma}^{(2)} = 1 + \sum_{v'}{}' \frac{4F|\langle v\,\sigma|p|v'\,\sigma\rangle|^2}{E_{v\sigma} - E_{v'\sigma}} \tag{9.37c}$$

as can be seen by comparing (9.35) with (9.36).

The form given in (9.36) has the advantage that extensive tabulations exist for the W coefficients. Hayashi and Pierce[112] have extended the original tables of Herschbach.[116] The most accessible copy of the more often used parts of the tables is to be found in the book by Wollrab,[343] (Appendix 12). We shall not go into the evaluation of the W coefficients but the following properties are useful.

(1) All coefficients of odd-order powers of π vanish for A torsional states, i.e. $W_{v0}^{(n)} \to 0$ for odd n.

(2) The W coefficients are a function only of s as long as V_6 and higher terms in the potential can be neglected.

(3) $W_{v0}^{(n)}(A)$ differs in both sign and magnitude from $W_{v\pm1}^{(n)}(E)$.

It is this property which together with (1) causes the energy levels to diverge and transmit barrier dependent splittings to the rotational spectrum.

(4) Several approximate relations exist among the W coefficients which become more valid with increasing s. One of the most useful is

$$W_{v0}^{(2)} \simeq -2W_{v\,\pm 1}^{(2)} \qquad (9.38)$$

which allows semi-quantitative conclusions to be drawn. Other relations are given by Herschbach.[117]

Using the W coefficients, the expanded Hamiltonian, correct to second order can be written as

$$\sum_{\alpha\beta} (\tfrac{1}{2}\mu_{\alpha\alpha}J_\alpha^2 + FW_{v\sigma}^{(0)} + FW_{v\sigma}^{(1)}[\rho_\alpha J_\alpha] + FW_{v\sigma}^{(2)}[\rho_\alpha\rho_\beta J_\alpha J_\beta] + \cdots) \qquad (9.39)$$

These terms are usually adequate as the higher terms which yield contributions of similar form to the centrifugal distortion corrections can often be neglected. This assumption becomes less valid as J increases and also as $b \to s$. The sum of the first and fourth terms in (9.39) yields a rigid-rotor type Hamiltonian. The off-diagonal elements which come from the fourth term can be eliminated by the procedures of Chapter 3, to yield the standard rigid-rotor Hamiltonian with inertial constants which have been modified from those of the reference configuration. It is however often sufficient to ignore the off-diagonal elements and only make the corrections required to the diagonal elements. Under these circumstances we get the convenient approximate expression for the Hamiltonian

$$[A_{v\sigma}J_A^2 + B_{v\sigma}J_B^2 + C_{v\sigma}J_C^2] + FW_{v\sigma}^{(0)} + FW_{v\sigma}^{(1)}[\rho_A J_A + \rho_B J_B + \rho_C J_C] + \cdots \qquad (9.40)$$

where $A_{v\sigma} = A + W_{v\sigma}^{(2)}F\rho_A^2$ etc. and as usual $\sigma = 0$ for the A state and ± 1 for the E states. If we neglect for the moment the terms involving $W_{v\sigma}^{(1)}$ then the energy-level pattern resolves itself into two pseudo-rigid-rotor stacks; one for the $A(\sigma = 0)$ levels and the other for the $E(\sigma = \pm 1)$ levels. The selection rule on σ is $\Delta\sigma = 0$ as rotation of the symmetric methyl group relative to the frame cannot affect the overall dipole moment ($\partial\mu/\partial\tau = 0$). Thus the spectrum will effectively consist of two sets of rigid-rotor transitions displaced to an extent determined by the barrier.

The linear term, $2Fp\pi$, is handled differently as it spoils the standard form of the rotational Hamiltonian. These terms are most important for low-J states and can usually be neglected for high-J lines when $s > 30$. At low values of J this term can, however, predominate over the even-order terms.

9.2d *Barrier Height Determination from Internal Rotation Splittings*

A simple calculation using the data for CH_3SiH_3[155] provides an example of the procedure to be followed in evaluating the barrier for a symmetric top molecule. It turns out, as we shall see, that for a symmetric top molecule the internal motion does not affect the *observed* spectral pattern and thus the commonly used method for getting round this is to introduce asymmetry by partial isotopic substitution. We can carry out an analysis for the $J = 2 \leftarrow 1$ lines of CH_3SiH_2D.

We can obtain a quite accurate F value by assuming that deuteration has not displaced the A-axis too far from the C—Si bond and that it still coincides reasonably well with the axis of the methyl group. Under these circumstances we have the situation that $\lambda_A \sim 1$, $\lambda_B \sim 0$ and $\lambda_C = 0$ and thus

(amu Å²)	(MHz)	
$I_A = 10{\cdot}810$	$A \sim 46\,750$	
$I_\tau = 3{\cdot}168$	$B = 10\,671$	$r \sim (I_A - I_\tau)/I_A = 0{\cdot}7069$
$rI_\tau = 2{\cdot}239$	$C = 10\,271$	
	$F \sim 225\,715$	

In this particular case we can neglect the effect of $W^{(2)}$ terms and thus the modified constants in (9.40) $A_{v\sigma}$ etc. are independent of σ, i.e. are the same for the A and E states and are essentially unchanged from their rigid-rotor value. The rotation dependent part of (9.40) can thus be written as two separate Hamiltonians for the A and E states.

$$H(A) = H_r, \qquad H(E) = H_r + \varepsilon J_A \tag{9.41}$$

where $\varepsilon = F\rho_A W_{01}^{(1)}$, H_r is just the usual rigid-rotor Hamiltonian and the associated matrix for $J = 1$ is (3.57). For the A states (3.57) is unchanged but for the E states it becomes

$$\begin{bmatrix} 2\alpha + \beta - \varepsilon & 0 & 2\gamma \\ 0 & 2\alpha & 0 \\ 2\gamma & 0 & 2\alpha + \beta + \varepsilon \end{bmatrix} \tag{9.42}$$

The three solutions of which are

$$2\alpha, \quad 2\alpha + \beta \pm \sqrt{4\gamma^2 + \varepsilon^2} \tag{9.43}$$

A very similar calculation can be carried out for the $J = 2$ levels. Note the sign dependence of the terms in ε (as they are linear in K) and also the interesting point that the $K = 0$ level is unperturbed. The energy-level expressions are thus:[155,117]

$$\begin{matrix} 1_{10} \\ 1_{11} \end{matrix} \qquad A + \tfrac{1}{2}(B + C) \pm \sqrt{\tfrac{1}{4}(B - C)^2 + \varepsilon^2} \tag{9.44a}$$

$$\begin{matrix} 2_{11} \\ 2_{12} \end{matrix} \qquad A + \tfrac{5}{2}(B + C) \pm \sqrt{\tfrac{9}{4}(B - C)^2 + \varepsilon^2} \tag{9.44b}$$

where $\varepsilon = 0$ for the A states and for the E states is an experimentally determinable parameter. In Figure 9.4 the observed spectrum of Kilb and Pierce[155] is shown schematically. Short circuiting the exact solution for ε let us note that $\varepsilon \ll (B - C)$ as indicated by the fact that the internal rotation splitting is $\sim$20 MHz, whereas the asymmetry splitting is $\sim$800 MHz. Thus we can re-arrange the square-root terms in (9.44) and write them as $\frac{1}{2}(B - C)\sqrt{1 + x}$ and $\frac{3}{2}(B - C)\sqrt{1 + \frac{1}{9}x}$ respectively where $x = 4\varepsilon^2/(B - C)^2 = \varepsilon^2/40\,000$.

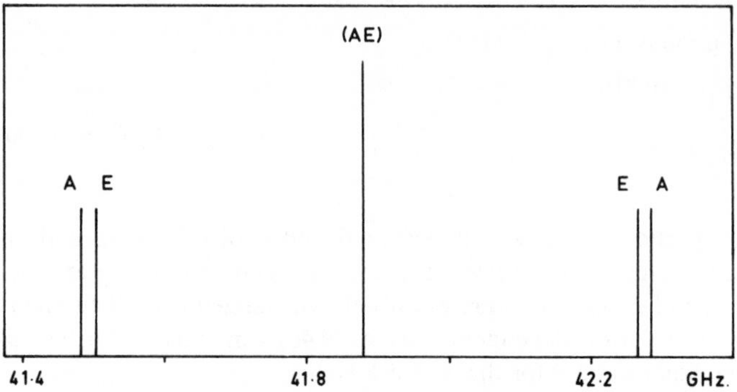

Figure 9.4 Schematic diagram of the spectrum of the $J = 2 \leftarrow 1$ transitions of SiH_2DCH_3. The frequencies observed by Kilb and Pierce[155] are: $2_{12} \leftarrow 1_{11}$, (A) 41 482·84, (E) 41 502·14; $2_{02} \leftarrow 1_{01}$, (A,E) 41 880·10; $2_{11} \leftarrow 1_{10}$ (E) 42 264·12, (A) 42 283·52 (MHz)

In this approximation we see that the separation of the A and E transitions depends only on the terms in x and as conveniently $B - C = 400$ MHz we obtain from Figure 9.4 and expansion of the square roots

$$\Delta E = 19{\cdot}3\,(\text{MHz}) = 100(\tfrac{2}{3}x - \tfrac{1}{4}x^2) \tag{9.45}$$

We thus see that:

(i) neglecting the x^2 term $x \cong 0{\cdot}3$, $|\varepsilon| = 110$
(ii) solving the quadratic $x = 0{\cdot}3305$, $|\varepsilon| = 115$
(iii) Kilb and Pierce[155] $x = 0{\cdot}331$, $|\varepsilon| = 115 \pm 2$.

As $|\varepsilon| = F\rho_A W_{01}^{(1)}$ and $\rho \cong I_\tau/I_A$ we see that

$$|W_{01}^{(1)}| = (115 \times 10{\cdot}8)/(225\,715 \times 3{\cdot}168) = 1{\cdot}738 \times 10^{-3}$$

using the Kilb and Pierce value for ε. If we refer to tables of W coefficients with $n = 1$, $v = 0$, $\sigma = 1$[112] (or Wollrab,[343] p. 433) we can interpolate over the values given there:

$$s = 32 \quad -0{\cdot}002778$$

$$s = 34 \quad -0{\cdot}002058$$

$$s = 36 \quad -0{\cdot}001536$$

$$s = 38 \quad -0{\cdot}001154$$

to yield a value for s of 35·2. From this we see that $V_3 = 596\ \text{cm}^{-1}$ ($\equiv 1703$ cal/mole). The value given by Kilb and Pierce is $V_3 = 570\ \text{cm}^{-1}$. In Figure 9.5 an explanatory energy-level diagram is drawn.

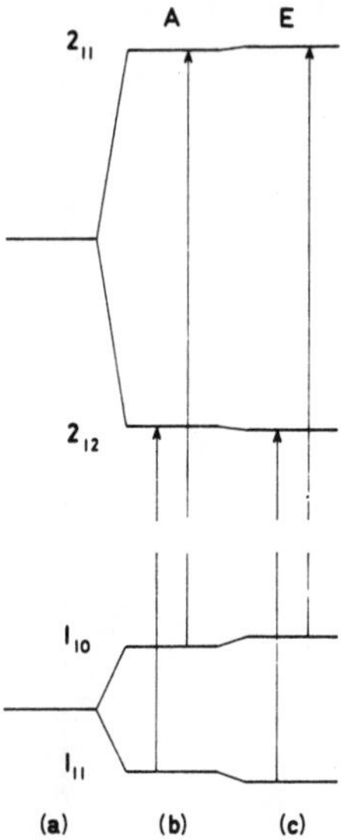

Figure 9.5 Energy levels of SiH_2DCH_3 with torsion–rotation coupling. (a) Symmetric rotor levels. (b) The *A* torsional levels affected only by asymmetry. (c) The *E* torsional levels affected both by asymmetry and internal rotation

In Figure 9.6 the $J = 3 \leftarrow 2$ transitions of thioacetaldehyde, CH_3CHS, are shown. The internal rotation causes the splitting of the outer $K = 1$ lines and the torsional satellites do not lie in a simple pattern. Compare this spectrum with that of cyanogen azide shown in Figure 5.6.

We can use the splittings observed in the spectrum of CH_3SiH_2F as an example of the use of the second-order coefficients $W^{(2)}$. Pierce[237] has observed several lines belonging to this molecule. It is particularly simple to use the Q_{01} branch lines for a barrier height determination. For instance the following lines were observed to be doublets.

$$5_{14} \leftarrow 5_{15} \quad 18040{\cdot}50 \quad (-1{\cdot}70)$$

$$6_{15} \leftarrow 6_{16} \quad 25113{\cdot}08 \quad (-2{\cdot}66)$$

$$7_{16} \leftarrow 7_{17} \quad 33178{\cdot}93 \quad (-3{\cdot}60)$$

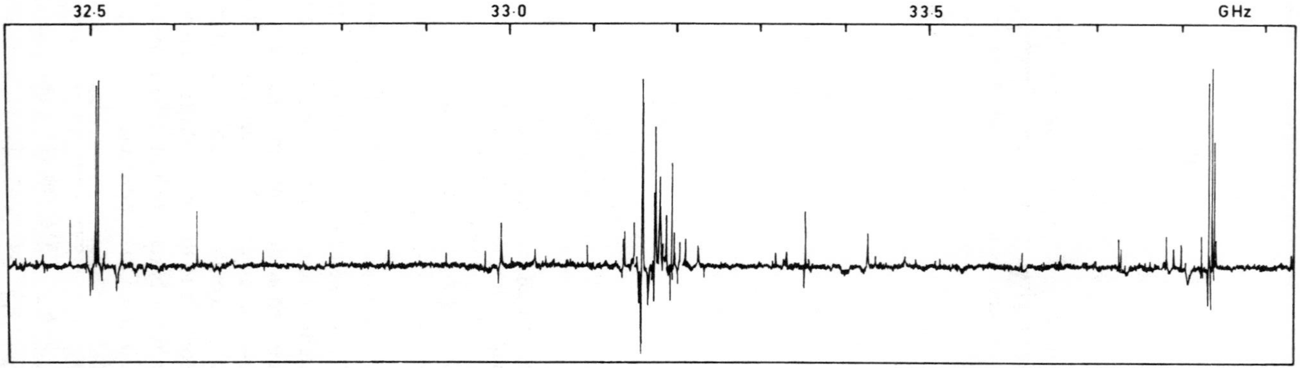

Figure 9.6 The $J = 3 \leftarrow 2$ transitions of thioacetaldehyde, CH_3CHS observed by Kroto, Landsberg and Suffolk.[166] Compare this spectrum with that of NCN_3 Figure 5.6. The internal rotation causes the outer $K_A = 1$ lines to be resolved into two approximately equally intense lines. The strongest line at the centre of the spectrum corresponds to the $K_A = 0$ line which shows some evidence of splitting, even on this wide-range scan. Other strong lines belong to the $K_A = 2$ transitions and the torsional satellites which do not lie in a simple regular sequence as do the satellites of NCN_3

these frequencies are in MHz and the splittings are given in brackets. For a near-prolate molecule the frequencies of these lines are given quite accurately by the expression

$$\Delta E = \tfrac{1}{2}(B - C)J(J + 1) \tag{9.46}$$

as shown in Chapter 5. In Table 9.2 the appropriate molecular parameters are listed. In this case it is reasonably valid to neglect the linear terms for the higher J lines, so we will choose the highest J, $7_{16} \leftarrow 7_{17}$ transition and use the following truncated form of expression (9.40)

$$H = A_{v\sigma}J_A^2 + B_{v\sigma}J_B^2 + C_{v\sigma}J_C^2 \tag{9.47}$$

where

$$B_{v\sigma} = B + W_{v\sigma}^{(2)}F\rho_B^2 \ \ldots \text{etc.} \tag{9.47a}$$

The C constant is the same for both A and E sublevels because $\lambda_c = 0$ by symmetry and thus also $\rho_C = 0$. The splitting in these lines is thus due only to the difference $B(E) - B(A)$, and as a result we can write

$$\Delta E_E - \Delta E_A = \tfrac{1}{2}\Delta BJ(J + 1) \tag{9.48}$$

and from the $7_{16} \leftarrow 7_{17}$ transition we find that

$$\Delta B = -0{\cdot}127 = [W_{00}^{(2)} - W_{01}^{(2)}]F\rho_B^2 \tag{9.49}$$

Table 9.2 Table of Molecular Parameters for CH_3SiH_2F

	MHz	λ_α	λ_α^2	I_α amu Å^2
A	20 050	0·85251	0·7268	25·2057
B	6 753·8	0·52271	0·2732	74·8282
C	5 545·3	0·0	0·0	91·1364
τ	161 410·0	—	—	3·131

From the definition of r (9.10) and F (9.12a) we can calculate

$$r = 0{\cdot}89829, \qquad rI_\tau = 2{\cdot}8125 \text{ (amu Å}^2\text{)}, \qquad F = 179\,689 \text{ (MHz)}$$

$$\rho_B^2 = (\lambda_B I_\tau / I_B)^2 = 4{\cdot}7837 \times 10^{-4}$$

Substituting these values into (9.49) we find that

$$[W_{00}^{(2)} - W_{01}^{(2)}] = -0{\cdot}001477 \simeq -3W_{01}^{(2)} \tag{9.50}$$

This last approximation in (9.50) follows from (9.38) which in this particular case holds very accurately. From tables[343] we can extract relevant data which is listed in Table 9.3. Graphical interpolation of this value (9.50)

Table 9.3

s	$W^{(2)}_{01} - W^{(2)}_{00}$
38	−0·002092
40	−0·001882
41	−0·001203
42	−0·000920

yields the result that $s = 40{\cdot}5$. As $V_3 = \frac{9}{4}Fs$ we see that $V_3 = 545{\cdot}8\ \text{cm}^{-1} \equiv$ 1560 cals/mole. The result obtained by Pierce is 1562 cals/mole· (29 979 MHz $\equiv 1\ \text{cm}^{-1}$ and $1\ \text{cm}^{-1} \equiv 2{\cdot}8592$ cals/mole).

9.2e *General Survey of Systems with Internal Rotation*

We have in the foregoing sections concentrated on giving a fairly detailed introduction to the theory which can be used to study molecules which contain a single symmetric internal rotor and a medium size threefold internal rotation barrier. This was the first and prototype case to be fully understood. There are however many different systems with internal rotation where this theory is not directly applicable. In this section we will give a brief descriptive survey of the various different cases, using a particular molecule as prototype example.

CH_2DCHO *α-monodeuteroacetaldehyde.* In this case the deuteration of methyl group has destroyed the C_3 symmetry of the internal-rotor inertia and the kinetic energy becomes a function of τ. The potential function however still retains its C_3 symmetry. Kilb, Lin and Wilson,[154] have studied this molecule and showed that rotational isomers exist. Quade and Lin have developed the relevant theory.[246]

CH_2FCHO *α-monofluoroacetaldehyde.*[264] In this case not only is the kinetic-energy expression dependent on τ but the barrier has also lost its C_3 symmetry. As a consequence (9.15) in general will contain several significant terms.

CH_3NO_2 *nitromethane.*[296] This molecule along with CH_3BF_2[212] has been studied in detail and shown to have a very low six-fold barrier. These systems are handled by using a basis Hamiltonian which corresponds to free internal rotation, and introducing the barrier potential as a small perturbation. The theory has been developed by Wilson, Lin and Lide.[337] The derived very small barrier heights, 6 cals/mole in CH_3NO_2 and 14 cals/mole in CH_3BF_2, have been used as the main justification for the general assumption that six-fold and higher terms in the potential (9.15) can often be neglected. One should note that in CF_3NO_2 the barrier is still rather high at 74 cals/mole.[300]

C_6H_5OH *phenol.* Quade[245,148] has considered molecules such as phenol and nitrobenzene which have two-fold barriers. In phenol the barrier height is 3·36 kcals/mole and the equilibrium configurations are planar.

H_2S_2 *hydrogenpersulphide.* The S—H bonds in this molecule lie at 91°20′ to the S—S bond and in the equilibrium configuration the dihedral angle is 90°36′. The molecule thus has a skew configuration. Winton and Winnewisser[341] have used the Lamb-dip technique to measure the small internal rotation splitting and have determined that $V_{trans} = 2373\ cm^{-1}$ and $V_{cis} = 2550\ cm^{-1}$. Oelfke and Gordy have studied H_2O_2.[220] D_2S_2 is perhaps most celebrated because it is essentially *the* accidental symmetric top.[338]

$(CH_3)_2CO$ *acetone.* Here there are two symmetric internal rotation tops. This molecule was studied by Swalen and Costain.[292] The two-top problem has been reviewed by Dreizler.[75]

In the case of acetone the lines are split into quartets due to the effects of internal rotation.[292,313] The four possible torsional states can be written as

$$|0\ 0\rangle, \qquad |0\ \pm1\rangle \pm |\pm1\ 0\rangle, \qquad |\pm1\ \pm1\rangle, \qquad |\pm1\ \mp1\rangle$$

The kets are labeled as $|\sigma_1\sigma_1\rangle$ where σ_1 and σ_2 are the torsional phase quantum numbers for each top, (9.22). In the case of thioacetone the effects of internal rotation are much smaller than in acetone. In Figure 9.7 the 4_{13}–3_{12} transition of thioacetone is depicted. Here only a triplet pattern is observed

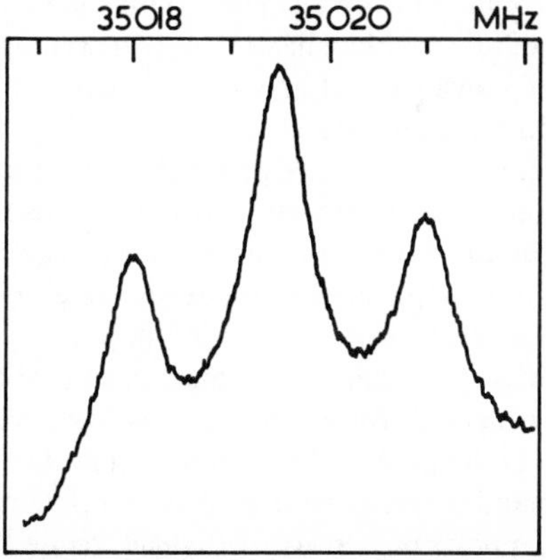

Figure 9.7 The $4_{13} \leftarrow 3_{12}$ transition of thioacetone, $(CH_3)_2CS$ observed by Kroto and Landsberg.[165a] This transition consists of a partially resolved quartet. The line assignments are: $|0\ 0\rangle$ at 35 021 MHz, $2^{-1/2}$ $\{|0\ \pm1\rangle \pm |\pm1\ 0\rangle\}$ at 35 019·5 and an *unresolved* doublet at 35 018 belonging to the states $|\pm1\ \pm1\rangle$ and $|\pm1\ \mp1\rangle$

because the splitting between the $|\pm 1 \ \pm 1\rangle$ and the $|\pm 1 \ \mp 1\rangle$ transitions at 35 018 MHz has not been resolved.[165a]

$CH_3CH_2CH_2F$ *n-propylfluoride*. Hirota[126] has studied this molecule in detail and has obtained detailed information about both three-fold barriers. Two distinct sets of transitions corresponding to the trans and gauche conformers were observed. For each of these two sets of spectra, splittings due to the methyl group internal rotation were observed in the torsional satellites.

$(CH_3)_3SiH$ *trimethyl silane*. Molecules such as trimethyl silane,[239] trimethyl phosphine[179] and trimethyl silylisocyanate[32] possess three equivalent methyl groups. Because they are essentially symmetric tops, the ground vibrational state gives rise to a straightforward symmetric rotor pattern (Figure 6.6). Lide and Mann[179] have studied the torsional satellites of this class of molecules and have formulated a procedure for assigning them.

9.3 Inversion of Ammonia

The dynamic properties of the ammonia molecule have been the subject of many experimental and theoretical investigations. The rotational and vibrational spectra indicate that the molecule has two pyramidal equilibrium structures and that tunneling occurs between these two mirror-image configurations via a planar intermediate (see (3.103)). The spectroscopic data can be satisfactorily accounted for by a barrier to inversion of about 2020 cm^{-1}. Several treatments of this problem have been reviewed by Wollrab[343] and as an example we shall consider that given by Swalen and Ibers[293] who use a Gaussian expression for the barrier.

A Hamiltonian for the inversion problem can be set up by adding a Gaussian hump potential contribution to the simple harmonic oscillator Hamiltonian (1.9). By adjusting the parameters which govern the shape of the hump, relative to the shape of the quadratic potential of the harmonic oscillator, one can generate a double minimum potential with the requisite form. One can then vary the shape parameters, so that the resulting eigenvalues match the observed vibrational energy-level pattern,[293] and thus determine the form of the potential and in particular the barrier height.

Chan, Zinn, Fernandez and Gwinn used such a potential in their original treatment of ring puckering in trimethylene oxide.[46,44] Thorson and Nakagawa[297] used a Lorentzian hump, and Dixon[68] a Gaussian hump, in their treatment of the rather more complicated problem which occurs in quasi-linear molecules. In this case the hump is in the *two-dimensional* bending potential.

With the assumption that a one-dimensional coordinate can adequately describe the inversion motion, the Hamiltonian can be written as

$$H = \tfrac{1}{2}\omega(p^2 + q^2 + \gamma\, e^{-cq^2}) \tag{9.51}$$

It is perhaps worth noting that there are really only two adjustable parameters γ and c; ω can be considered as a scale factor. To see what this means, consider the harmonic oscillator Hamiltonian; there is no flexibility in fitting more than one transition in this case because this Hamiltonian generates eigenvalues which are equidistantly spaced and it is in the *variations* in the spacings that the anharmonic information lies. The problem can be solved by adding the matrix elements of the Gaussian term (in a harmonic oscillator basis) to the diagonal matrix of the harmonic oscillator (1.23). The resulting matrix is truncated at some convenient value of v and then diagonalized. Swalen and Ibers found that they could truncate at around $v = 20$ and calculate the lowest eight vibrational levels without significant truncation error. Their results are given in Table 9.4 and compared with the observed energy-level pattern. The generation of Gaussian matrix elements in a convenient way for computational purposes has been discussed by Bell.[15] Bell also considers the two- and three-dimensional problems.

If the barrier is very high then the lowest eigenstates are essentially doubly degenerate. As the barrier decreases the degeneracy is lifted and the resulting splitting frequency can be related to the tunneling frequency, the levels near the top of the barrier being the most widely split. A second and complementary way of looking at the effect is to start at the limiting case where the molecule is planar and the barrier is negligible. In this case the levels are those of the harmonic oscillator and are equidistantly spaced. As a small perturbing hump is introduced and increased in magnitude, it has the effect of pushing the vibrational levels upwards. The hump pushes the levels with v, even, more strongly than those with v, odd, and as a result the odd and even levels pair up. This occurs because the levels with v, even, are symmetric and possess maxima in the coordinate distribution function at the planar configuration where the hump maximum lies. The odd levels are less strongly perturbed because they have nodes at this point.

Table 9.4 Calculated[293] and Observed[16] Inversion Energy-level Pattern for the Ammonia Molecule

Level	Observed Energy, cm^{-1}	Calculated energy, cm^{-1}
3^-	2895·5	2893
3^+	2383·5	2387
2^-	1910	1884
2^+	1597·6	1601
1^-	968·3	967
1^+	932·5	928
0^-	0·794	0·98
0^+	0·0	0·0

9.3a Spectra of Ammonia

As a result of the flexibility of ammonia, the ground-state level is split into two states, the 0^- state lying 0·794 cm^{-1} ($\sim$24·0 GHz) above the lowest state, 0^+. This splitting is an order of magnitude smaller than the rotational-energy level spacings. As a result we can consider the rotational-energy levels to be split apart by $\pm\frac{1}{2}E_{inv}$ from the oblate symmetric rotor pattern as shown in Figures 9.8 and 9.9. Due to the exclusion principle, when $K = 0$ only 0^+ states with J odd exist whereas for the 0^- state only the J even levels exist.

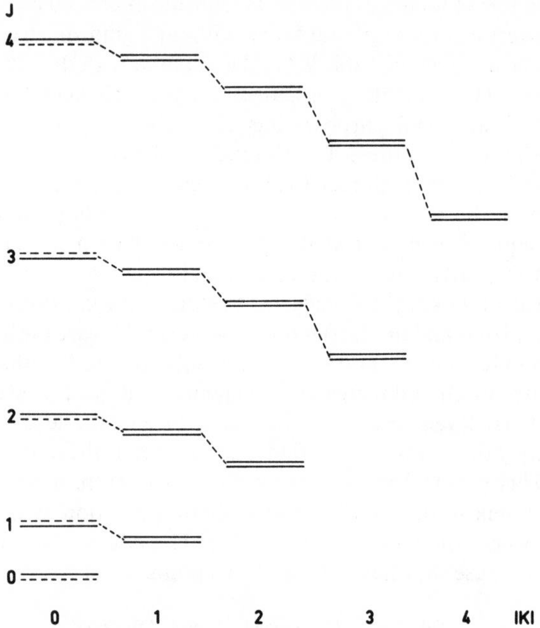

Figure 9.8 The rotation–inversion energy levels of ammonia

Electric dipole transitions between the inversion levels can occur according to the selection rules $v^+ \leftrightarrow v^-$, $\Delta J = 0$ and $\Delta K = 0$ ($K \neq 0$). This set of transitions gives rise to a large number of lines in the 24 GHz region. These transitions were in fact the first ones detected using microwave techniques.[48]

Costain[51] has fitted the frequencies of the inversion spectrum to the exponential function.

$$\Delta E(JK) = 23\,785{\cdot}88 \exp[A_1 J(J+1) + A_2 K^2 + A_3 J^2(J+1)^2 + A_4 J(J+1)K^2 + A_5 K^4] \tag{9.52}$$

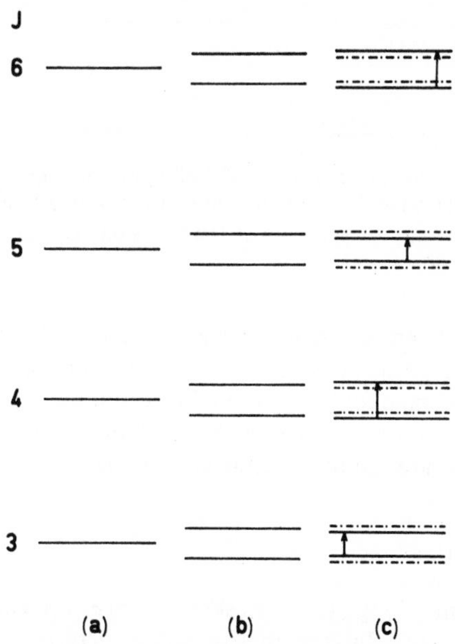

Figure 9.9 The rotation–inversion levels of ammonia with $|K| = 3$. (a) The levels assuming inversion is negligible. (b) The levels including the inversion splitting. (c) The levels including K-doubling[219,303]

which predicts the lines to $\sim 1{\cdot}3$ MHz. An extended form of this treatment has been given by Schnabel, Törring and Wilke.[269,100] There is a fourth-order K-doubling interaction which Nielsen and Dennison have discussed,[219] it involves matrix elements of the type $\Delta K = 6$ and splits the A_1 and A_2 levels with $|K| = 3$ apart. When $|K| = 3$ the Exclusion Principle indicates that for the 0^+ state only the A_2 rotational (or A_2' overall) levels (upper K-doublet when J odd, lower when J even) levels can occur. For the 0^- state, only the A_1 (or A_2'' overall) levels (lower K-doublet when J odd, upper when J even) occur. This situation is shown in Figure 9.9.

Helminger and Gordy[114] have observed the $J = 2 \leftarrow 1$ and $J = 1 \leftarrow 0$ transitions in the millimetre wave and Dowling has observed the far infra-red transitions with $\Delta J = +1$ from $J = 1$ to $J = 11$.[72] A computer simulation of the far infra-red spectrum based on Dowling's line measurements is shown in Figure 9.10. For these $\Delta J = +1$ transitions a change of inversion state occurs simultaneously, i.e. $0^- \leftrightarrow 0^+$. As a result the infra-red spectrum consists of two R-branches, with K-substructure, separated by about twice the inversion splitting.

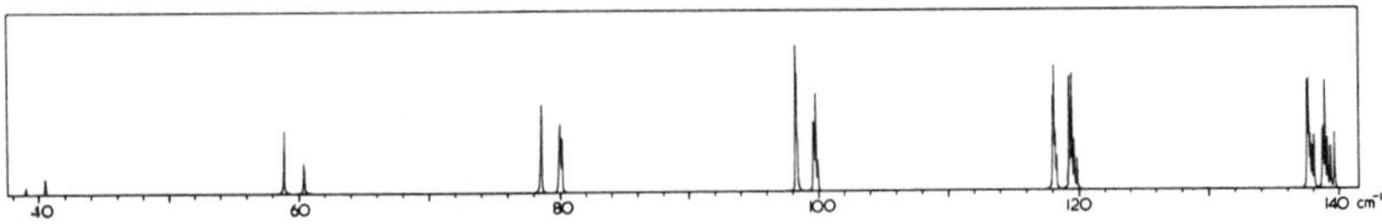

Figure 9.10 Part of a computer synthesis of the far infra-red spectrum of NH_3 using the line frequencies of Dowling.[72] The spectrum consists of two branches split apart by approximately twice the inversion frequency

The Raman spectrum has been investigated by a number of workers.[285,122] In this case the selection rules restrict the transitions such that they occur within the 0^+ or 0^- manifolds ($\pm \leftrightarrow \pm$). As a result the two sets of transitions overlap, because the dependence of the rotational constants on inversion state is small compared to the resolution available.

9.4 Ring Puckering

A number of small ring molecules show an inversion effect related to that observed in ammonia. In trimethylene oxide there is an out-of-plane deformation vibration which has a small hump at the planar configuration. The spectra and theory of this type of motion have been the subject of a number of papers by Chan, Gwinn and coworkers[43–47] who have shown that the potential has a strong quartic component, as predicted by Bell.[14] A large number of these types of molecule have now been studied by both infra-red and microwave techniques. The infra-red gives the potential surface information which indicates that in general the potential can be described with good accuracy by

$$V = aq^4 + bq^2 \tag{9.53}$$

Depending on the molecule, b may vary in relative magnitude and sign to a. When b is zero the potential is essentially pure quartic and when b has a negative value a double minimum potential is generated. Depending on the type of potential the lower vibrational levels might lie in a rather irregular pattern.[305]

Sometimes the inversion splitting is small enough to occur in the microwave range.[111] In general one observes a rotational microwave spectrum where each line is accompanied by several vibrational satellites. In Figure 9.11 transitions of trimethylene oxide are shown. Because the potential function is complicated, the dependence of the rotational constants on the inversion coordinates is also complicated. As a result the vibrational satellites may depart significantly (Figure 9.11) from the usual regular spacing (Figure 5.6).

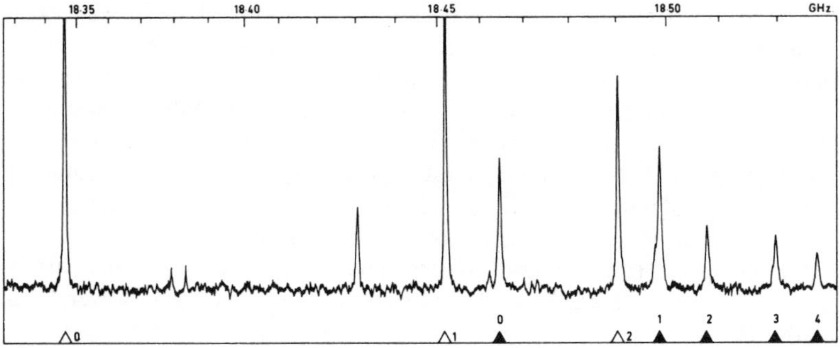

Figure 9.11 The microwave spectrum of trimethylene oxide in the region 18·32 to 18·53 GHz. The lines shown belong to the $1_{01} \leftarrow 0_{00}$ (▲) and $5_{51} \leftarrow 5_{32}$ (△) and their associated vibrational satellites. The uneven spacing of the satellites is characteristic of a molecule with puckering and contrasts with the even spacing associated with a harmonic vibration Figures 5.6 and 5.9. The relevant data are given in Table 9.5

The observed rotational constants can be compared to the theoretically calculated inertial constants[47,111] using a relation such as

$$\langle v|B_{\alpha\alpha}|v\rangle = B^0_{\alpha\alpha} + \beta_2\langle v|q^2|v\rangle + \beta_4\langle v|q^4|v\rangle + \cdots \tag{9.54}$$

In Table 9.5 are given the rotational constants for trimethylene oxide obtained by Chan, Zinn, Fernandez and Gwinn[46,47,43] for the various excited puckering vibrational states. The values of $(B_{v+1} - B_v)$ etc. are also given in Table 9.5. The haphazard variation in the rotational constants indicated by these differences is characteristic of molecules which exhibit puckering motion.[43] The largest irregularities tend to occur for levels close to the maximum. In this case $v = 0$. The frequencies of the $1_{01} \leftarrow 0_{00}$ and $5_{51} \leftarrow 5_{32}$ transitions are also given in Table 9.5. Centrifugal distortion has been neglected in these calculations. In Figure 9.11 certain of these transitions are shown and the uneven satellite pattern is quite evident, particularly between $v = 0$ and $v = 1$ for the $1_{01} \leftarrow 0_{00}$ transition. The frequency of this transition is given by $B_v + C_v$.

Table 9.5 Rotational Constants and Transition Frequencies of Trimethylene Oxide[46,47,43]

v	0	1	2	3	4
A_v	12 045·2 (12·8)	12 058·0 (0·9)	12 058·9 (1·3)	12 060·2 (−2·2)	12 058·0
B_v	11 734·0 (−8·0)	11 726·0 (−7·2)	11 718·8 (−8·8)	11 710·0 (−11·3)	11 698·7
C_v	6 730·7 (41·9)	6 772·6 (16·5)	6 789·1 (20·5)	6 809·6 (18·0)	6 827·6
1_{01} 0_{00}	18 464·7	18 498·6	18 507·9	18 519·6	18 526·3
5_{51} 5_{32}	18 343·75	18 457·31	18 490·66	18 534·17	18 566·2

Five-membered rings have two puckering vibrations which in symmetric systems, such as cyclopentane, are degenerate. The resulting puckering motion can be considered to rotate round the ring. This type of motion has been called pseudorotation. In non-symmetric rings the vibrations are no longer degenerate but they may still interact if the associated energies are not too different. In these cases the potential governing the phase of the rotation has maxima and minima. The theory of this type of motion has been developed and applied to tetrahydrafuran[110] and other molecules.[171]

Six-membered rings tend not to show puckering behaviour as the barrier appears to be too high. Several conformation studies have however been made.[171,203]

9.5 Quasi-linear Molecules and Quasi-symmetric Tops

In some molecules the distinction between a bent and a linear structure is blurred. Some aspects of this problem have already been discussed in Section 6.11c. Consider a molecule such as $Si(CH_3)_3NCO$ where the potential governing the bending of the $\widehat{Si{-}N{=}C}$ angle is highly anharmonic. There may be a small potential maximum at the linear configuration which may lie close to or below the vibrational ground state. There is evidence for such a potential in SiH_3NCO[94] and in HCNO a similar situation appears to apply to the $\widehat{H{-}C{\equiv}N}$ angle. Qualitatively the rotational structure of the ground state of $Si(CH_3)_3NCO$ is consistent with that expected for a symmetric top. Similarly that of SiH_3NCO is consistent with that of a symmetric top and that in HCNO consistent with a linear molecule. Quantitative study of the ground-state structures on the assumption of linearity tends to yield certain bond lengths shorter than expected, as discussed in Section 6.11c.

In such systems the quasi-linear character causes the bending vibrational levels to be close together and well populated. As a consequence the ground-state spectrum is often accompanied by many vibrational satellites. This is the case in $Si(CH_3)_3NCO$, and in Figure 9.12 the set of transitions $J = 16 \leftarrow 15$ are shown. Analysis of the vibrational satellites in these molecules is potentially a direct source of information about the anharmonic character of the bending potential.

The theory of vibration in quasi-linear molecules has been developed by Thorson and Nakagawa[297] and Dixon[8] and further developed by Johns.[142] The vibration–rotation problem has been studied by Hougen, Bunker and Johns.[135] Lide and Matsumura[180] have analysed vibration–rotation parameters in terms of a force field in curvilinear coordinates which are more realistic than the conventional linear coordinates when a molecule bends.[242]

The problem of molecules such as CH_3NCO is slightly different from that in SiH_3NCO. The spectrum is rather complicated and appears to be consistent with a molecule with a predominantly bent configuration for the $\widehat{C{-}N{=}C}$ angle of 140–150°.[175]

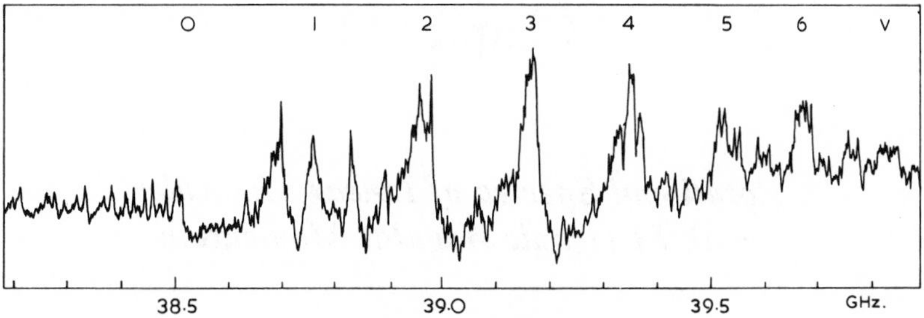

Figure 9.12 The $J = 16 \leftarrow 15$ transition of trimethylsilyl isocyanate and its associated vibrational satellites observed by Careless, Green and Kroto.[32] The fine structure associated with the ground state line is shown in Figure 6.6 in detail. Note the l-doublets when $v = 1$

In general, the symmetric top problem can be considered in terms of a correlation between two limiting cases. On the one extreme the system behaves like an asymmetric rotor with internal rotation and the motion of the essentially non-linear chain can be described using the standard internal rotation theory. As the chain straightens the C_3 character of the potential gradually becomes unimportant and the motion can be treated by the standard theory for degenerate vibrations. CH_3NCO is closer to the asymmetric rotor limit whereas SiH_3NCO and $Si(CH_3)_3NCO$ are closer to the symmetric top limit.

Chapter 10

Rotational Spectra of Linear Molecules with Electronic Angular Momentum

10.1 Electronic Angular Momentum

The theory developed so far in this book requires quite substantial modification when the species under study possesses electronic angular momentum. The best known examples are the $^3\Sigma_g^-$ ground state of O_2 which possesses resultant spin angular momentum and the $^2\Pi$ ground state of NO which has both orbital and spin angular momentum. Until recently only a small number of such systems had been studied where circumstances were favourable. For instance, O_2 and NO are stable and in the case of OH an unusual regenerative process effectively prolongs the lifetime. Recent improvements in techniques have enabled the microwave spectra of many more species to be detected.[203] Gas phase magnetic resonance techniques for detecting these species have also been successfully developed as discussed in Section 10.6.

We must now take into account the fact that the total angular momentum $\boldsymbol{J}$, devoid of nuclear spin, will in general be a sum of nuclear rotational angular momentum, $\boldsymbol{R}$, electronic angular momentum, $\boldsymbol{L}$, and spin angular momentum, $\boldsymbol{S}$. Thus, if we neglect vibrational angular momentum,

$$\boldsymbol{J} = \boldsymbol{R} + \boldsymbol{L} + \boldsymbol{S} \tag{10.1}$$

Note that for linear molecules in their ground vibrational state $R_c = 0$ where the c-axis lies along the internuclear axis.

10.2 Classification of Electronic States of Linear Molecules

In Chapter 2 the overall *spin free* Hamiltonian was written as the sum of an electronic part, H_e, and a nuclear part, H_n, a separation justified by the Born–Oppenheimer approximation. The procedure for calculating electronic energy-level patterns in general resolves itself into finding eigenfunctions, Ψ, which diagonalize H_e. These techniques have been discussed by many authors.[210,188,140] For our purposes we need only consider those basic properties of Ψ which are important for rotational calculations. Thus the overall form of Ψ is not necessary as many integrals over these functions occur as constants which parametrize the expressions we wish to develop. For instance the molecule-fixed dipole moment is such an integral, and in Chapter 7 this quantity is the constant which parametrizes the Stark energy and can be determined by analysis of the Stark shift pattern.

For a non-rotating linear molecule we must consider the motion of the electrons in the molecular field. The field has cylindrical symmetry about the internuclear axis and the resulting potential energy function, if spin–orbit coupling can be neglected, will be independent of the angle χ about the c-axis from a plane through the axis. As a consequence χ will appear only in the kinetic-energy expression. The momentum conjugate to χ, the electronic angular-momentum component $L_c \equiv -i\partial/\partial\chi$, commutes with H_e.

$$[H_e, L_c] = 0 \tag{10.2}$$

This enables us to use the quantum number $\Lambda(\equiv L_c')$ defined by the relation

$$L_c\Psi_e = \Lambda\Psi_e \tag{10.3}$$

to specify the basis, *non-rotating*, electronic eigenfunctions. The commutation of an operator with the Hamiltonian signifies that the associated quantity, here L_c, is conserved. Such, so-called, constants of motion are often represented by an associated quantum number, here Λ, which is often called a 'good' quantum number. The electronic eigenfunctions can thus be specified by Λ and from (10.3) can be written in the usual exponential form (see Chapter 3) $\Psi_e = F\exp(\pm i\Lambda\chi)$, where F is a function of all other electronic variables and independent of χ. Using the C_∞ operation of the point group $C_{\infty v}$ it is usual to classify the electronic functions as $\Sigma, \Pi, \Delta, \ldots$, etc. as $\Lambda = 0, \pm 1, \pm 2, \ldots$, etc. respectively. If there is a resultant spin then the spin multiplicity ($\equiv 2S + 1$) is added as a superscript to these state signatures. Although spin operators do not appear explicitly in the basis Hamiltonian H_e (2.3) they must be carefully taken into account when H_e is solved by ensuring that the Ψ_e have a form consistent with the Pauli exclusion principle. When second-order magnetic interactions are considered it is necessary to add to H_e, perturbation terms which include spin operators and correspond to interactions involving the spin magnetic moment. When these operators are important (10.2) may no longer hold and Λ will cease to be a good quantum number. There are however degrees of goodness which depend on the magnitude of the perturbation terms. If these are not too large the quantum number may still be useful to specify the eigenfunctions. In the non-rotating molecule $J_c(\equiv L_c + S_c)$ commutes with the Hamiltonian which includes the perturbation terms and thus Ω, defined by the equation

$$J_c\Psi_e = \Omega\Psi_e \tag{10.4}$$

will be a good quantum number. $\Omega = \Lambda + \Sigma$ where $\Sigma \equiv S_c'$. Thus the ground state of NO can be specified by Ω, and $|\Omega|$ is usually appended as a subscript.

There are two possible values of $|\Omega|$ which in this case are $\frac{1}{2}$ and $\frac{3}{2}$. The resulting states, $^2\Pi_{1/2}$ and $^2\Pi_{3/2}$, which are degenerate in the absence of the spin-dependent perturbation, are split apart in its presence.

The symmetry of the functions with respect to the σ_v operation of $C_{\infty v}$ should also be considered. $\sigma_v f(\chi) \rightarrow f(-\chi)$ and thus the eigenfunctions specified by (10.3) with $\Lambda \neq 0$ occur in degenerate pairs, whereas Σ states with $\Lambda = 0$ are non-degenerate. Whether the function is symmetric or anti-symmetric, with respect to σ_v, is signified by a superscript + or − on the term signature. The majority of stable molecules such as CO and N_2 have $^1\Sigma^+$ ground states. The ground state of O_2 on the other hand is $^3\Sigma^-$. When $\Lambda \neq 0$, the + and − states are degenerate in the absence of rotation and unless terms which can lift this degeneracy are significant, the $\pm$ specification is usually dropped.

We can see how a Σ^- state can occur by considering the lowest electronic states of O_2. In a molecular orbital description, the lowest electron configuration is $(1s\sigma)^2(1s\sigma^*)^2(2s\sigma)^2(2s\sigma^*)^2(2p\sigma)^2(2p\pi)^4(2p\pi^*)^2$. The angular momentum contribution from electrons in the closed shells is zero and thus we need only consider the contributions from the two equivalent π^* electrons. As they are π electrons, their individual momentum components l_{c1} and l_{c2} have eigenvalues $\lambda_1 = \pm 1$ and $\lambda_2 = \pm 1$ which must be added to give $\Lambda = \lambda_1 + \lambda_2$. As a result there are four states, two with $\Lambda = 0$ and two with $\Lambda = \pm 2$. We obtain the correct symmetry by applying σ_v to the angular portions of the linear combination of orbital products. Thus

$$\Psi(\Sigma^\pm) = F[e^{+i\chi_1} e^{-i\chi_2} \pm e^{-i\chi_1} e^{+i\chi_2}] \tag{10.5a}$$

$$\Psi(\Delta^\pm) = F[e^{+i\chi_1} e^{+i\chi_2} \pm e^{-i\chi_1} e^{-i\chi_2}] \tag{10.5b}$$

The Σ^- state changes sign under σ_v, and is unchanged under rotation, C_∞. This physically paradoxical behaviour does not occur for single orbitals. The Exclusion Principle, when applied to the total eigenfunction *including spin*, allows only the terms $^3\Sigma^-$, $^1\Sigma^+$ and $^1\Delta$ to arise from a π^2 configuration.

For $D_{\infty h}$ molecules such as H_2 and C_2H_2 the inversion operation i exists and allows us to specify the electronic wavefunctions as g or u depending on whether Ψ_e is invariant or changes sign on application of i. Thus the ground state of O_2 is completely specified by $^3\Sigma_g^-$.

10.3 Interactions Involving Electronic Angular Momentum

In the previous section the basic angular-momentum properties of the electronic wavefunction were outlined assuming that the *nuclei* were stationary and that electron orbital motion was independent of spin for a given electron state. There is of course the Pauli Exclusion Principle which dictates the allowed forms of the wave functions and thus also the energy through exchange terms for rather obscure reasons. We now consider some interactions which can be introduced in a formal way into the overall Hamiltonian.

10.3a Spin–Orbit Coupling

Associated with the resultant electron spin angular momentum $\boldsymbol{S}$ is a magnetic moment, $\boldsymbol{\mu} = g\beta\boldsymbol{S}$, where β is the Bohr Magneton and g is called the gyromagnetic ratio. If the molecule also has a resultant orbital angular momentum (as is the case when $\Lambda \neq 0$) then the circulating charge will generate a magnetic field component along the axis, B_c. As in an electron spin resonance experiment the two possible orientation spin states for a single electron will possess different energies. The energy of interaction $H = -B_c\mu_c$ can be written in terms of $L_c(\propto B_c)$ and $S_c(\propto \mu_c)$ as

$$H_{LS} = AL_cS_c \tag{10.6}$$

where A is the spin–orbit coupling constant. We will consider A as a parameter which can be obtained from experiment, though in some cases it can be calculated from the electronic wavefunctions with fair accuracy.[139] A simple energy-level diagram is given in Figure 10.1 for the $^2\Pi$ ground state of NO where $\Lambda = \pm 1$ and $\Sigma = \pm\frac{1}{2}$. The basis states can be represented by the ket $|\Lambda\ \Sigma\rangle$. We see that the four states, $|\pm 1 \pm \frac{1}{2}\rangle$ and $|\pm 1 \mp \frac{1}{2}\rangle$, degenerate in the absence of H_{LS}, are split into two sets which can be specified by the quantum number $|\Omega|$. The constant A in NO is, as we shall see, $\sim 124\ \text{cm}^{-1}$.

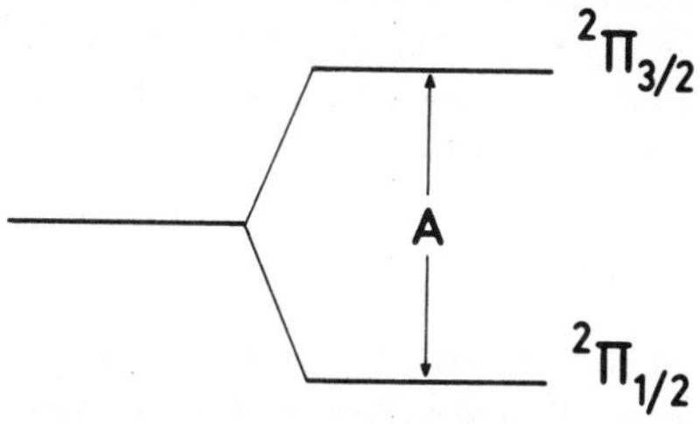

Figure 10.1 The splitting of a $^2\Pi$ state by spin–orbit coupling

10.3b Spin–Rotation Interactions

Another type of interaction which involves spin is often important and can be considered to originate in a coupling between rotational angular momentum and spin. Two mechanisms can give rise to magnetic-field components along $\boldsymbol{R}$. One of these is the rotation of the molecule as a rigid charge distribution. The second is less direct, but more important, and occurs because the electrons do not follow the molecular frame exactly and one consequence is that rotationally dependent magnetic fields are generated. Thus we can expect a term of the form

$$H_{RS} = \gamma \boldsymbol{R} \cdot \boldsymbol{S} \tag{10.7}$$

to contribute to the energy at least in high rotational levels. Hougen has pointed out that this physical way of introducing this term has some unsatisfactory consequences at least as far as $^4\Sigma$ and higher multiplicity states are concerned and thus as usual prefers a more phenomenological approach to the problem.[130,134]

10.3c Spin–Spin Interactions

In states where $S \geqslant 1$ such as in the $^3\Sigma_g^-$ state of O_2 there are, in addition to the γ dependent term just discussed, spin–spin coupling terms which are most important at low J. These types of terms are discussed in nuclear magnetic resonance texts where they are of prime importance. They give rise to the terms involving the coefficient λ in the equations (10.17) for a $^3\Sigma$ state.

10.3d Λ-doubling

One more phenomenon must be considered and this is called Λ-doubling. We have so far tacitly assumed that the orbital angular momentum is strongly coupled to the axis and in a particular electronic eigenstate there is no orbital angular momentum at right angles to the axis. As rotational motion occurs however, the electronic motion may not quite follow and Λ will no longer exactly specify the state. A splitting called Λ-type doubling between the Λ^+ and Λ^- states can thus occur which increases with increasing J. This effect is discussed in detail in Section 10.5.

10.4 Hund's Coupling Cases

The importance of the various interactions we have just introduced in any particular situation is reflected in the resulting energy-level pattern. Hund recognized that the energy-level patterns could be classified according to the magnitude of the combined splittings due to angular momentum–axis coupling and spin–orbit coupling *relative to rotational spacings*. The overall classification must be swallowed whole and to this end Hund's cases *a*, *b*, *c* and *d* are summarized in Table 10.1.

As implied in Table 10.1 an initial classification is carried out, into states for which the axial field causes the various $\Lambda\ (\equiv L_c')$ components of a particular $\boldsymbol{L}$ to be well separated in energy (relative to BJ), and cases in which the interaction is so weak that the separations are small and the states are nearly degenerate. A further subclassification can now be carried out according to the magnitude of spin–orbit coupling splittings, again relative to BJ.

Table 10.1 Summary of Hund's Cases[a]

Coupling of L to axis	Spin–orbit coupling
Strong, states with different $\|\Lambda\|$ well separated	*Case a*. Strong, states with different $\|\Omega\|$ well separated
Strong, states with different $\|\Lambda\|$ well separated	*Case b*. Weak, states with different $\|\Omega\|$ close.
Weaker than spin–orbit coupling	*Case c*. Strong, states with different $\|\Omega\|$ well separated
Weak, states with different $\|\Lambda\|$ are close	*Case d*. Weak, states with different $\|\Omega\|$ close.

[a] The criterion of well separated or close is defined by $\Delta E \gg BJ$ and $\Delta E \ll BJ$ respectively.

In *Hund's case a* not only is $\boldsymbol{L}$ strongly coupled to the axis but $\boldsymbol{S}$ is strongly coupled to $\boldsymbol{L}$ and therefore *also* to the axis. The spin–orbit coupling constant is large, causing states with different values of $|\Omega|$ to be well separated; $\Delta E \gg BJ$. Such a case is depicted in Figure 10.2 at least at low J.

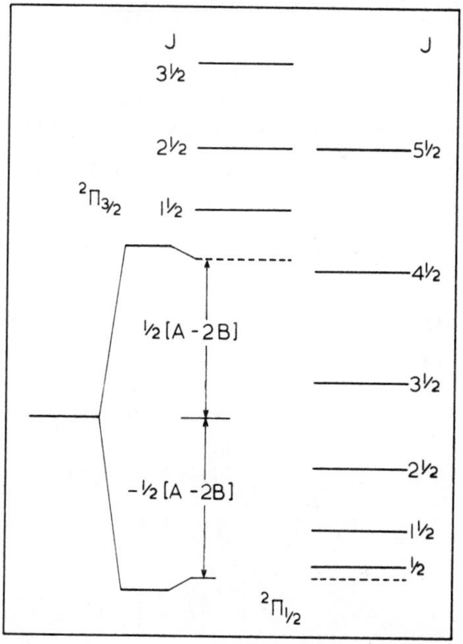

Figure 10.2 The lower rotational energy-level pattern for a molecule is a $^2\Pi$ state in Hund's case *a* governed by (10.14)

In *Hund's case b* $\boldsymbol{L}$ is still strongly coupled to the axis but spin–orbit splittings are small. Σ states usually belong to this category as do the high J levels of a state for which A is small; $\Delta E \ll BJ$.

The case *intermediate between a and b* often occurs. As J increases a state initially in case a may pass through a situation where $\Delta E \sim BJ$, to case b at high J.

In *Hund's case c* the coupling of $\boldsymbol{L}$ to the axis is weaker than that of $\boldsymbol{L}$ with $\boldsymbol{S}$, a situation which can occur when heavy atoms are present. In this case it is $\boldsymbol{L} + \boldsymbol{S} = \boldsymbol{J}$ which couples to the axis.

In Hund's case d not only is the coupling of $\boldsymbol{L}$ with the axis weak but so also is spin–orbit coupling. Such a situation occurs for highly excited Rydberg states of light molecules such as H_2. The electrons giving rise to the angular momentum are so highly excited that the internuclear separation is small compared to the size of the occupied orbital and again atomic or essentially spherical fields prevail.

10.5 Energy Levels and Spectra; Hund's Cases *a*, *b* and *a–b*

In setting up the basic Hamiltonian for the problem we must note that in general the total angular momentum $\boldsymbol{J}$ has components

$$J_a = R_a + L_a + S_a, \qquad J_b = R_b + L_b + S_b, \qquad J_c = L_c + S_c \tag{10.8}$$

where $\boldsymbol{R}$ is the angular momentum of the nuclear frame. The treatments of rotational problems in the previous chapters have essentially assumed that $\boldsymbol{J} \equiv \boldsymbol{R}$. When this—as here—is not the case, the rotational Hamiltonian for a diatomic molecule becomes

$$H_r = B\boldsymbol{R}^2 = B[(J_a - L_a - S_a)^2 + (J_b - L_b - S_b)^2] \tag{10.9}$$

To H_r must be added those coupling terms, considered in the previous section, which may yield a significant contribution and the resulting Hamiltonian solved. Solution is often not too difficult, as the Hamiltonian matrix involves a small finite basis set in the electronic functions. All other electronic states are far away and can often be neglected to a first approximation. As an example we consider a doublet spin state in which both spin–orbit and spin–rotation interactions are important. The Hamiltonian can be written

$$H = B\boldsymbol{R}^2 + \gamma \boldsymbol{R} . \boldsymbol{S} + AL_cS_c \tag{10.10}$$

On expansion this expression becomes

$$\begin{aligned} H = {} & B[(\boldsymbol{J}^2 - J_c^2) + (\boldsymbol{L}^2 - L_c^2)] - (B - \tfrac{1}{2}\gamma)(J^+S^- + J^-S^+) \\ & + (B - \tfrac{1}{2}\gamma)(L^+S^- + L^-S^+) - B(J^+L^- + J^-L^+) \\ & + (B - \gamma)(\boldsymbol{S}^2 - S_c^2) + AL_cS_c \end{aligned} \tag{10.11}$$

For a doublet state such as the $^2\Pi$ state of NO we can set up the Hamiltonian in the case *a* basis $|J\,\Lambda\,\Sigma\rangle$. J is a 'good quantum number' even when the interactions are included, i.e. the Hamiltonian (10.1) is diagonal in J—and thus we can deal with each J family separately using $|J\,\Lambda\,\Sigma\rangle$. In setting up the matrix we shall consider only the first-order energies for the states $|\Lambda\,\tfrac{1}{2}\rangle$, $|\Lambda\,-\tfrac{1}{2}\rangle$, $|-\Lambda\,\tfrac{1}{2}\rangle$ and $|-\Lambda\,-\tfrac{1}{2}\rangle$. We can drop the terms involving $L^{\pm}$ as they can only contribute to second order involving expressions such as $\langle\Pi|L^+|\Sigma\rangle\langle\Sigma|L^-|\Pi\rangle/\Delta E(\Pi\Sigma)$ as multiplier and are thus small. They can be important if there is a nearby Σ or Δ state ($\Delta\Lambda = \pm 1$) causing splittings called Λ-type doubling.

We must note a rather subtle point about the matrix elements of the various type of momenta which appear in (10.11). In Section 3 the matrix elements of $\boldsymbol{J}$ in terms of molecule-fixed component J_α were shown to obey the commutation relation (3.30) 'with the negative sign of i'. The 'internal' angular momenta S_α and L_α must however be considered as vectors *relative* to the molecule-fixed axes and thus obey commutation relations with the so-called 'normal!' positive sign of i (3.5). The overall angular momentum $\boldsymbol{J}$ is to be considered as a vector *relative to spatial axes and then referred to molecular axes* (Van Vleck[310] and Chapter 3).

We thus obtain the matrix

	$\|J\,\Lambda\,\tfrac{1}{2}\rangle$	$\|J\,\Lambda\,-\tfrac{1}{2}\rangle$
$\langle J\;\Lambda\;\tfrac{1}{2}\|$	$B[(J+\tfrac{1}{2})^2 - \Lambda^2] - \tfrac{1}{2}\gamma + \tfrac{1}{2}(A-2B)\Lambda$	$-(B-\tfrac{1}{2}\gamma)[(J+\tfrac{1}{2})^2 - \Lambda^2]^{1/2}$
$\langle J\,\Lambda\,-\tfrac{1}{2}\|$	$-(B-\tfrac{1}{2}\gamma)[(J+\tfrac{1}{2})^2 - \Lambda^2]^{1/2}$	$B[(J+\tfrac{1}{2})^2 - \Lambda^2] - \tfrac{1}{2}\gamma - \tfrac{1}{2}(A-2B)\Lambda$

(10.12)

In setting up this matrix we have dropped out the diagonal term in the Hamiltonian, $B(\boldsymbol{L}^2 - L_\zeta^2)$ to yield a somewhat more symmetric final expression. This type of term is considered in more detail in Section 10.5c. It is only necessary to set up half the matrix as the blocks with Λ positive are identical to those with Λ negative yielding degenerate solutions. The solutions are, with a little rearrangement,

$$B[(J+\tfrac{1}{2})^2 - \Lambda^2] - \tfrac{1}{2}\gamma \pm \sqrt{(B-\tfrac{1}{2}\gamma)^2(J+\tfrac{1}{2})^2 + \tfrac{1}{4}\Lambda^2(A-\gamma)(A-4B+\gamma)} \tag{10.13}$$

This is essentially the result obtained by Hill and Van Vleck[125] with an additional spin–rotation γ contribution. This expression describes the energy-level pattern as increasing rotational motion (represented by the increasing magnitude of the off-diagonal term) 'uncouples' the spin from the axis. Thus the molecule's energy-level pattern changes from case *a* at low *J* to case *b* at high *J*.

10.5a Hund's Case a Limit

In the case *a* limit the spin–orbit coupling term is large and at low *J* we can usually neglect spin–rotation interactions and set $\gamma = 0$. We can generate a very convenient approximate solution of the Hamiltonian which is often as accurate as experiment by using the second-order perturbation solution of (10.12) given by the general expression (1.48)

$$E(^2\Pi_{3/2}) = +\tfrac{1}{2}\bar{A} + B_{\text{eff}+}[(J+\tfrac{1}{2})^2 - \Lambda^2] \qquad (10.14a)$$

$$E(^2\Pi_{1/2}) = -\tfrac{1}{2}\bar{A} + B_{\text{eff}-}[(J+\tfrac{1}{2})^2 - \Lambda^2] \qquad (10.14b)$$

$$\bar{A} = A - 2B, \qquad B_{\text{eff}\pm} = B[1 \pm B/\bar{A}\Lambda]$$

Thus we obtain the energy-level pattern shown in Figure 10.2. If *A* is positive, the $^2\Pi_{3/2}$ state lies higher than the $^2\Pi_{1/2}$ state, and the situation is called regular. A subscript *r* is sometimes appended to the term signature, i.e. $^2\Pi_r$. When *A* is negative the reverse occurs and we get a $^2\Pi_i$, inverted, state. We see that we have in this case essentially two distinct electronic states with slightly different *B* values and separated by $\sim A$. For NO, Hall and Dowling have obtained the far infra-red spectrum[107] shown in Figure 10.3, where the two resulting sets of transitions are quite clearly observed. Each set can be analysed separately to obtain the following rotational constants (in cm^{-1})[107]

$$B_{\text{eff}}(^2\Pi_{1/2}) = 1{\cdot}67185 \pm 0{\cdot}00008$$

$$B_{\text{eff}}(^2\Pi_{3/2}) = 1{\cdot}72018 \pm 0{\cdot}00006$$

$$\Delta B_{\text{eff}} = 0{\cdot}04832 \pm 0{\cdot}00006$$

$$B_{\text{av}} = 1{\cdot}6960$$

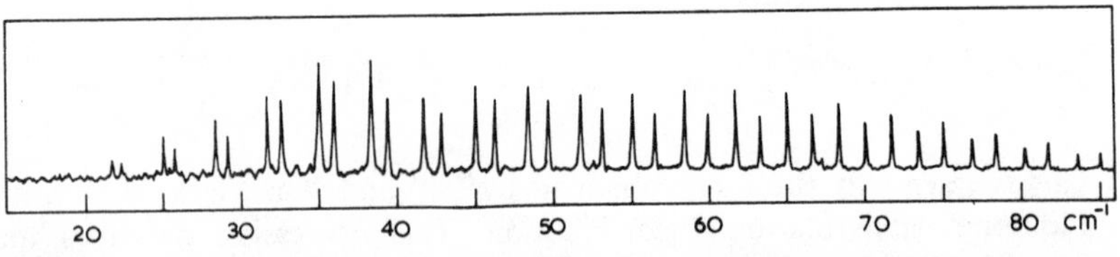

Figure 10.3 The far infra-red spectrum of NO after Hall and Dowling.[107] Transitions with $\Delta J = +1$ occur within the $^2\Pi_{3/2}$ and $^2\Pi_{1/2}$ manifolds giving rise to two branches

From (10.14) we see that $2B^2/\bar{A} = \Delta B_{\mathrm{eff}}$ yielding the result $A = 122{\cdot}44\ \mathrm{cm}^{-1}$. A more accurate result can be obtained from the vibrational spectrum[141] and yields the value $A = 123{\cdot}16\ \mathrm{cm}^{-1}$.

10.5b Hund's Case b Limit

The doublet-state problem has been formulated in such a way that case *b* expressions are also simply obtained. If Λ is set $= 0$, then we obtain the energy-level expression for a $^2\Sigma$ state which cannot belong to case *a*. In this case the solution of (10.12) is easily obtained using (1.40) with $\delta = 0$ as

$$E(J) = B(J + \tfrac{1}{2})^2 - \tfrac{1}{2}\gamma \pm (B - \tfrac{1}{2}\gamma)(J + \tfrac{1}{2}) \qquad (10.15)$$

This expression can be usefully rephrased as

$$F_1(N) = BN(N + 1) + \tfrac{1}{2}\gamma N \qquad \begin{cases} N = J - \tfrac{1}{2} \\ J = N + \tfrac{1}{2} \end{cases} \qquad (10.16a)$$

$$F_2(N) = BN(N + 1) - \tfrac{1}{2}\gamma(N + 1) \qquad \begin{cases} N = J + \tfrac{1}{2} \\ J = N - \tfrac{1}{2} \end{cases} \qquad (10.16b)$$

where in a rather formal way the quantum number N has been introduced such that the quantum number for overall angular momentum $J = N \pm \tfrac{1}{2}$. N thus can be considered to be the quantum number associated with the rotational angular momentum of the nuclear framework to which the spin can add, to yield $J = N + \tfrac{1}{2}$, or subtract, to yield $J = N - \tfrac{1}{2}$.

The O_2 molecule has a $^3\Sigma_g^-$ ground state and has been studied in detail by several workers in both the microwave[186,299] and far infra-red[91] regions. Spin–spin coupling can occur when $S \geqslant 1$ and is particularly important at low rotational energies. If this interaction is included then the energy-level expressions obtained by Schlapp[268,303] for $^3\Sigma$ states can be generated

$$\begin{aligned} F_1(N) &= BN(N + 1) - \lambda + B(2N + 3) \\ &\quad - B[a^2 - 2a + (2N + 3)^2]^{1/2} + \gamma(N + 1) + \tfrac{1}{2}\gamma \end{aligned} \qquad (10.17a)$$

$$F_2(N) = BN(N + 1) \qquad (10.17b)$$

$$\begin{aligned} F_3(N) &= BN(N + 1) - \lambda - B(2N - 1) \\ &\quad + B[a^2 - 2a + (2N - 1)^2]^{1/2} - \gamma N + \tfrac{1}{2}\gamma \end{aligned} \qquad (10.17c)$$

In these expressions λ is the spin–spin interaction constant and $a = \lambda/B$. For F_1, $J = N + 1$; F_2, $J = N$ and for F_3, $J = N - 1$. Hougen has developed these expressions by a technique which avoids vector coupling arguments which appear to lead to error in the case of Σ states of higher multiplicity than three.[130,134] Microwave[186,3] and magnetic resonance measurements[299,306,61] have been made on O_2 and SO which both have $^3\Sigma^-$ ground electronic states. The resulting constants are given in Table 10.2.

Table 10.2

GHz	B	λ	γ
O_2	43·1	59·5	−0·252
SO	21·5	158·2	−0·164

As can be seen from (10.17) and Table 10.2, at low J values the energy-level pattern is dominated by the effects of spin–spin coupling. In the case of O_2 we must also realize that as the oxygen atoms have zero spin and the ground state is an antisymmetric ${}^3\Sigma_g^-$ electronic state, only rotational states with N *odd* may occur. The lowest state in the rotational manifold thus has $N = 1$. In this case we see that the pattern of the $N = 1$ multiplet is approximately given by

$$F_1(1) \simeq 2B - \lambda, \qquad J = 2$$

$$F_2(1) = 2B, \qquad J = 1$$

$$F_3(1) = 2B - 2\lambda, \qquad J = 0$$

where the γ term has been neglected. The three levels are roughly equidistantly spaced. At higher values of J where $N \gg a$ or equivalently $BN \gg \lambda$ the square-root terms in (10.17) can be expanded to yield the following more easily digested approximate relations[122]

$$F_1(N) = BN(N+1) - \frac{2\lambda(N+1)}{2N+3} + \gamma(N+1) + \tfrac{1}{2}\gamma \qquad (10.18a)$$

$$F_2(N) = BN(N+1) \qquad (10.18b)$$

$$F_3(N) = BN(N+1) - \frac{2\lambda N}{2N-1} - \gamma N + \tfrac{1}{2}\gamma \qquad (10.18c)$$

For the F_1 and F_3 states, if we neglect the γ term which is small at low J, we see that the spin–spin coupling terms in λ converge to the same limiting value of $-\lambda$ as N increases. As a consequence the levels tend to fall into a pattern in which the F_1 and F_3 levels are nearly degenerate and lie approximately λ below the F_2 level which is unaffected. As N increases to very high values ~30 the spin–rotation term becomes significant and pushes them apart again. In the microwave region transitions have been observed[186] among the various $F(N)$ components with $\Delta N = 0$ and $\Delta J \pm 1$ which are allowed by magnetic dipole selection rules. There are also a set of very weak transitions which have been observed in the far infra-red by Gebbie, Burroughs and Bird[91] for which the selection rule is $\Delta N = +2$ and $\Delta J = 0, \pm 1$. The $N = 3 \leftarrow 1$, $J = 2$ transition has been observed by McKnight and Gordy in the submillimetre region.[186] Tinkham and Strandberg have carried out a very complete theoretical and experimental study of the magnetic resonance spectrum of O_2.[299]

10.5c Λ-doubling Energy

In the introductory analysis of the energy levels of a molecule in a $^2\Pi$ electronic state we chose to neglect Λ-doubling because it is usually a small effect. In the case of NO the associated splitting is too small to show up in the far infra-red spectrum depicted in Figure 10.3. In the case of OH the splitting is ~1660 MHz in the lowest rotational level and transitions can occur between the two Λ-doubling components. This particular transition is used by radio astronomers as a powerful probe of the conditions in interstellar molecular clouds. Λ-doubling, though rather small, can be readily detected in the gas-phase magnetic resonance spectrum of NO which is discussed in Section 10.6.

To take this effect into account, we must set up a much more complete matrix representation of the Hamiltonian (10.11) than that given in (10.12). We shall also take this opportunity to apply the Van Vleck transformation (Section A8) which is particularly useful in this problem.

Λ-doubling can be understood by considering the matrix elements which involve the electronic state shift operators $L^{\pm}$ which are associated with the selection rule $\Delta\Lambda = \pm 1$. In the case of a Π state, which has $\Lambda = \pm 1$, these off-diagonal terms mix the Π state with $\Sigma\,(\Lambda = 0)$ and $\Delta\,(\Lambda = \pm 2)$ states which may be quite far away in energy. To take the interaction into account we will use a basis set which not only includes the $^2\Pi_{3/2}$ and $^2\Pi_{1/2}$ states but also a 'nearby' $^2\Sigma$ state. The cumulative effect of other interacting states can be included in a similar manner.

We will use the ket

$$|J\,L\,S\,\Omega\,\Lambda\,\Sigma\rangle \tag{10.19}$$

to represent the six basis functions which arise, i.e. with $\Omega = \pm 3/2$ and $\pm 1/2$ for the $^2\Pi$ states and $\Omega = \pm 1/2$ for the $^2\Sigma$ state. Before generating the various matrix elements we must consider a little more carefully the validity of the set of quantum numbers specifying the basis states in (10.19). J will always be a good quantum number in our field-free problem whether the molecule is rotating or not. The orbital quantum number L will not in general be a good quantum number, even in the non-rotating system, except in the case *d* limit where the energy is independent of the orientation of $\boldsymbol{L}$ relative to the axis. In the non-rotating molecule S will only be a good quantum number as long as spin–orbit coupling is small. We are going to set up our problem in the case *a* limit where Ω, Λ and Σ are all good quantum numbers in the non-rotating limit. In Section 10.5 we only considered off-diagonal interactions which involved Ω and Σ and in this approximation the goodness of Ω and Σ was spoiled. We tacitly assumed in Section 10.5 that S was a good quantum number and that in evaluating the matrix elements of $S^{\pm}$ which involve S we could use the usual values $\pm 1/2$. This problem is discussed in detail by Hougen[134] with particular reference to diatomic molecules. The dilemma can be circumnavigated because as long as Σ ($\equiv S_c'$) is a good quantum number the selection rules on Σ implied by the matrix elements of $\mathbf{S}$ still hold. Also one can allow for this discrepancy by introducing an adjustable parameter, γ. Thus in this case

$$S^{\pm}|J\ L\ S\ \Omega\ \Lambda\ \Sigma\rangle = (1-\gamma)\sqrt{(S\mp\Sigma)(S\pm\Sigma+1)}|J\ L\ S\ \Omega\ \Lambda\ \Sigma\pm 1\rangle \quad (10.20)$$

where γ is a number usually much smaller than unity. L is in general a rather poorer quantum number than S but as Λ is usually very good it can be treated in a somewhat similar manner. As the reservations which apply to L are quite great it is quite common not to evaluate the associated matrix elements and leave them as experimentally determinable parameters.[134] Van Vleck calls the situation when L is good, and the matrix elements can be evaluated, the 'case of pure precession!'

Using the basis vector (10.19) and Table 3.1 we can generate the following set of matrix relations.

$$\mathbf{J}^2 - J_c^2|J\ L\ S\ \Omega\ \Lambda\ \Sigma\rangle = [J(J+1)-\Omega^2]|J\ L\ S\ \Omega\ \Lambda\ \Sigma\rangle$$

$$\mathbf{L}^2 - L_c^2|J\ L\ S\ \Omega\ \Lambda\ \Sigma\rangle = [L(L+1)-\Lambda^2]|J\ L\ S\ \Omega\ \Lambda\ \Sigma\rangle$$

$$\mathbf{S}^2 - S_c^2|J\ L\ S\ \Omega\ \Lambda\ \Sigma\rangle = [S(S+1)-\Sigma^2]|J\ L\ S\ \Omega\ \Lambda\ \Sigma\rangle$$

$$J^{\pm}S^{\mp}|J\ L\ S\ \Omega\ \Lambda\ \Sigma\rangle = \sqrt{(J\pm\Omega)(J\mp\Omega+1)(S\pm\Sigma)(S\mp\Sigma+1)} \times |J\ L\ S\ \Omega\mp 1\ \Lambda\ \Sigma\mp 1\rangle$$

$$J^{\pm}L^{\mp}|J\ L\ S\ \Omega\ \Lambda\ \Sigma\rangle = \sqrt{(J\pm\Omega)(J\mp\Omega+1)(L\pm\Lambda)(L\mp\Lambda+1)} \times |J\ L\ S\ \Omega\mp 1\ \Lambda\mp 1\ \Sigma\rangle$$

$$L^{\pm}S^{\mp}|J\ L\ S\ \Omega\ \Lambda\ \Sigma\rangle = \sqrt{(L\mp\Lambda)(L\pm\Lambda+1)(S\pm\Sigma)(S\mp\Sigma+1)} \times |J\ L\ S\ \Omega\ \Lambda\pm 1\ \Sigma\mp 1\rangle$$

$$L_cS_c|J\ L\ S\ \Omega\ \Lambda\ \Sigma\rangle = \Lambda\Sigma|J\ L\ S\ \Omega\ \Lambda\ \Sigma\rangle \quad (10.21)$$

Note the subtle way in which operator products such as $J^{\pm}S^{\mp}$ simultaneously shift the Ω and Σ quantum numbers. As discussed in Chapter 3, J^+ shifts $\Omega \equiv K$ DOWN by one unit whereas S^+, as described in Section 10.5, shifts Σ UP by one unit.

Table 10.3 Matrix for Λ-doubling

(a)

	$\lvert 1\ \tfrac{1}{2}\rangle$	$\lvert 1\ -\tfrac{1}{2}\rangle$	$\lvert -1\ \tfrac{1}{2}\rangle$	$\lvert -1\ -\tfrac{1}{2}\rangle$	$\lvert 0\ \tfrac{1}{2}\rangle$	$\lvert 0\ -\tfrac{1}{2}\rangle$	
$\langle 1\ \tfrac{1}{2}\rvert$	α'	γ'			a		$^2\Pi_{3/2}$
$\langle 1\ -\tfrac{1}{2}\rvert$	γ'	β'			b	c	$^2\Pi_{1/2}$
$\langle -1\ \tfrac{1}{2}\rvert$			β'	γ'	c	b	$^2\Pi_{1/2}$
$\langle -1\ -\tfrac{1}{2}\rvert$			γ'	α'		a	$^2\Pi_{3/2}$
$\langle 0\ \tfrac{1}{2}\rvert$	a	b	c		ζ		$^2\Sigma$
$\langle 0\ -\tfrac{1}{2}\rvert$		c	b	a		ζ	$^2\Sigma$

(b)

α	γ	δ	
γ	β	ε	δ
δ	ε	β	γ
	δ	γ	α

(c)

$\dfrac{1}{\sqrt{2}}$

-1			1
	-1	1	
	1	1	
1			1

(d)

α	$\gamma - \delta$			$^2\Pi^-_{3/2}$
$\gamma - \delta$	$\beta - \varepsilon$			$^2\Pi^-_{1/2}$
		$\beta + \varepsilon$	$\gamma + \delta$	$^2\Pi^+_{1/2}$
		$\gamma + \delta$	α	$^2\Pi^+_{3/2}$

$$\alpha' = B[J(J+1) - \tfrac{3}{4}] + \tfrac{1}{2}A + Ba_{\Pi}$$

$$\beta' = B[J(J+1) - \tfrac{1}{4}] - \tfrac{1}{2}A + Ba_{\Pi}$$

$$\gamma' = -B\sqrt{(J + \tfrac{3}{2})(J - \tfrac{1}{2})}$$

$$\zeta = E(\Sigma) - E(\Pi) = \Delta$$

$a = -Ba_{\Pi\Sigma}\sqrt{(J + \tfrac{3}{2})(J - \tfrac{1}{2})}$ $\quad \alpha = \alpha' + a^2/\Delta$ $\quad \delta = ac/\Delta$

$b = (B + \tfrac{1}{2}A)a_{\Pi\Sigma}$ $\quad \beta = \beta' + (b^2 + c^2)/\Delta$ $\quad \varepsilon = 2bc/\Delta$

$c = -Ba_{\Pi\Sigma}(J + \tfrac{1}{2})$ $\quad \gamma = \gamma' + ab/\Delta$

In this set, (10.21), we have neglected the reservations we have discussed about S and L. In the case of S this is quite valid for our purposes. The assumption for L that 'pure precession' holds is less valid and in fact we will not assume this is so in general. It is common practice to use expressions of the type

$$L^2 - L_c^2|J\ L\ S\ \Omega\ \Lambda\ \Sigma\rangle = a_\Lambda|J\ L\ S\ \Omega\ \Lambda\ \Sigma\rangle \tag{10.22a}$$

$$L^{\pm}|J\ L\ S\ \Omega\ \Lambda\ \Sigma\rangle = a_{\Lambda\Lambda\pm1}|J\ L\ S\ \Omega\ \Lambda \pm 1\ \Sigma\rangle \tag{10.22b}$$

When $\Lambda = \pm 1$ the parameters can be written as a_Π and $a_{\Pi\Sigma}$ or $a_{\Pi\Delta}$.

Using this set of matrix elements we can set up the matrix shown in Table 10.3(a) where all but the most necessary indices have been suppressed from the basis set specification. The spin rotation, γ, interaction term has also been set to zero. Previously we solved the top left-hand 4×4 submatrix neglecting the interference caused by distant states. The inclusion of the $^2\Sigma$ state in our basis set introduces numerous off-diagonal elements represented by a, b and c which link the Σ and Π states, Table 10.3(a). It is convenient to use the Van Vleck procedure (Section A8) which in effect removes these off-diagonal terms and takes their effect into account by introducing correction terms *within* the Π and Σ blocks. Using the prescription for carrying out the transformation (A8.3) we obtain the matrix in Table 10.3(b) for the Π states.

In this basis representation Π^+ and Π^- states are degenerate in the absence of terms in δ and ε which mix the states with $\Lambda = +1$ with those for which $\Lambda = -1$. These terms spoil the 'goodness' of the Λ quantum number and cause L_c to 'uncouple' from the internuclear axis and thus lift the $\pm$ degeneracy. To keep our Quantum Mechanical conscience clear we shall transform to the correct representation in which the basis functions possess parity. This can be accomplished by the matrix U, Table 10.3(c), according to $U^{-1}\tilde{H}U = \tilde{\tilde{H}}$. U is essentially the 4×4 equivalent of the Wang matrix (3.59). When U has even $\times$ even dimension it differs from (3.59) only in the absence of the central element $\sqrt{2}$. The resulting transformed matrix, Table 10.3(d), is block diagonalized into two 2×2 matrices which are very similar to (10.12) but differ in that they now contain the Λ-doubling information. The solutions are thus

$$E(^2\Pi^-_{1/2}) = \beta - \varepsilon + (\gamma - \delta)^2(\beta - \varepsilon - \alpha)^{-1} \tag{10.23a}$$

$$E(^2\Pi^+_{1/2}) = \beta + \varepsilon + (\gamma + \delta)^2(\beta + \varepsilon - \alpha)^{-1} \tag{10.23b}$$

$$E(^2\Pi^-_{3/2}) = \alpha + (\gamma - \delta)^2(\alpha - \beta + \varepsilon)^{-1} \tag{10.23c}$$

$$E(^2\Pi^+_{3/2}) = \alpha + (\gamma + \delta)^2(\alpha - \beta - \varepsilon)^{-1} \tag{10.23d}$$

To the energy which we have already calculated (10.14) we must thus add the following Λ-doubling corrections:

$$\Delta E(^2\Pi^{\pm}_{1/2}) = \pm[\varepsilon + 2\gamma\delta(\beta - \alpha)^{-1}] \tag{10.24a}$$

$$\Delta E(^2\Pi^{\pm}_{3/2}) = \pm 2\gamma\delta(\alpha - \beta)^{-1} \tag{10.24b}$$

where we have dropped the small ε term in the denominator. The Λ-doubling *splitting* is thus evaluated as[309,303]

$$\Delta E(^2\Pi_{1/2}) = a(J + \tfrac{1}{2}) + b(J + \tfrac{3}{2})(J - \tfrac{1}{2})(J + \tfrac{1}{2}) \tag{10.25a}$$

$$\Delta E(^2\Pi_{3/2}) = b(J + \tfrac{3}{2})(J - \tfrac{1}{2})(J + \tfrac{1}{2}) \tag{10.25b}$$

where

$$a = 2B(B + \tfrac{1}{2}A)a^2_{\Sigma\Pi}(\Delta E_{\Sigma\Pi})^{-1} \tag{10.26a}$$

$$b = 2B^3 a^2_{\Sigma\Pi}(A\Delta E_{\Sigma\Pi})^{-1} \tag{10.26b}$$

In the pure precession approximation $a^2_{\Sigma\Pi} = L(L + 1) = 2$ and thus

$$a \sim 4AB/\Delta E_{\Sigma\Pi}, \qquad b_\Lambda \sim 8B^3/A\Delta E_{\Sigma\Pi} \tag{10.27}$$

These final simplified expressions are the result of liberally diluting the calculation to taste with numerous approximations. Note also that as J increases the J^3 dependent term, which is often neglected, will become important for the $^2\Pi_{1/2}$ state.

10.6 Electron Resonance Spectra of Diatomic Molecules

In 1949 Beringer and Castle observed the magnetic resonance spectrum of NO[19] and the theoretical treatment was carried out by Margenau and Henry.[191] The theory was further developed by Frosch and Foley[89] and Lin and Mizushima.[182] Further experimental microwave studies on NO were carried out by Burrus and Gordy[31] and Gallagher, Bedard and Johnson.[90]

The application of magnetic resonance techniques to the general study of free radicals has been developed by Radford[250,251] and Carrington, Levy and Miller.[40,38] A large number of diatomic molecules have now been studied[35] as have two linear triatomic species NCO and NCS.[37]

We shall here restrict ourselves to the salient features which apply to diatomic molecules in $^2\Pi$ electronic states, taking the spectrum of NO as our example. The Hamiltonian for the *free* molecule in a $^2\Pi$ electronic state was discussed in Section 10.5. We shall consider the NO molecule in its lowest J levels where it is close to the Hund's case *a* limit. In this limit we can effectively consider the molecule as having two separate electronic states, a $^2\Pi_{3/2}$ state lying ~124 cm^{-1} above the $^2\Pi_{1/2}$ state. The $^2\Pi_{1/2}$ state corresponds to a situation in which the individual magnetic moments associated with orbital angular momentum and spin angular momentum cancel, yielding an essentially diamagnetic species and is not in this case accessible by spin resonance techniques. In the $^2\Pi_{3/2}$ state, on the other hand, the magnetic moments reinforce each other and as a result this state is paramagnetic and exhibits a strong Zeeman effect. As increasing rotation mixes the two states (Section 6.5) the $^2\Pi_{1/2}$ will begin to exhibit some paramagnetic behaviour. In OH the rotational spacings are large relative to the spin–orbit coupling energy ($A = 139{\cdot}7$, $B = 18{\cdot}9\ \text{cm}^{-1}$) and as a result the $^2\Pi_{3/2}$ and $^2\Pi_{1/2}$ states are strongly mixed and both have been studied.[250,251]

Finally one might note that one does not need a spin magnetic moment to apply this technique. The $^1\Delta$ states of O_2[84] and SO[39], which have only orbital angular momentum, have also been studied.

10.6a The Hamiltonian for Electron Resonance

Many interaction terms must be taken into account in the analysis of spin resonance spectra. Because small energy separations are involved, non-field-dependent terms have a significant effect on the spectral pattern. It is convenient to write the Hamiltonian as the sum

$$H = H_0 + H_Z + H_{hf} + H_Q \tag{10.28}$$

H_0, *The Basis Hamiltonian.* H_0 corresponds to the Hamiltonian for the free molecule exclusive of nuclear-spin and field-dependent terms. The basic form is given in (10.10), though we do not need to consider all the individual terms in the following simplified treatment. If for instance Λ-doubling is small, we can neglect the contribution from the term $-B(J^+L^- + J^-L^+)$. Λ-doubling is large in OH and must be taken into account.[250] In NO it splits the Π^+ and Π^- states by $\sim 0{\cdot}9$ MHz[20,27] and we can take it into account at the end. We shall thus use as our Hamiltonian the reduced form whose matrix is (10.12) in the case *a* representation. We can write our basis functions as

$$|\Lambda\, \Sigma\, \Omega\, J\, M_J\, I\, M_I\rangle \tag{10.29}$$

where we have taken into account the possibility of nuclear spin and have also specified the space-fixed components of $\boldsymbol{J}$ and $\boldsymbol{I}$.

H_Z, *The Zeeman Hamiltonian.* There are several terms which contribute to the energy of the molecule under the influence of a magnetic field. The main one which governs the overall position of the spectrum is that between the combined orbital and spin magnetic moment and the external magnetic field $\boldsymbol{H}$. We can write this as

$$H_{Z_e} = \beta \boldsymbol{H}\cdot(g_l \boldsymbol{L} + g_e \boldsymbol{S}) \tag{10.30}$$

where g_l is the electronic orbital g factor, g_e the electron spin g factor and β is the Bohr Magneton. g_l is essentially unity and g_e is $2{\cdot}002322$. A second term which corresponds to the interaction energy between the nitrogen atom's nuclear spin magnetic moment and the external field can be written as

$$H_{Z_n} = g_I \beta_N \boldsymbol{H} \cdot \boldsymbol{I} \tag{10.31}$$

We should also include a term which represents the interaction between the external field and that generated by overall rotation of the molecule. This we can write as

$$H_{Z_r} = g_r^N \beta \boldsymbol{H} \cdot (\boldsymbol{J} - \boldsymbol{L} - \boldsymbol{S}) = g_r^N \beta \boldsymbol{H} \cdot \boldsymbol{R} \tag{10.32}$$

H_{hf}, *The Hyperfine Interaction Hamiltonian.* This contribution essentially originates from the expression

$$g_e \beta \sum_{ne} g_n \beta_n \left\{ r_{ne}^{-5}[3(\boldsymbol{S}_e \cdot \boldsymbol{r}_{ne})(\boldsymbol{I}_n \cdot \boldsymbol{r}_{ne}) - r_{ne}^2 \boldsymbol{I}_n \cdot \boldsymbol{S}_e] + \frac{8\pi}{3} \delta(\boldsymbol{r}_{ne}) \boldsymbol{I}_n \cdot \boldsymbol{S}_e \right\} \tag{10.33}$$

where the summation is carried out over nuclei, n, and electrons, e. g_n and β_n are the nuclear spin g factor and Bohr magneton respectively, $\boldsymbol{r}_{ne}$ is the vector between nucleus n and electron e. The square bracketed term represents the electron-dipole–nuclear-dipole magnetic interaction and the third term represents the Fermi contact energy.[41] We can drop the summation in the case of NO where we have only one unpaired electron and only one atom (nitrogen) with a non-zero nuclear spin.

To (10.33) we must also add the interaction energy between the nitrogen's nuclear magnetic moment and the magnetic field generated by the electron's orbital motion. This we can write as

$$a\boldsymbol{L} \cdot \boldsymbol{I} \tag{10.34}$$

If we now use the Wigner–Eckart theorem discussed in Chapter 8 and relate $\boldsymbol{r}_{ne}$ to $\boldsymbol{L}$, and add (10.33) to (10.34) we obtain the hyperfine Hamiltonian as

$$H_{hf} = a\boldsymbol{L} \cdot \boldsymbol{I} + b\boldsymbol{I} \cdot \boldsymbol{S} + c(\boldsymbol{S} \cdot \boldsymbol{L})(\boldsymbol{I} \cdot \boldsymbol{L}) \tag{10.35}$$

where we can consider the constants a, b and c as molecular parameters determinable from experiment.[89]

H_Q, *The Quadrupole Hamiltonian.* If the magnetic field is strong enough to break down the $\boldsymbol{I} \cdot \boldsymbol{J}$ coupling, the quadrupole energy must be slightly modified from the field-free value (8.39). We need to determine the diagonal matrix elements of H_Q given by either (8.16) or (8.14) in the representation (10.29). In (8.16) only the $m = 0$ term yields a diagonal contribution and as we are only going to determine the quadrupole energy to first order we see with the aid of Table 8.1 that we obtain

$$E_Q^{(1)} = \langle \Lambda\, \Sigma\, \Omega\, J\, M_J\, I\, M_I | \tfrac{1}{2} V_0^{(2)} F_0^{(2)} | \Lambda\, \Sigma\, \Omega\, J\, M_J\, I\, M_I \rangle \tag{10.36}$$

10.6b The Energy of a Free Radical in a Magnetic Field

Having determined the form of the electron resonance Hamiltonian we shall evaluate the most important terms for the $^2\Pi_{3/2}$ electronic state of NO for $J = 3/2$. The main features of the spectrum can be understood with the aid of first- and second-order perturbation theory. For an accurate fitting of the spectrum it is preferable to set up the complete Hamiltonian matrix[36] and diagonalize it numerically.

The first-order energy is thus given by

$$E^{(1)} = \langle \Lambda\, \Sigma\, \Omega\, J\, M_J\, I\, M_I | H_{Z_e} + H_{hf} + H_Q | \Lambda\, \Sigma\, \Omega\, J\, M_J\, I\, M_I \rangle \quad (10.37)$$

We shall neglect H_{Z_n} and H_{Z_r} because the associated g factors are several orders of magnitude smaller than the electronic factors, in fact of the order of $g_e m/M$ where m is the electron mass and M is the nuclear mass. The most important term is H_{Z_e}, the Zeeman energy, as this governs the general magnitude of the field strength required for a given resonance frequency. We can rewrite H_{Z_e} in index notation as

$$H_{Z_e} = \beta H_i S_{\alpha i}(L_\alpha + g_e S_\alpha) \quad (10.38)$$

where $S_{\alpha i}$ is the αith element of the direction cosine matrix. If, as is usually the case, the field has only one non-zero component, say H_z, and we neglect the off-diagonal contribution for $\boldsymbol{L}$ and $\boldsymbol{S}$ we can simplify (10.38) as

$$H_{Z_e} = \beta H_z S_{cz}(L_c + g_e S_c) \quad (10.39)$$

which on evaluation, using Table 4.1, yields the result

$$E_{Z_e}^{(1)} = \beta H_z \left[\frac{(\Lambda + g_e\Sigma)\Omega}{J(J+1)}\right] M_J \quad (10.40)$$

For the $J = 3/2$ level of the $^2\Pi_{3/2}$ state we see that this reduces to

$$E_{Z_e}^{(1)} = \tfrac{4}{5}\beta H_z M_J \quad (10.41)$$

The g value for this level should thus be 0·8 if we assume perfect Hund's case *a*. We have also taken $g_e = 2$.

We can evaluate the diagonal matrix elements of the hyperfine interaction term (10.35) in a very similar way. For instance in index notation we see that

$$aL_i I_i = aS_{\alpha i} L_\alpha I_i \quad (10.42)$$

for which only the term $aS_{zc}L_c I_z$ will yield a diagonal contribution, which is

$$a\frac{\Omega M_J \Lambda M_I}{J(J+1)} \quad (10.43)$$

using the results of Table 4.1, and setting $K = \Omega$ and $m = M_J$. The term $\boldsymbol{I} \cdot \boldsymbol{S}$ can be evaluated in a similar way. As $\boldsymbol{S} \cdot \boldsymbol{L}$ is a scalar and thus independent of the axis representation it can be evaluated in either space-fixed components ($\boldsymbol{S} \cdot \boldsymbol{L} \equiv S_i L_i$) or more conveniently for our purposes in terms of molecule-fixed components ($\boldsymbol{S} \cdot \boldsymbol{L} \equiv S_\alpha L_\alpha$). As a result we see that the diagonal term is simply $S_c L_c$ which yields the matrix element $\Lambda\Sigma$. As a result we see that

$$E_{hf}^{(1)} = (a\Lambda + b\Sigma + c\Sigma\Lambda^2)\frac{\Omega}{J(J+1)}M_I M_J \tag{10.44}$$

which for this particular case becomes

$$E_{hf}^{(1)}(^2\Pi_{3/2}, J = 3/2) = \tfrac{2}{5}(a + \tfrac{1}{2}b + \tfrac{1}{2}c)M_I M_J \tag{10.45}$$

as given by Frosch and Foley.[89]

It is not too difficult to evaluate the diagonal quadrupole contribution represented by (10.36) by combining the relations (8.18), (8.20), (8.22), (8.23), (8.24) and (8.35) to obtain the result

$$E_Q^{(1)} = \frac{eQq}{4}\frac{[3M_I^2 - I(I+1)][3M_J^2 - J(J+1)]}{I(2I-1)(2J-1)(2J+3)}\left[\frac{3\Omega^2}{J(J+1)} - 1\right] \tag{10.46}$$

where $K \rightarrow \Omega = \Sigma + \Lambda$.

There are several off-diagonal terms that must be considered in an accurate treatment, however we will consider only the most important one. It comes from (10.39) which has terms in S_{zc} which can mix states differing by $\Delta J = \pm 1$. As a consequence we can take a second-order perturbation term between the $J = 3/2$ and $J = 5/2$ states which differ in energy by $5B_{\text{eff}}(^2\Pi_{3/2})$ according to (10.14). The required matrix element is

$$\langle \Lambda\, \Sigma\, \Omega\, J\, M_J\, I\, M_I | \beta H_z S_{zc}(L_c + g_e S_c) | \Lambda\, \Sigma\, \Omega\, J{+}1\, M_J\, I\, M_I \rangle$$

$$= \beta H_z \left[\frac{\sqrt{(J+\Omega+1)(J-\Omega+1)(J+M_J+1)(J-M_J+1)}}{(J+1)\sqrt{(2J+1)(2J+3)}}\right](\Lambda + 2\Sigma) \tag{10.47}$$

The energy separation between the $J = 3/2$ and $J = 5/2$ state is only $\sim 5 \times 1{\cdot}7\ \text{cm}^{-1}$ and thus the energy denominator in the second-order perturbation term is not large compared to the first-order magnetic field perturbation energy which is usually chosen to be of the order of 9·3 GHz or $\sim 1/3\ \text{cm}^{-1}$. This gives an idea of the magnitude of the second-order contribution to the Zeeman energy. If we take $g_e = 2$ we obtain the result that

$$E_{Z_e}^{(2)} = \frac{8}{75}\beta^2 H_z^2 M_J^2 \left[\frac{1}{\Delta E}\right] \tag{10.48}$$

This term is quadratic in the magnetic field strength. There is also a term independent of M_J, which we can neglect. In this case ΔE is 258·0 GHz ($\equiv 5 \times 1{\cdot}720 = 8{\cdot}60\ \mathrm{cm}^{-1}$) and the first-order Zeeman energy is approximately 9·3 GHz as given by the resonance frequency. The resulting second-order energy contribution is thus quite significant. There are off-diagonal matrix elements which arise from terms in (10.39) involving the molecule-fixed *a* and *b* angular momentum components of $\boldsymbol{S}$. They connect states with $\Delta\Sigma = \pm 1$ and thus effectively mix the $^2\Pi_{3/2}$ state with the $^2\Pi_{1/2}$ state. As the energy denominator is $\sim 124\ \mathrm{cm}^{-1}$ in this case, they yield similar contributions to the quadratic field term but an order of magnitude smaller than (10.48), and we shall neglect them in this simplified treatment.

Collecting all the results together, and inserting the appropriate quantum numbers the approximate expression

$$E(M_I M_J) = A_1 M_J + A_2 M_J^2 + A_3 M_I M_J + A_4(3M_I^2 - 2)(M_J^2 - \tfrac{5}{4}) \qquad (10.49)$$

is obtained for the energy of the $J = 3/2$ level of a $^2\Pi_{3/2}$ state. The coefficients are given by

$$\begin{aligned} A_1 &= g_0 \beta H_z \simeq 0{\cdot}8\,\beta H_z \\ A_2 &\simeq (0{\cdot}002133) B^{-1} \beta^2 H_z^2 \\ A_3 &= \tfrac{2}{5}[a + \tfrac{1}{2}(b + c)] \\ A_4 &= eQq/20 \end{aligned} \qquad (10.49a)$$

It is possible to determine the values of the individual hyperfine coefficients in A_3 if the $\Delta J = +1$ microwave transitions are also studied.[90] The energy-level diagram which results from (10.49) for NO is given semi-quantitatively in Figure 10.4 for a given field strength. Λ-doubling has not been included.

10.6c Electron Resonance Spectra

Magnetic resonance studies are, for reasons discussed in Section 11.5, carried out using a fixed frequency and a *swept* field. As a result analysis is not quite as straightforward as it would be if the conditions were reversed and the frequency were swept at fixed field. This is a consequence of the nonlinearity of the relation between the field and frequency, indicated in (10.49). It is usual, in accurate studies, to determine numerically the set of A coefficients (or their equivalents in a more complete matrix diagonalization procedure) of (10.28) which most satisfactorily predicts the resonance frequency from the resonant field strengths. The non-linear term is however quite small ($\sim 1\%$) at 8000–9000 Gauss and $\sim$9·3 GHz which are the field and frequency values often employed; it also tends to cancel out in this case. As a result, in our simplified study of NO, we shall neglect this problem and assume that field and frequency are linearly related over a small range. The error introduced by this device is not significant in our treatment.

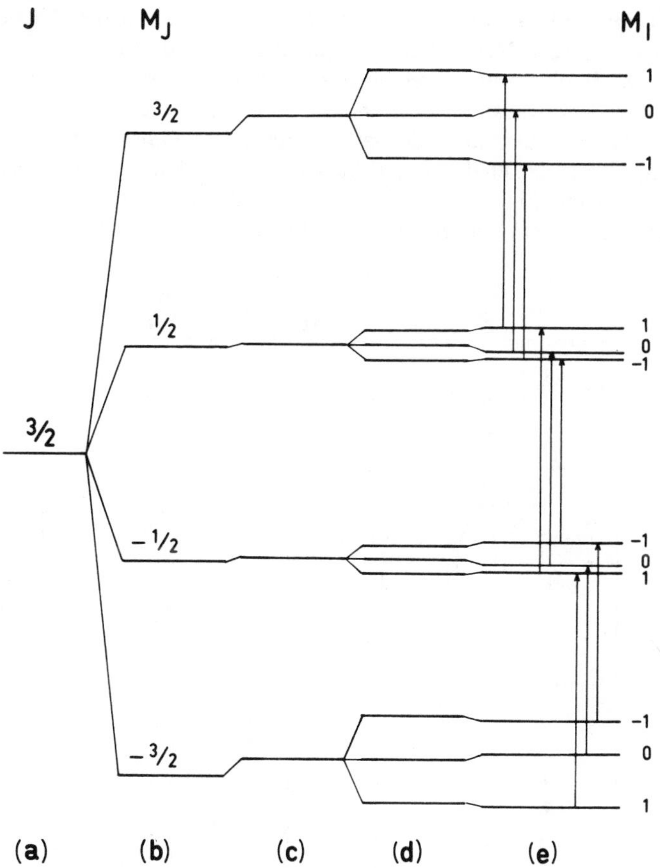

Figure 10.4 Energy-level diagram for the $J = 3/2$ level of the $^2\Pi_{3/2}$ state of NO. (a) In the absence of the field and hyperfine interactions, (b) first-order Zeeman splitting, (c) second-order Zeeman shifts, (d) hyperfine splittings due to spin of N nucleus and (e) the quadrupole shifts. (b), (c), (d) and (e) indicate the effects associated with the terms in A_1, A_2, A_3 and A_4 respectively (10.49). Λ-doubling has been omitted. The selection rules are $\Delta M_J = \pm 1, \Delta M_I = 0$

Beringer and Castle[19] originally observed the magnetic dipole transitions of NO which are governed by the selection rules $\Delta J = 0$, $\Delta M_J = \pm 1$, $\Delta M_I = 0$ and $\pm \leftrightarrow \pm$ which take place *within* the two Λ-doublet manifolds. The electric dipole transitions which are subject to the selection rules $\Delta J = 0$, $\Delta M_J = \pm 1$, $\Delta M_I = 0$ and $\pm \leftrightarrow \mp$ have also been observed.[20,27] These take place *between* the two Λ-doublet manifolds.

The origin and form of the Λ-type doubling is discussed in detail in Section 10.5c. In the case of NO each level is split apart by ~0·9 MHz. The two sets of *magnetic* dipole transitions coincide and cannot be resolved. The simplified energy-level and transition diagram Figure 10.4 in which this splitting has been neglected is therefore adequate for an understanding of the magnetic dipole spectrum depicted in Figure 10.5(a). On the other hand, in the electric dipole spectrum Figure 10.5(b) the upper and lower stacks do combine, resulting in two sets of transitions, split apart by twice the Λ-doubling energy.

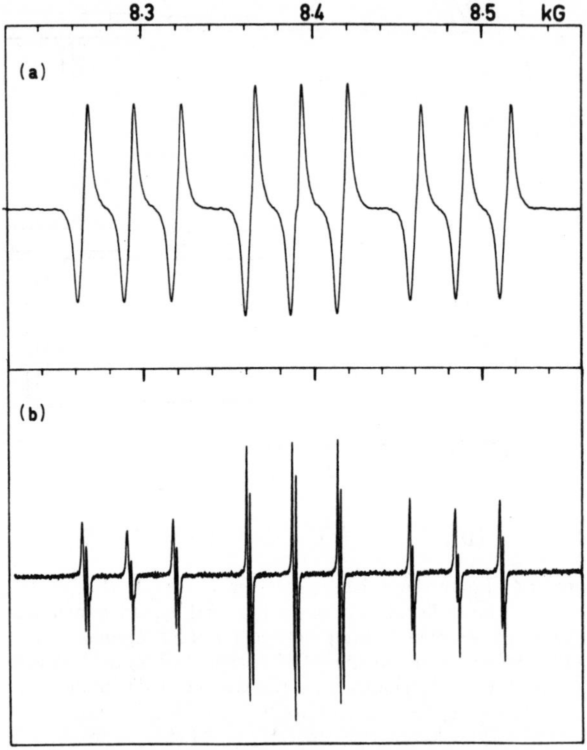

Figure 10.5 Magnetic resonance spectrum of NO. (a) Magnetic dipole spectrum. (b) Electric dipole spectrum

If a molecule possesses both magnetic and electric dipole transitions, experimental considerations determine which ones are detected as discussed in Section 11.8. In general it is expedient to seek electric dipole transitions because they are significantly stronger. The intensity ratio will be of the order of $\mu_m^2/\mu_e^2 \sim 10^{-4}$ when μ_m is 1 Bohr magneton and μ_e is 1 Debye.

The energy-level diagram in Figure 10.4 facilitates the assignment and analysis of the magnetic dipole spectrum of NO. We must however always bear in mind that the field required to bring a particular transition into resonance is essentially inversely proportional to the frequency, or energy separation, indicated in Figure 10.4. In Table 10.4 is given a convenient break-down scheme which shows how the various contributions combine to generate the observed spectrum. The characteristic pattern which results depends on the relative magnitudes of the various A constants and in this case $|A_1| \gg |A_2| > |A_3| \gg |A_4|$. The term A_2 is the one which depends quadratically on the field strength. Thus if the experiment were carried out at say half the frequency given in Table 10.4, this term would be a quarter of its present value and the order of $|A_3|$ and $|A_2|$ would be reversed.

Table 10.4 Field Strengths (in Gauss) of the Electron Resonance Spectrum of NO $^2\Pi_{3/2}$ ($J = 3/2$) State at 9270·65 MHz[a]

		$M'_J \leftarrow M''_J$	
M_I	$3/2 \leftarrow 1/2$	$1/2 \leftarrow -1/2$	$-1/2 \leftarrow -3/2$
1	8399·26[a]	8501·49	8601·75[a]
0	8425·71	8528·28	8629·71
−1	8453·44	8555·84	8656·44
Average	8426·14	8528·54	8629·30

[a] These measurements, which are the means of the Λ-doublets, are listed by Beringer, Rawson and Henry[21,182] who quote the above resonance frequency for all but two lines. The resonance fields for these two lines have been adjusted to the predicted values at 9270·65 MHz.

In the approximation represented by the energy-level expression (10.49) we see that the overall pattern is symmetric about the $M_J = \frac{1}{2} \leftarrow -\frac{1}{2}, M_I = 0$ line which lies at the centre of the spectrum. This line indicates the value of A_1 exactly. Symmetrically placed about this central line at $A_1 + A_3$ and $A_1 - A_3$, unperturbed by first-order quadrupole interactions, lie the $M_I = +1$ and $M_I = -1$ satellites of the $M_J = 1/2 \leftarrow -1/2$ transition. Similar triplet patterns, whose origins occur at fields equivalent to $A_1 + 2A_2$ and $A_1 - 2A_2$, they correspond to the $M_J = 3/2 \leftarrow 1/2$ and $M_J = -1/2 \leftarrow -3/2$ transitions. These two outer triplets are however affected by quadrupole shifts as can be seen from Table 10.5 and Figure 10.4. This interaction manifests itself as an unequal splitting of the outer triplet patterns. This effect is not obvious in Figure 10.5 as eQq is rather small. In OCl, however, eQq is much larger and this effect is nicely observed.[36,38]

Table 10.5 Breakdown of Contributions to the Transition Energy

$M'_J \leftarrow M''_J$	M_I	$E_Z^{(1)}$	$E_Z^{(2)}$	E_{hf}	E_Q[a]
	1	A_1	$+2A_2$	$+A_3$	$(+)2A_4$
$3/2 \leftarrow 1/2$	0	A_1	$+2A_2$	·	$(-)4A_4$
	-1	A_1	$+2A_2$	$-A_3$	$(+)2A_4$
	1	A_1	·	$+A_3$	·
$1/2 \leftarrow -1/2$	0	A_1	·	·	·
	-1	A_1	·	$-A_3$	·
	1	A_1	$-2A_2$	$+A_3$	$(-)2A_4$
$-1/2 \leftarrow -3/2$	0	A_1	$-2A_2$	·	$(+)4A_4$
	-1	A_1	$-2A_2$	$-A_3$	$(-)2A_4$

[a]The signs in brackets of the terms in A_4 are correct according to expression (10.49). In Figure 10.4 the sign of A_4 has been *reversed* to account for the negative sign of eQq in NO.

10.6d Analysis

An accurate analysis is, as we have noted, most accurately carried out by numerically diagonalizing the Hamiltonian matrix. The simple analysis which we will carry out is however a necessary preliminary to such an exact treatment and gives a good understanding of the significance of the various terms.

The position of the central line yields the value of A_1, from which we can determine g_0—the g value of the state directly. From Tables 10.4 and 10.5 we see that $A_1 \equiv 9270{\cdot}65$ MHz at 8528·28 Gauss and therefore

$$g_0 = \frac{A_1}{\beta H_z} = \frac{1}{\beta}\frac{\omega_0}{H_z} = 0{\cdot}7766 \qquad (10.50)$$

The ratio $[\omega_0/H_z]$ is here 1·08705 MHz/Gauss and the Bohr magneton $\beta = 1{\cdot}39968$ MHz/Gauss. The value of g_0 given by Brown and Radford[27] is 0·777246. Note that the simple theory for a 'pure' $^2\Pi_{3/2}$ state indicates that g_0 should be 0·8 exactly.

We can average the field values given in the columns of Table 10.4 and, as indicated in Table 10.5, this yields the *origins* of the three triplets. The values are listed in the last row of Table 10.4. This averaging procedure eliminates the effect of hyperfine and quadrupole coupling. The separation of the two outer *origins* yields the field equivalent of $4A_2$ and thus gives the coefficient of the term which is quadratic in the field. The average value of A_2 is 50·79 Gauss, which when multiplied by the average conversion factor, $[\omega_0/H_z] =$ 1·08705, yields the frequency equivalent of A_2 as 55·21 MHz. We have neglected other non-linear field-dependent terms which can make contributions of the order of a few per cent to A_2 at these field strengths. Lin and

Mizushima[182] give A_2 as $3{\cdot}915 \times 10^{-7}\beta^2 H_z^2$ which is equivalent to 55·8 MHz in this case. This term is particularly useful when pure microwave transitions have not been detected, as careful analysis can yield the B value to an accuracy of ~0·1%. The simple, single expression for A_2 given in (10.48a) is only correct to about 5%. This is because we have neglected the further second-order interactions which correspond to field-induced mixing between the $^2\Pi_{1/2}$ and $^2\Pi_{3/2}$ states which also contribute to the quadratic field dependence. Using this relation we obtain an estimate of B as 55·1 GHz. The value obtained by Hall and Dowling was 51·605 GHz[107] as is given in Section 10.5a.

Many free radicals tend to be more easily detected by electron resonance techniques than by straight microwave methods. As a consequence an estimate of the rotational constant from a careful analysis of the non-linear Zeeman term can aid the search for pure microwave transitions.

The average separations of the $M_I = +1$ and -1 components for the three triplets yields, according to Table 10.5, the value of $2A_3$. As a result we find that $A_3 = 27{\cdot}20$ Gauss which is equivalent to 29·57 MHz. Lin and Mizushima give the function $\frac{2}{5}[a + \frac{1}{2}(b + c)]$ as the coefficient of $M_I M_J$, i.e. A_3. Using the values of a, b and c obtained by Brown and Radford[27] this function is 29·8 MHz.

We can determine the sign of eQq as negative by noting that the observed resonance lines for the two outer $M_I = 0$ lines, Table 10.4, lie outside their respective origins, i.e. $8425{\cdot}71 < 8426{\cdot}14$ and $8629{\cdot}71 > 8629{\cdot}30$. The energy-level diagram in Figure 10.4 has been drawn in accordance with this negative value. By permuting the various relations in Table 10.4 the value of eQq can be derived as −2·25 MHz. Using the more accurate data of Brown and Radford[27] the same method yields an average value of −1·72 MHz.

Finally we can turn our attention to the electric dipole spectrum shown in Figure 10.5(b). This spectrum differs from the magnetic dipole spectrum in that each line has been split apart by the field equivalent of twice the Λ-doubling frequency. In this case the average splitting of the lines in Figure 10.5(b) is 1·67 Gauss[27] which yields a Λ-doubling frequency of 0·908 MHz. The frequency given by Brown and Radford is 0·906 MHz for the $J = 3/2$ state. They have also studied the $J = 5/2$ level whose Λ-doubling frequency is 3·601 MHz. From the expressions developed in Section 10.5c we see that the ratio should be 1:4, as is observed. It is worth noting that according to (10.26b) the Λ-doubling parameter b is proportional to B^3 in the case of a $^2\Pi_{3/2}$ state. Thus in a light molecule, such as OH where $B \sim 565{\cdot}8$ GHz compared with 51·6 GHz for NO, the splitting frequency is much larger and in this case is ~1667 MHz. This large splitting dominates the electric dipole spectrum of OH[250, Figure 6] whereas in many other cases the Λ-doubling is too small to be resolved.

Chapter 11

Experimental Techniques

Numerous methods have been developed to detect the spectra associated with rotational transitions. For many molecules transitions occur in the microwave region, Table 11.1, and can conveniently be studied using the Stark spectrometer introduced by Hughes and Wilson,[138] which has been the most powerful all-purpose device for studying rotational spectra. The rotational transitions of light molecules tend to lie in the millimetre, submillimetre and far infra-red regions. Gordy and co-workers[99,101] have developed frequency multipliers for generating the high frequencies required for millimetre wave work. Far infra-red grating spectrometers as well as Fourier transform interferometers have also been used to study these spectra. We shall discuss most of the basic techniques as well as some of the more specialized ones which have been developed to overcome specific problems.

Table 11.1 Approximate Frequency and Wavelength Ranges of the Microwave, Millimetre Wave and Far Infra-Red Regions

	Wavelength (cm)	Wavenumber (cm^{-1})	Frequencies (GHz)
Microwave	0·3–30·0	3·3–0·033	100–1
Millimetre wave	0·05–0·3	20–3·3	600–100
Far infra-red	0·005–0·1	200–10	6000–300

There are a number of general problems which must be overcome when studying rotational spectra. The quanta are of fairly low energy and are thus more difficult to detect than, say, quanta at optical frequencies. The absorption coefficients are very small in the microwave region and the loss in power due to absorption is usually less than 0·1 %. Because detector noise swamps signals as small as this, it is usual to use phase-sensitive detection techniques to rescue the signal. The energy levels which take part in the transition are often relatively close and the Boltzmann population difference is small. As a result high powers are not always useful because saturation occurs readily, decreasing the line height and increasing the line width. Raising the pressure may not be useful when pressure broadening becomes important. The increased collision frequency at higher pressures decreases the lifetime of the states and thus the line widths are increased, limiting the resolution. As discussed in Chapter 4, $\gamma(\omega_0)$ is essentially independent of pressure over the useful range. At microwave and higher frequencies Doppler broadening can be a limiting factor in the resolution if the pressure is low. Apart from saturation, pressure and Doppler broadening, there is also the possibility of modulation broadening which introduces a line width of the order of the modulation frequency or side bands may appear.[303]

11.1 Microwave Spectrometers

In most microwave spectrometers monochromatic radiation is passed through a sample cell and the transmitted radiation detected. The modification of the transmitted radiation due to the sample is electronically amplified and plotted as a function of the radiation frequency to yield the spectrum. In Figure 11.1 is depicted a block diagram for a typical Stark spectrometer and the individual units and their modes of operation are described in the following subsections.

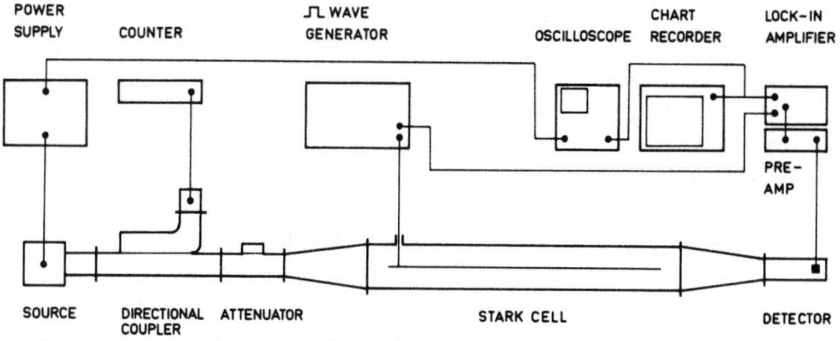

Figure 11.1 Block diagram of a Stark modulation microwave spectrometer

11.1a Microwave Sources

The majority of spectrometers have in the past used reflex-klystrons as microwave radiation sources. These devices are essentially valves in which a beam of electrons is accelerated through a cavity and is reflected back through the cavity by a high-voltage electrode (reflector). Electromagnetic oscillations build up in the cavity and modulate the electron beam causing the electrons to bunch together. The bunching occurs because fast electrons are retarded by the resonant field whereas the slower ones are accelerated.

If the dimensions and voltages are propitious the oscillations of the bunches can deliver power to the cavity which can be coupled into the waveguide.

The resonant frequency can be tuned over a wide range by mechanically deforming the cavity *and* simultaneously adjusting the reflector potential to maintain resonant oscillations. The frequency can also be tuned over a small range by varying the reflector potential alone. The electronic tuning range may only be a few tens of MHz.

Backward wave oscillators are related to klystrons but have the advantage that they can be tuned electronically over a whole band. In these devices the beam of electrons passes through a series of consecutive holes in a spiral cavity in which the radiation is generated. They tend to be rather unstable (frequency-wise) and need to be phase-locked to a stable low-frequency sweep oscillator.

In the phase-lock technique a small amount of source power is mixed with a harmonic of a stable low frequency sweeper generated by a non-linear crystal, which can also serve as the mixer element. A beat between the two frequencies at a convenient intermediate frequency, say 30 MHz, is used to drive a feedback circuit which adjusts the voltage governing the frequency of the source. As the low frequency is swept the feedback circuit maintains the 30 MHz interval and the high-frequency source follows the low-frequency sweep unit.

Gunn diodes can also be used. They have two advantages, one is that they need only a 20 volt power supply compared with the 300–4000 volts required for klystrons and backward wave oscillators. They can be mechanically tuned over a whole range without requiring any voltage adjustment as is the case for klystrons.

Up to 18 GHz it is possible to measure the frequency directly using counters which are commercially available. Above this frequency it is usual to use a beat technique. If the source is phase locked to a low-frequency sweeper one can simply measure this low frequency with a counter, multiply by the harmonic number, and add (or subtract) the intermediate frequency. A common measurement procedure uses a frequency multiplier chain to generate a frequency of the order of 500 MHz from a 5 MHz crystal controlled oscillator which is stable to ~ 1 part in 10^7. This 500 MHz frequency is then used to generate harmonics in a non-linear crystal which can also be used to mix the harmonic with the source frequency. An intermediate beat frequency is detected with a tunable radio receiver at a frequency of $\sim 30 \pm 15$ MHz. This allows a series of measurements to be carried out within the ± 15 MHz tuning range of the receiver.

Cavity wavemeters which are accurate to between ± 1 to 5 MHz are usually used to make a rough estimate of the frequency before setting up the tedious multiplier chain measurement procedure. These devices consist of a cavity whose resonance frequency is governed by a calibrated plunger.

11.1b Microwave Cells

Several different types of cell have been developed and which one is most convenient depends on the particular experiment. Each has its own advantages and disadvantages, though perhaps the Hughes–Wilson Stark cell is the one which is most generally applicable. The main types can be listed as

(1) Plain waveguide cells,
(2) Stark waveguide cells,
(3) Parallel plate cells,
(4) Free space cells,
(5) Cavity cells.

Plain waveguide cells. These essentially consist of lengths of standard waveguide, of the order of 10 feet long.[287] They are usually vacuum sealed by mica windows, as are most cells. They can be coiled into a spiral shape so that they can be placed in a vessel when careful temperature control is important. The main disadvantage of such cells is that they usually require source-modulation techniques and the detected signal is modified by cell transmission characteristics which are usually quite unpleasant. Below 50 GHz the Stark cell is often to be preferred. Such cells are usually used in the millimeter range where the small waveguide dimensions, required for efficient propagation, make Stark cells impractical.

Stark cells. The neatest way of eliminating cell background characteristics is by using the Stark modulation cell.[138] In this case a flat metal strip (septum) is mounted half way between the broad faces of the waveguide cell.[287] The septum is located in this position by two flat grooved strips of teflon which lie against the narrow inside faces of the guide, Figure 11.2. As the insulating teflon is flush against the narrow face where the electromagnetic radiation has a node, its dielectric properties are not too important. The Hewlett–Packard cell has eliminated most of the dielectric by cutting the locating

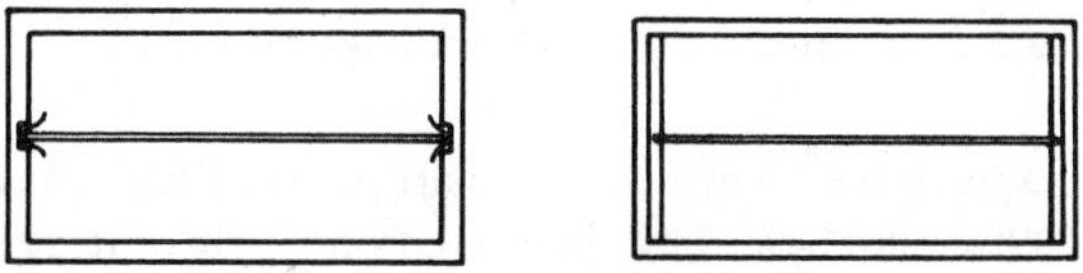

(a) (b)

Figure 11.2 Cross-sections of Stark cells. (a) The septum is located in a groove machined in the wall of the wave guide and insulated with a thin strip of teflon film. (b) Conventional Stark cell in which the septum is located in a groove machined in a flat strip of teflon which lies flush with the inside (narrow) wall of the waveguide groove in the wall of the guide itself and using a thin film to insulate the septum. These cells are usually of X-band (8–12·4 GHz) or K-band (12·4–18 GHz) dimensions. If they are much smaller accurate machining becomes difficult and also problems start to occur because of electrical breakdown of the sample gas.

By injecting a 0–2000 Volt zero-based square-wave potential between the septum and the guide, Figure 11.1, the resonance frequencies of the molecules can be modulated via the Stark effect. The modulation is followed by a phase lock detector and will thus only respond to molecular resonances and unmodulated background characteristics are ignored.

In the arrangement depicted in Figure 11.2 the applied field is parallel to the electric vector of the propagated radiation in the guide. As a result only $\Delta M = 0$ transitions are detected.

In some cells the guide has been split in half by cutting across the broad face, and the two half-sections held apart by insulators and placed inside a containing vessel. The voltage which is applied across the two halves of the cell is now perpendicular to the electric vector so allowing $\Delta M = \pm 1$ transitions to be studied.[178,303]

Parallel plate cells. Two parallel metal plates can be used to propagate microwave radiation. As such plates can be accurately machined flat and accurately separated with spacers they can be used for careful dipole moment measurements.[209] The plates are usually placed inside a suitable vacuum system and microwave radiation is fed in and out using horns. This type of cell is also useful for working with unstable species as a very high pumping rate can be maintained across the cell between the plates.[143]

Free space cells. In such cells the radiation is emitted from a horn at one end of the free-space vacuum cell and collected by a second matching horn which feeds the transmitted radiation to a suitable detector.[153] Such systems are useful when fast pumping of unstable species is required or when the species under study undergoes metal catalyzed decomposition.

Cavity cells. Spectrometers which have narrow band tuned cavities can be used in specialized studies. They allow a very high effective sensitivity over a small volume. It is usual to define the number Q (called the Quality factor) to specify the properties of such cavities. Q is essentially 2π (average energy stored)/(energy lost per cycle). Resonant radiation is trapped in such a cavity undergoing many reflections and thus yielding very high effective path lengths. Q essentially gives the path length in wavelengths and the effective path length is $Q\lambda/2\pi$. Such cells are useful for magnetic resonance studies as discussed in Section 11.8.

11.1c Detection and Amplification

In general the transmitted radiation is detected by a diode crystal rectifier. In such a unit a tungsten whisker makes a point contact with a piece of semiconductor such as silicon. Such a system represents a rectifier which because of the small contact area has a very low capacitance, enabling it to operate at the very high frequencies of microwave and millimetre wave radiation. The detector itself consists of a suitable piece of waveguide in one face of which is mounted the diode crystal cartridge. In the opposite face of the guide is mounted a coaxial socket the central pin of which extends across the guide to make contact with the whisker. The case of the cartridge makes contact with the guide and the central pin acts as the antenna for receiving the transmitted radiation. The forward and backward impedances are different and thus the incoming radiation gives rise to a d.c. current with a typical output of ~ 1 milliamp per milliwatt of power.

In a Stark spectrometer the transmitted radiation is only modulated when a molecular resonance occurs. On resonance a small square wave is imprinted on the top of the d.c. signal. This a.c. component is then amplified by a suitable preamplifier which can match the ambiguous impedance of the diode to the input of a phase sensitive detector (PSD). Such a device uses a reference signal from the square-wave generator to select and average only signals which oscillate at the modulation frequency. This amplifier averages the signal to give a d.c. output suitable for chart recorder or oscilloscope display, Figure 11.1.

The optimum modulation frequency is about 30–100 KHz. The use of lower frequencies is hampered by detector noise levels. Detector noise has two main contributions one is the Johnson (or thermal) noise caused by thermal electron fluctuations in the crystal and is proportional to the temperature. The second contribution which is called crystal noise is proportional to the square of the current, inversely proportional to the modulation frequency and essentially independent of temperature.

Frequencies higher than 100 KHz may not be used because modulation broadening becomes a line width determining factor. In the Stark spectrometer this is caused by the inability of the collisional process to relax molecules fast enough to follow the field.

11.2 Frequency Multiplier Millimetre Wave Techniques

Gordy and co-workers[99,101] have developed techniques for generating harmonics of a low frequency using non-linear crystals. Using these methods transitions as high as 813 GHz have been detected. A review of the techniques has been given by Baker.[11] The multiplier consists of a crossed-waveguide structure with a suitable non-linear crystal mounted in the high-frequency guide opposite a hole which interconnects the two guides. A tungsten whisker is mounted on a tuning stub so that the whisker interconnects the two guides making a point contact with the crystal. Radiation at the lower frequency generated by a suitable high-power source can now be used to excite harmonics of the crystal which can propagate in the high frequency guide.

Source modulation is usually used as the small size makes the high fields necessary for modulating spectra difficult to achieve. Absorption coefficients are however much higher than in the microwave region because of the strong frequency dependence (Chapter 4). This is important because the harmonic generated power is frequently small. Diode and other far infra-red detectors are used in this technique.[11,156]

11.3 Far Infra-red Fourier Transform Interferometers

High-resolution grating instruments have been used for the detection of rotational transitions in the far infra-red. Fourier transform interferometric techniques have certain advantages and have been used effectively down to $5\ \mathrm{cm}^{-1}$. In Figure 11.3 is shown the layout of the lamellar grating spectrometer used by Dowling[73,74,108] to obtain the spectra of CO and NO given in Figures 5.1 and 10.3 respectively is shown. This spectrometer works on a similar principle to that of the Michelson interferometer which uses a semitransparent beamsplitter. Several reviews on Fourier transform techniques have been written.[50,307,266]

The lamellar grating splits the radiation falling on it into two beams which are phase shifted relative to one another. By varying the grid depth the phase lag can be varied. If the two beams are subsequently recombined then they will interfere so that the intensity at a particular wavelength is given as a function of the path difference, x, by

$$I(x) = 2A^2\left(1 + \cos 2\pi\frac{x}{\lambda}\right) \tag{11.1}$$

where A is the amplitude (here assumed to be the same for both beams) of each beam and λ is the wavelength of the radiation. Integrating this function over all frequencies the total intensity is

$$I(x) = \int_{-\infty}^{+\infty} I(\omega)\,d\omega + \int_{-\infty}^{+\infty} I(\omega)\cos(2\pi\omega x)\,d\omega \tag{11.2}$$

where $\omega = 1/\lambda$. Now

$$I(0) = 2\int_{-\infty}^{+\infty} I(\omega)\,d\omega$$

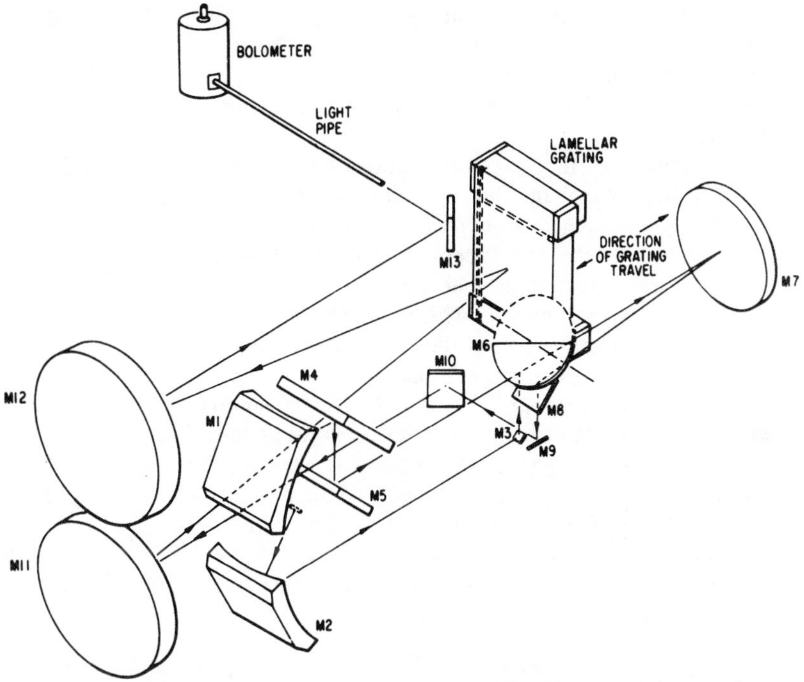

Figure 11.3 Schematic diagram of the layout of the lamellar grating interferometer used by Dowling to obtain rotational spectra in the far infra-red (from [73]). The sample cell lies underneath the grating. M6 is a rotating sector which modulates the signal. The lamellar grating itself consists of a reflecting grid with a depth which can be varied

and if $F(x) = I(x) - \frac{1}{2}I(0)$ we see that

$$F(x) = \int_{-\infty}^{+\infty} I(\omega) \cos(2\pi\omega x)\, d\omega \tag{11.3}$$

whose Fourier transform is

$$I(\omega) = \int_{-\infty}^{+\infty} F(x) \cos(2\pi\omega x)\, dx \tag{11.4}$$

This expression is usually approached by the practically accessible function

$$I(\omega) \simeq \sum_{-a}^{+a} F(x) \cos(2\pi\omega x)\, \Delta x \tag{11.5}$$

The combined signal intensity at a point after passage through the sample is monitored as a function of the path difference. The path difference is usually stepped, the signals recorded digitally and the data transformed using a computer. The difference between $I(\omega)$ with and without the sample yields the spectrum.

One advantage that these techniques have over conventional grating spectrometers is that the radiation does not need to be focused through a slit. This interface is a frequent point at which radiation is lost in conventional spectrometers. In this region sensitivity is noise limited and this point gives rise to a second advantage that the Fourier transform technique has over conventional spectrometers which sample the frequencies sequentially. Fellgett[85] noted that whereas the signal integrates in direct proportion to time, the noise integrates as the square root. As a result, noise $\propto \sqrt{\text{signal}}$ and there is a consequent multiplex gain, or Fellgett advantage, in these techniques where the radiation over the whole range is continually detected.

Resolution is limited by the mechanical accuracy with which the path differences can be varied, the planarity of the mirrors etc., which are more critical at higher frequencies though here the multiplex gain is greater. The resolution is of the order of 0·06 cm^{-1} for interferometers which operate between 25 and 200 cm^{-1}. Connes has developed near infra-red interferometers with a resolution of ~0·005 cm^{-1}.[50,266]

11.4 Lamb Dip Techniques

The Lamb dip technique has been applied by Costain in the microwave region[54] and by Winton and Gordy in the millimetre wave region[340] in order to eliminate Doppler broadening of absorption lines. In this method the radiation is focused by a paraboloid mirror into a parallel beam which propagates through a free space cell to be reflected back through the cell by a flat mirror to the detector. A standing wave field is thus set up. Consider a frequency ω_0 exactly at the center of a Doppler broadened line. This radiation will only be absorbed by those molecules whose random velocities lie perpendicular to the direction of propagation. For a frequency $\omega_0 + \delta\omega$ (still within the Doppler profile), the set of molecules with the appropriate velocity component $-v_z$ in this direction will absorb *on the first transit*. The reflected radiation is however absorbed by a new set of molecules with component $+v_z$. As a consequence more molecules are available to absorb in the wings than at the center frequency. If enough power is available to saturate the transition, saturation will occur at ω_0 first and under these conditions a dip appears at ω_0. With this technique Costain has obtained dip half-widths of 3 kHz. The method can be used for fairly strong transitions which can be saturated when resolution is limited by Doppler broadening. This technique is discussed by Shimoda and Shimizu.[273]

11.5 Molecular Beam Techniques, Maser Spectrometers

As discussed in previous sections the detection of rotational spectra requires several problems to be solved. Molecular beam techniques can to some extent overcome some of these such as detector noise, Doppler broadening and pressure broadening.

It is convenient to describe first the celebrated ammonia beam maser developed by Gordon, Zeiger and Townes.[98] In this device a beam of ammonia molecules diffuse from a high-pressure zone through a nozzle into a high-vacuum chamber along the axis of a cylindrically symmetric inhomogeneous electric field produced by four parallel long electrodes. The electrodes are constructed such that in cross section the adjacent faces form the four segments of a hyperbola. A high static voltage is applied to two opposite poles relative to the other two. The resulting potential can be described by $V = V_0 + axy$ and the field $dV/dr = ar\,(r = \sqrt{x^2 + y^2})$. The field is thus proportional to the distance from the axis of symmetry of the rods.[17]

As discussed in Chapter 9, the ammonia molecule in its ground state has two inversion states, each with its own set of rotational levels. The 0^+ state lies $\sim$24 GHz below the 0^- state. If we neglect the rotational energy, for simplicity, we can describe the Stark interaction by a matrix of the form (1.39) where $\Delta E = \Delta E_{\text{inv}}$ and $\alpha = \mu EMK/J(J + 1)$. A similar type of Stark interaction is discussed in Section 7.6. As a result the two states push each other apart under the influence of the field. As the molecules in the beam progress along the path between the electrodes the upper 0^- states whose energy decreases with decreasing field are focused towards the axis. The lower 0^+ states are defocused and deflected out of the region between the electrodes. A resonant cavity placed at a suitable position along the axis is entered only be molecules in the upper state. Population inversion is thus achieved and if radiation of the correct frequency is admitted into the cavity stimulated emission can be detected. By stimulating emission at right angles to the axis, Doppler effects can be eliminated. The technique can be used for high-resolution spectroscopy yielding line widths of the order of 5 kHz. It can be used when the species is volatile and has a large enough dipole moment that differential focusing can be achieved.

Molecular-beam techniques which employ similar electric and magnetic field focusing methods have been used to carry out high-resolution spectroscopy.[254,304] The type of experiments which can be carried out can be described with the aid of Figure 11.4. Consider a beam of molecules produced by admitting the gas from a high-pressure region through a nozzle into a very low-pressure tank. The lower the value of J, the higher in general is the magnitude of the effective dipole moment (space-fixed) and the more strongly will that state be deflected by an electric field. By using an appropriate A-field one can arrange that only certain states will pass through the collimator into the C-field region *together* with states whose μ_{eff} is so small that they are not deflected by the fields. The unfocused states are eliminated by the stop-wire which is bypassed by the selected state. The B-field can be adjusted to focus a selected state onto a suitable detector. In Figure 11.4 the detector is an electron bombardment ionizer followed by a mass analyzer.[323]

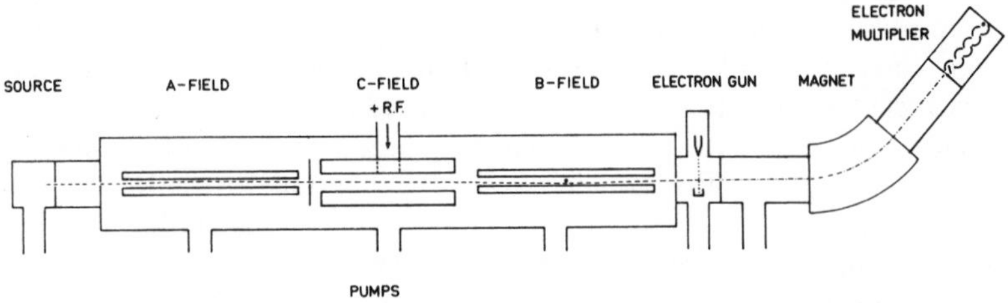

Figure 11.4 Schematic diagram of a molecular beam electric resonance spectrometer. The path ---- is that of a molecule under study. After passage through the electron beam ····· the ionized fragments –·–·– are detected by the mass spectrometer

By irradiating the beam in the C-field region transitions can be stimulated so that fewer molecules reach the detector. As a consequence a molecular resonance is detected in this so-called 'flop out' experiment by a drop in the number of molecules reaching the detector. A 'flop in' experiment can also be carried out by setting the B-field to refocus only the upper states and a resonance will be observed by noting the frequency at which molecules are detected.

In the molecular-beam electric-resonance experiment a homogeneous static C-field is applied which can be varied. In this type of experiment the splitting between the various Stark components of a particular J state can be adjusted till a resonance takes place.[136,147] In this case one can observe $\Delta J = 0$, $\Delta M = \pm 1$ transitions.

In these experiments not only are pressure and Doppler broadening reduced but a significant factor is that losses in sensitivity associated with power-induced detector noise are avoided. At low frequencies the absorption coefficient is low and high radiation densities are necessary to cause transitions. The modulation, by a weak absorption, of the transmitted radiation would in a conventional spectrometer be swamped by current noise.

The techniques have been used in many different ways and the field has been reviewed by Ramsey[254] and Trischka.[304] Klemperer and co-workers have studied some interesting quasi-molecules using a supersonic nozzle to produce beams of associated species such as $(HF)_2$.[80]

11.6 Acoustic Detection Spectrometers

A technique which neatly bypasses the detector noise problem, as do the molecular beam methods, is the acoustic detection method discussed by Krupnov.[167] This method takes advantage of the fact that on absorption of resonant radiation, the excited molecules of a gas relax via collisions, so giving rise to radiation-induced pressure fluctuations. These pressure fluctuations are detected using a sensitive microphone. A similar technique is used in infra-red detectors for specific gases. Again the technique avoids power-induced noise. It has the disadvantage that the vapour pressure must be fairly high to produce large enough pressure fluctuations. The technique has become effective with the development of millimeter wave backward wave oscillator sweepers. The specialized detectors necessary for detecting millimeter waves are also eliminated.

11.7 Multiple Resonance Experiments

Many types of double and multiple resonance experiments have been carried out. The appearance of tuned lasers will ensure that infra-red and optical frequencies will be used more and more in such experiments. Shimoda and Shimizu[273] and Carrington[35] have reviewed the field.

The method involves a simultaneous irradiation with two frequencies. In a particular experiment a particular transition frequency is pumped, resulting in a decrease in the concentration of molecules in the lower state and an increase in the concentration of the upper state—compared with the Boltzmann distribution. By monitoring the intensity of a second transition, which connects with either of these two levels as the pump frequency is varied, it is possible to obtain many different types of information. There are many variations of this basic double resonance scheme. For microwave–microwave methods Oka has given detailed technical data.[225]

As there are so many different types of problem which can be studied by double resonance (DR) we will just mention a few of the main ones:

(1) A straightforward application of DR is to pump a particular transition with modulated radiation and sweep the spectrum using phase-sensitive detection. Only transitions directly connected with the states between which pumping occurs will be detected. Thus it is a powerful method of assigning lines in a complex spectrum.[58]

(2) Oka has used DR to obtain detailed information about energy-transfer processes in small molecules.[223] By pumping a particular transition and noting how the pump energy is relaxed he has found that certain selection rules apply to collisional energy transfer processes.

(3) A weak transition can be stimulated by pumping it with very high power and noting the intensity variation in an allowed transition which is

connected with one of the pumped states. As an example Oka has detected $\Delta J = 3$ transitions.[222]

(4) By pumping an infra-red or optical transition and simultaneously sweeping the microwave frequency it is possible to study the rotational spectrum of the excited vibrational or electronic state. A typical example of the type of experiment which will become more general with the development of tuned frequency lasers is the study of magnetic resonance in excited states. German and Zare[92] have studied the $^2\Sigma^+$ state of OH. Silvers, Bergman and Klemperer[277] have studied the $^1\Pi$ state of CS by an optical–microwave DR technique which involved monitoring the variation of the fluorescence as a function of microwave frequency. It is important to note that the line width is that associated with microwave transitions.

(5) Pure rotational transitions in CH_4 have also been detected using DR techniques (p. 150).[60a]

11.8 Magnetic Resonance Spectroscopy

In Section 10.6 the theory of magnetic resonance spectroscopy has been discussed. The experimental technique was first used by Beringer and Castle[19] and has been further developed by Radford,[250] and Carrington, Levy and Miller.[40] The technique takes advantage of the high sensitivity and small volume of a high-Q tuned cavity to detect unstable radicals. Such cavities have small frequency ranges; however a resonance can be tuned in by placing the cavity in a magnetic field as in standard electron spin resonance and nuclear magnetic resonance techniques. The small cavity size allows adequate filling by an unstable species produced by various discharge techniques. Because of the high effective path length, associated with the high value of Q, small amounts of the radicals can be detected.

The design of the cavity determines the type of transitions which can be detected. In one type of cavity[40] both magnetic and electric dipole transitions can be detected. It uses Zeeman modulation coils in a similar way to that used in conventional nuclear magnetic resonance and electron spin resonance methods. In a second type of cavity[40] Stark modulation is used and the cavity is so oriented that only electric dipole transitions can be detected. It has the advantage over the first type of cavity that the resonance spectrum of O_2 can be eliminated. The electric dipole spectrum of NO shown in Figure 10.5(b) was taken with the first type of cavity.

11.9 Raman Spectra

In 1928 Raman and Krishnan[252] observed the vibrational Raman effect in liquids at the same time that Landsberg and Mandelstam[170] observed the effect in crystalline quartz. Essentially free rotational Raman transitions were observed in liquid hydrogen by McLennan and McLeod[187] and Rasetti[257] observed the pure rotational transitions in gaseous H_2, N_2 and

O_2. A few simple polyatomic molecules such as NH_3 and C_2H_2[66,176] were also studied at this time. Interest in Raman spectroscopy waned somewhat until about 1951 when experimental techniques were improved.[325,326] Lasers have also been used as sources of intense monochromatic radiation, though not as extensively for gases as in the case of condensed phases.

Raman spectroscopy is in general only used for studying molecules whose rotational spectra are inaccessible by standard microwave techniques as is the case for example in CO_2. Stoicheff has given a detailed review of the results of high-resolution Raman studies together with experimental details.[285,286] In a typical experiment a gas, at a fairly high pressure ($\sim$1 atm), in a cylindrical cell is irradiated by a batch of arc lamps which are designed to emit the Hg 4358 Å line strongly.[325,326] A White mirror system[328] is used to gather as much of the scattered radiation as possible, and focus it on the slit of a high-resolution grating spectrograph. The spectrum is usually detected photographically as exposure times of several hours are often necessary.

11.10 Other Experimental Techniques

Many variations of the techniques already discussed have been used in specific cases and further ingenious experiments will undoubtedly be devised in the future. We shall list here some of the techniques together with references which one should be aware of.

(1) *Bridge microwave spectrometers.* These are useful for the accurate measurement of microwave intensities.[329]

(2) *High-temperature microwave cells.* Special cells devised for high-temperature work have been effectively used.[184a]

(3) *Zeeman effect.* By placing the waveguide between the poles of a large electromagnet, Flygare and his associates have carried out a complete investigation of the Zeeman effect in diamagnetic molecules. The results are reviewed by Flygare and Benson.[88]

(4) *Laser–Stark spectroscopy.* Several workers have used the Stark effect to bring a molecular transition into resonance with infra-red and optical laser lines.[272,273,79]

(5) *Radio telescopes.* The rotational spectra from molecules associated with interstellar dust clouds can be detected. Using a parametric or maser amplifier it is possible to observe the transitions in either absorption or emission. Several reviews of this exploding field have been given.[303,283,281,255,226] It is likely that molecules which are almost impossible to study in the laboratory will be studied by this technique. There is evidence that a particular interstellar line belongs to HNC[282] whose spectrum has not so far been observed in the laboratory.

Appendices

A1 Notation

Notation is always a problem in quantum mechanics and it is doubtful if there is any convention which will satisfy more than a small number of people. The problem arises because the ambivalent nature of many quantities requires the use of several techniques when their properties are to be developed. In general the convention which follows has been used: vectors and vector operators are in bold face italics ($\boldsymbol{J}$); vector components and other tensor elements in light face italics with subscripts (J_z, $S_{\alpha i}$); quantum numbers, parameters and other quantities in light face italic (j, B_v) or greek (α). Occasionally it will be expedient to manipulate matrices and in such cases bold face roman notation for a second rank tensor will be used. Thus the term $\frac{1}{2}\omega_\alpha I_{\alpha\beta}\omega_\beta \equiv \frac{1}{2}\boldsymbol{\omega}^\dagger \mathbf{I}\boldsymbol{\omega}$ (A1.5) in matrix notation. The rotation transformation tensor $S_{\alpha i}^{-1}$ has been written as $\mathbf{S}^{-1}$ in relations such as (2.9). The symbol H has been used for the Hamiltonian which is a scalar operator and J^2 (also really a scalar operator) for the square of $\boldsymbol{J}$.

A1.1 Units of Angular Momentum Operators and Parameters

The correct Quantum Mechanical operator corresponding to the observable $\bar{p}_i$ is $-i\hbar\partial/\partial q_i$ where $\hbar(\equiv h/2\pi)$. If the momenta are defined so that they are in units of $\hbar$ then the resulting calculations are simplified and in particular we see that

$$[\bar{q}_i, \bar{p}_j] = i\hbar\delta_{ij} \quad \equiv \quad [q_i, p_j] = i\delta_{ij} \tag{A1.1}$$

$$[\bar{J}_i, \bar{J}_j] = i\hbar e_{ijk}\bar{J}_k \quad \equiv \quad [J_i, J_j] = ie_{ijk}J_k \tag{A1.2}$$

This allows us to define parameters such as B in the Hamiltonian (3.48) to have the same units as the resulting energy $E(J)$ in (3.49). This will usually be MHz or cm^{-1} in rotational spectroscopy.

A1.2 Angular Momentum Quantum Numbers

In general the following convention is adhered to:

$j\,k\,m$ for the quantum numbers associated with generalized angular momenta.
$J\,K\,M$ for the quantum numbers which apply directly to molecular problems.

Thus in this text K and M both are signed and the standard modulus sign will be used when necessary, $|K|$ and $|M|$. Note that Herzberg[123] uses K and M to represent $|K|$ and $|M|$.

A1.3 Zero Point and Equilibrium Quantities

Here again it is difficult to set up a consistent and neat notation which can be used theoretically as well as experimentally.

$A_0\,B_0\,C_0$ represent the observed rotational constants for the zeroth vibrational level.
$A\,B\,C$ represent instantaneous rotational constants.
$I_A^0\,I_B^0\,I_C^0$ represent the principal moments of inertia derived from $A_0\,B_0\,C_0$.

$I_{\alpha\beta}^e$ is an element of the equilibrium moment of inertia tensor.
$I_{\alpha\beta}$ is an element of the instantaneous moment of inertia tensor (2.25).
$I'_{\alpha\beta}$ is defined in (6.40).
$I''_{\alpha\beta}$ is defined in (6.41).
$\mu_{\alpha\beta}$ is an element of the inverse of $\mathbf{I}'$, i.e. $\boldsymbol{\mu} = \mathbf{I}'^{-1}$, (6.43).
I_α^s is a calculated substitution moment of inertia obtained by using the bond lengths r_s derived using Kraitchman's equations.
I_α^m are the mass dependent estimates of I_α^e, (Section 6.11).
$r_{n\alpha}$ instantaneous molecule-fixed coordinate of atom n, (2.23).
$r_{n\alpha}^e$ molecule-fixed equilibrium coordinate (2.23).
r_α^0 general molecule-fixed coordinate obtained directly from $I_A^0\,I_B^0\,I_C^0$.
r_α^s molecule-fixed coordinate obtained from $I_A^0\,I_B^0\,I_C^0$ using Kraitchman's equations.
r_α^m mass dependent estimate of r_α^e (Section 6.11).
r_α^e *The* equilibrium coordinate.
Bond *length* quantities will be represented by $r_s(AB)$, $r_0(AB)$, $r_m(AB)$, $r_c(AB)$, etc.

In experimental papers it is standard to assume that the equilibrium quantities I_α^e and r_α^e can be obtained from (6.73a) if all the necessary values of $A_0B_0C_0$ and the αs are known. This is, of course, not strictly correct but is a good approximation (Section 6.11).

A1.4 Cartesian Indices

The following letters are used as cartesian subscripts.

$x\,y\,z$	represent space-fixed axes.
$a\,b\,c$	represent molecule-fixed axes (*unordered*).
$A\,B\,C$	represent molecule-fixed axes, *ordered* according to the convention that $I_A \leqslant I_B \leqslant I_C$ and thus $A \geqslant B \geqslant C$.
$i\,j\,k$	represent space-fixed summation indices.
$\alpha\,\beta\,\gamma$	represent molecule-fixed summation indices.

A1.5 Summation Convention

According to this convention any expression in which a *cartesian* index is repeated is to be interpreted as the sum of all three cartesian values. As an example the scalar product of the vectors $\boldsymbol{A}$ and $\boldsymbol{B}$, often written

$$\boldsymbol{A}\cdot\boldsymbol{B} \equiv A_xB_x + A_yB_y + A_zB_z \equiv \sum_i^{xyz} A_iB_i \equiv \boldsymbol{A}^\dagger\boldsymbol{B} \tag{A1.3}$$

in summation convention, is simply

$$A_iB_i \tag{A1.4}$$

In summation convention the tensor product in (2.29) becomes

$$\tfrac{1}{2}\boldsymbol{\omega}^\dagger\mathbf{I}\boldsymbol{\omega} \equiv \tfrac{1}{2}\omega_iI_{ij}\omega_j \tag{A1.5}$$

Examples of summation convention. Consider the relation between scalar product $\boldsymbol{a}\cdot\boldsymbol{b}$ and the scalar product $\boldsymbol{A}\cdot\boldsymbol{B}$ where $\boldsymbol{A} = \mathbf{S}^{-1}\boldsymbol{a}$ and $\boldsymbol{B} = \mathbf{S}^{-1}\boldsymbol{b}$. In summation convention

$$S_{ij}^{-1}a_j = A_i \quad \text{and} \quad S_{ik}^{-1}b_k = B_i \tag{A1.6}$$

$$\boldsymbol{A}\cdot\boldsymbol{B} \equiv A_iB_i = S_{ij}^{-1}S_{ik}^{-1}a_jb_k \tag{A1.7}$$

Now $\mathbf{S}^{-1} = \mathbf{S}^\dagger$ where $\mathbf{S}^\dagger$ is the transpose of $\mathbf{S}$, i.e. $S_{ij}^{-1} = S_{ji}$ and therefore we can write (A1.7) as

$$S_{ji}S_{ik}^{-1}a_jb_k = (\mathbf{SS}^{-1})_{jk}a_jb_k = \delta_{jk}a_jb_k \tag{A1.8}$$

The last step follows because $\mathbf{SS}^{-1} = \mathbf{1}$ and thus we see that

$$\boldsymbol{A}\cdot\boldsymbol{B} \equiv a_jb_j \equiv \boldsymbol{a}\cdot\boldsymbol{b} \tag{A1.9}$$

$$\boldsymbol{A}\cdot\boldsymbol{B} \equiv \boldsymbol{A}^\dagger\boldsymbol{B} = (\mathbf{S}^\dagger\boldsymbol{a})^\dagger\mathbf{S}^\dagger\boldsymbol{b} = \boldsymbol{a}^\dagger\mathbf{SS}^\dagger\boldsymbol{b} = \boldsymbol{a}^\dagger\boldsymbol{b} \equiv \boldsymbol{a}\cdot\boldsymbol{b} \tag{A1.9a}$$

Thus we can write

$$(\mathbf{S}^{-1}\boldsymbol{a})\cdot(\dot{\mathbf{S}}^{-1}\boldsymbol{b}) = (\mathbf{SS}^{-1}\boldsymbol{a})\cdot(\mathbf{S}\dot{\mathbf{S}}^{-1}\boldsymbol{b}) \equiv (\mathbf{S}\dot{\mathbf{S}}^{-1})_{ij}a_ib_j \tag{A1.10}$$

A1.6 Permutation Symbol e_{ijk}

In quantum mechanics we often have to deal with quantities which do not commute and to evaluate vector product expressions, account must be taken of this. A neat way of handling them is to make use of summation convention together with the symbol e_{ijk} which goes under various titles: permutation symbol, unit antisymmetric tensor, cartesian tensor symbol or Levi–Cività symbol. It is defined by the following set of conditions

$$\begin{aligned} e_{xyz} &= e_{yzx} = e_{zxy} = 1 \quad \text{cyclic order } xyz\text{, etc.} \\ e_{xzy} &= e_{yxz} = e_{zyx} = -1 \quad \text{anticyclic order } yxz\text{, etc.} \end{aligned} \tag{A1.11}$$

Zero for other permutations such as e_{xzz}, etc.

From these defining relations we see that

$$e_{ijk} \equiv [\delta_{ix}\delta_{jy}\delta_{kz} + \delta_{iy}\delta_{jz}\delta_{kx} + \delta_{iz}\delta_{jx}\delta_{ky} - \delta_{iz}\delta_{jy}\delta_{kx} - \delta_{ix}\delta_{jz}\delta_{ky} - \delta_{iy}\delta_{jx}\delta_{kz}] \tag{A1.12}$$

This symbol is conveniently written as a determinant[318]

$$\begin{vmatrix} \delta_{ix} & \delta_{iy} & \delta_{iz} \\ \delta_{jx} & \delta_{jy} & \delta_{jz} \\ \delta_{kx} & \delta_{ky} & \delta_{kz} \end{vmatrix} \tag{A1.13}$$

The product of two such symbols becomes

$$e_{ijk}e_{lmn} = \begin{vmatrix} \delta_{il} & \delta_{im} & \delta_{in} \\ \delta_{jl} & \delta_{jm} & \delta_{jn} \\ \delta_{kl} & \delta_{km} & \delta_{kn} \end{vmatrix} \tag{A1.14}$$

This result is most easily obtained by multiplying the determinant (A1.13) times the equivalent determinant for e_{lmn}. Note that transposing the determinant of e_{lmn} (which does not alter the value) makes life easier.

When certain of the indices are identical (A1.14) reduces to

$$e_{ijk}e_{imn} = \begin{vmatrix} \delta_{jm} & \delta_{jn} \\ \delta_{km} & \delta_{kn} \end{vmatrix} = \delta_{jm}\delta_{kn} - \delta_{jn}\delta_{km} \tag{A1.15}$$

$$e_{ijk}e_{ijn} = 2\delta_{kn} \tag{A1.16}$$

$$e_{ijk}e_{ijk} = 6 \tag{A1.17}$$

The power of this notation can be seen for instance when we evaluate a triple cross-product $(\boldsymbol{A} \times \boldsymbol{B}) \times \boldsymbol{C}$

$$\boldsymbol{F} = (\boldsymbol{A} \times \boldsymbol{B}) \times \boldsymbol{C} \quad \text{and let} \quad \boldsymbol{D} = \boldsymbol{A} \times \boldsymbol{B} \tag{A1.18}$$

according to (A1.11) $\boldsymbol{A} \times \boldsymbol{B} \equiv e_{ijk}A_jB_k$ and thus

$$D_i = e_{ijk}A_jB_k, \qquad F_l = e_{lmn}D_mC_n \tag{A1.19}$$

matching $m \rightarrow i$ and cycling the indices such that $e_{lin} \rightarrow e_{inl}$ we see that

$$F_l = e_{ijk}e_{inl}A_jB_kC_n = (\delta_{jn}\delta_{kl} - \delta_{jl}\delta_{kn})A_jB_kC_n \tag{A1.20}$$

We can evaluate this as

$$F_l = A_jB_lC_j - A_lB_nC_n \tag{A1.21}$$

When $\boldsymbol{A}$, $\boldsymbol{B}$ and $\boldsymbol{C}$ commute this can be written in the usual notation as

$$\boldsymbol{F} = \boldsymbol{B}(\boldsymbol{A} \cdot \boldsymbol{C}) - \boldsymbol{A}(\boldsymbol{B} \cdot \boldsymbol{C}) \tag{A1.22}$$

When they *do not* commute it is often worth while to introduce the commutators and thus (A1.21) becomes

$$F_l = [A_j, B_l]C_j + B_lA_jC_j - A_lB_nC_n \tag{A1.23}$$

These results are essentially the same as those deduced in section 12^2 of the book of Condon and Shortley.[49] In essentially the same way we can show that

$$\boldsymbol{F} = \boldsymbol{J} \times (\boldsymbol{J} \times \boldsymbol{T}) = i\boldsymbol{J} \times \boldsymbol{T} + \boldsymbol{J}(\boldsymbol{J} \cdot \boldsymbol{T}) - \boldsymbol{J}^2\boldsymbol{T} \tag{A1.24}$$

$$\equiv F_i = ie_{ijk}J_jT_k + J_iJ_kT_k - J_kJ_kT_i \tag{A1.24a}$$

The determinant det A of the 3×3 cartesian tensor $A_{i\alpha}$ where $i = xyz$ and $\alpha = abc$ can be written out as

$$\begin{aligned}\det A = {} & A_{ax}A_{by}A_{cz} + A_{bx}A_{cy}A_{az} + A_{cx}A_{ay}A_{bz} \\ & - A_{az}A_{by}A_{cx} - A_{bz}A_{cy}A_{ax} - A_{cz}A_{ay}A_{bx}\end{aligned} \tag{A1.25}$$

With a little clear thought we see that this can also be written as

$$\det A = \tfrac{1}{6}e_{ijk}e_{\alpha\beta\gamma}A_{\alpha i}A_{\beta j}A_{\gamma k} \tag{A1.26}$$

multiplying both sides by $e_{\alpha\beta\gamma}$ and using (A1.17) indicates that

$$\det A e_{\alpha\beta\gamma} = e_{ijk}A_{\alpha i}A_{\beta j}A_{\gamma k} \tag{A1.27}$$

Consider this relation (A1.27) when the matrix **A** is the rotation matrix **S**. In this case one can post-multiply both sides by $S_{\gamma k}^{-1}$ ($\equiv S_{k\gamma}$) and note that the determinant of **S** is unity and thus obtain the result

$$e_{\alpha\beta\gamma}S_{\gamma k} = e_{ijk}S_{\alpha i}S_{\beta j} \tag{A1.28}$$

which is equivalent to Van Vleck's statement[310] that one element of the **S** tensor is equal to its cofactor (see Section 3.4).

A2 Angular Velocities, ω_α

The angular velocities, ω_α, can be defined in terms of χ, θ and φ, and $\dot{\chi}$, $\dot{\theta}$ and $\dot{\varphi}$. For instance

$$\omega_\alpha = (\dot{\boldsymbol{\chi}})_\alpha + (\dot{\boldsymbol{\theta}})_\alpha + (\dot{\boldsymbol{\varphi}})_\alpha \tag{A2.1}$$

Study of Figure 2.3 together with this relation yields the results

$$\omega_a = \dot{\theta} \sin \chi - \dot{\varphi} \cos \chi \sin \theta \tag{A2.2a}$$

$$\omega_b = \dot{\theta} \cos \chi + \dot{\varphi} \sin \chi \sin \theta \tag{A2.2b}$$

$$\omega_c = \dot{\chi} + \dot{\varphi} \cos \theta \tag{A2.2c}$$

In the case of ω_b for instance, we see that $(\dot{\boldsymbol{\chi}})_b = 0$, $(\dot{\boldsymbol{\theta}})_b = \dot{\theta} \cos \chi$ and $(\boldsymbol{\varphi})_b = \dot{\varphi} \sin \chi \sin \theta$.

The form of the tensor can now be proven. If we write $\mathbf{S} = \mathbf{ABC}$ where $\mathbf{A}$, $\mathbf{B}$ and $\mathbf{C}$ are the three rotation matrices defined in (2.12) then we can write ($\mathbf{A}^\dagger$ is the transpose of $\mathbf{A}$)

$$\dot{\mathbf{S}} = \mathbf{AB\dot{C}} + \mathbf{A\dot{B}C} + \mathbf{\dot{A}BC} \tag{A2.3}$$

$$\dot{\mathbf{S}}^\dagger = \dot{\mathbf{C}}^\dagger\mathbf{B}^\dagger\mathbf{A}^\dagger + \mathbf{C}^\dagger\dot{\mathbf{B}}^\dagger\mathbf{A}^\dagger + \mathbf{C}^\dagger\mathbf{B}^\dagger\dot{\mathbf{A}}^\dagger \tag{A2.4}$$

If we note that $\mathbf{CC}^\dagger = \mathbf{1}$, etc., we can therefore write

$$\mathbf{S}\dot{\mathbf{S}}^\dagger = \mathbf{ABC}\dot{\mathbf{C}}^\dagger\mathbf{B}^\dagger\mathbf{A}^\dagger + \mathbf{AB}\dot{\mathbf{B}}^\dagger\mathbf{A}^\dagger + \mathbf{A}\dot{\mathbf{A}}^\dagger \tag{A2.5}$$

On evaluating the three terms in (A2.5) we obtain

$$\begin{bmatrix} 0 & -\cos\theta & \sin\theta \sin\chi \\ \cos\theta & 0 & \sin\theta\cos\chi \\ -\sin\theta\sin\chi & -\sin\theta\cos\chi & 0 \end{bmatrix} \dot{\varphi} + \begin{bmatrix} 0 & 0 & \cos\chi \\ 0 & 0 & -\sin\chi \\ -\cos\chi & \sin\chi & 0 \end{bmatrix} \dot{\theta} + \begin{bmatrix} 0 & -1 & 0 \\ 1 & 0 & 0 \\ 0 & 0 & 0 \end{bmatrix} \dot{\chi} \tag{A2.6}$$

which on summation and comparison with (A2.2) indicates that

$$S_{\beta\varepsilon}\dot{S}^{-1}_{\varepsilon\gamma} \equiv (\mathbf{S}\dot{\mathbf{S}}^{-1})_{\beta\gamma} = -e_{\alpha\beta\gamma}\omega_\alpha \tag{A2.7}$$

A3 Reduction of the Double Commutator $[\boldsymbol{J}^2, [\boldsymbol{J}^2, \boldsymbol{T}]]$

The double commutator $[\boldsymbol{J}^2, [\boldsymbol{J}^2, \boldsymbol{T}]]$ can be reduced according to (4.39) using the method outlined by Dirac[67] and rephrased by Condon and Shortley.[49] The reduction is a good exercise in summation technique and the use of the permutation symbol e_{ijk}.

First we evaluate $[\boldsymbol{J}^2, \boldsymbol{T}]$:

$$[J_i J_i, T_j] = J_i[J_i, T_j] + [J_i, T_j]J_i = ie_{ijk}(J_i T_k + T_k J_i) \tag{A3.1}$$

$$= ie_{ijk}(2J_i T_k - ie_{ilk}T_l) = -2i(e_{jik}J_i T_k - iT_j) \tag{A3.2}$$

$$\equiv [\boldsymbol{J}^2, \boldsymbol{T}] = -i(\boldsymbol{J} \times \boldsymbol{T} - \boldsymbol{T} \times \boldsymbol{J}) = -2i(\boldsymbol{J} \times \boldsymbol{T} - i\boldsymbol{T}) \tag{A3.2a}$$

In this reduction we have made use of (3.22) and (A1.16). If we note that $[\boldsymbol{J}^2, \boldsymbol{J} \times \boldsymbol{T}] = \boldsymbol{J} \times [\boldsymbol{J}^2, \boldsymbol{T}]$ we can reduce the double commutator

$$[\boldsymbol{J}^2, [\boldsymbol{J}^2, \boldsymbol{T}]] = -2i\{-2i\boldsymbol{J} \times (\boldsymbol{J} \times \boldsymbol{T}) - 2\boldsymbol{J} \times \boldsymbol{T} - i(\boldsymbol{J}^2\boldsymbol{T} - \boldsymbol{T}\boldsymbol{J}^2)\} \tag{A3.3}$$

$$[\boldsymbol{J}^2, [\boldsymbol{J}^2, \boldsymbol{T}]] = -4\boldsymbol{J}(\boldsymbol{J} \cdot \boldsymbol{T}) + 2(\boldsymbol{J}^2\boldsymbol{T} + \boldsymbol{T}\boldsymbol{J}^2) \tag{A3.4}$$

where we have made use of the result (A1.24).

The relation (A3.4) can now be used to derive the selection rule on Δj. Confining our attention to transitions for which $\Delta j \neq 0$ we can equate the off-diagonal matrix elements of the left-hand side of (A3.4) with those of the right-hand side. Because the scalar product of two T class operators commutes with $\boldsymbol{J}$, the term in $\boldsymbol{J}(\boldsymbol{J} \cdot \boldsymbol{T})$ does not contribute to the off-diagonal elements in j. Thus

$$\langle j\,k\,m|\boldsymbol{J}^4\boldsymbol{T} - 2\boldsymbol{J}^2\boldsymbol{T}\boldsymbol{J}^2 + \boldsymbol{T}\boldsymbol{J}^4|j'\,k'\,m'\rangle = 2\langle j\,k\,m|\boldsymbol{J}^2\boldsymbol{T} + \boldsymbol{T}\boldsymbol{J}^2|j'\,k'\,m'\rangle \tag{A3.5}$$

which indicates that

$$\begin{aligned} &[j^2(j+1)^2 - 2j(j+1)j'(j'+1) + j'^2(j'+1)^2]\langle j\,k\,m|\boldsymbol{T}|j'\,k'\,m'\rangle \\ &\qquad = 2[j(j+1) + j'(j'+1)]\,\langle j\,k\,m|\boldsymbol{T}|j'\,k'\,m'\rangle \end{aligned} \tag{A3.6}$$

The algebraic factors on the left- and right-hand sides in this expression can be combined in the following, not immediately obvious, way[49,67]

$$[j(j+1) - j'(j'+1)]^2 = (j - j')^2(j + j' + 1)^2 \tag{A3.7}$$

The right-hand side factor on the other hand can be rearranged:

$$2[j(j+1) + j'(j'+1)] = (j + j' + 1)^2 + (j - j')^2 - 1 \tag{A3.8}$$

Using these expressions (A3.6) can be rearranged to give

$$[(j + j' + 1)^2 - 1][(j - j')^2 - 1]\langle j\,k\,m|\boldsymbol{T}|j'\,k'\,m'\rangle = 0 \tag{A3.9}$$

A4 Second-order Perturbation Contributions

Consider the term $H' = aq\boldsymbol{J}^2$ given in (6.33) the basis functions in which this term is to be evaluated are the product functions $|vJ\rangle = |v\rangle|J\rangle$. $\boldsymbol{J}^2$ is diagonal in this basis whereas q is off-diagonal in v as indicated in (1.28) and Table 1.1. If we evaluate the perturbation contribution to second order as given by (1.46c) we see that

$$E^{(2)} = a^2\left[\frac{\langle v|q|v+1\rangle\langle v+1|q|v\rangle}{E^0_{v\,J} - E^0_{v+1\,J}} + \frac{\langle v|q|v-1\rangle\langle v-1|q|v\rangle}{E^0_{v\,J} - E^0_{v-1\,J}}\right]\langle J|\boldsymbol{J}^4|J\rangle \tag{A4.1}$$

In the rigid-rotor approximation the rotational energy is independent of v and thus the denominators in (A4.1) are $-\omega$ and $+\omega$ respectively and thus

$$E^{(2)} = a^2\left\{-\frac{1}{2\omega}\right\}J^2(J+1)^2 \tag{A4.2}$$

A5 The Two-dimensional Harmonic Oscillator

The matrix elements of the two-dimensional isotropic oscillator have been given by several authors using various phase conventions.[335,218,201] Starting with the Hamiltonian

$$H = \tfrac{1}{2}\omega(p_1^2 + p_2^2 + q_1^2 + q_2^2) \tag{A5.1}$$

one can calculate the matrix elements according to the procedure such as that given by Moffit and Liehr[201] ($p_\pm = p_1 \pm ip_2, q_\pm = q_1 \pm iq_2$)

$$\langle v+1\; l+1|q_+|v\; l\rangle = \alpha\sqrt{(v+l+2)} \tag{A5.2a}$$

$$\langle v-1\; l+1|q_+|v\; l\rangle = \alpha\sqrt{(v-l)} \tag{A5.2b}$$

$$\langle v-1\; l-1|q_-|v\; l\rangle = \alpha\sqrt{(v+l)} \tag{A5.2c}$$

$$\langle v+1\; l-1|q_-|v\; l\rangle = \alpha\sqrt{(v-l+2)} \tag{A5.2d}$$

$$\langle v+1\; l+1|p_+|v\; l\rangle = i\alpha\sqrt{(v+l+2)} \tag{A5.3a}$$

$$\langle v-1\; l+1|p_+|v\; l\rangle = -i\alpha\sqrt{(v-l)} \tag{A5.3b}$$

$$\langle v-1\; l-1|p_-|v\; l\rangle = -i\alpha\sqrt{(v+l)} \tag{A5.3c}$$

$$\langle v+1\; l-1|p_-|v\; l\rangle = i\alpha\sqrt{(v-l+2)} \tag{A5.3d}$$

where $\alpha = 1/\sqrt{2}$. In these equations l is the quantum number associated with the vibrational angular momentum.

$$\boldsymbol{L} = q_1p_2 - q_2p_1$$

In the above phase convention all of the matrix elements of q are real.

A6 Kraitchman's Equations

These equations first published in a general form by Kraitchman[165] are a generalization of the procedure discussed in Section 6.11b for the determination of the coordinates of an atom in a linear molecule relative to the center of mass. The derivation of the equations is discussed in several texts.[100,343] The equations are exact for rigid molecules only, and the results of applying them to semi-rigid molecules were first discussed by Costain.[57] They have been further discussed by Watson.[321]

Suppose that one atom of mass m is replaced by one of mass $m + \Delta m$ and all internuclear distances remain unchanged. Let us define

$$\mu = \frac{M\Delta m}{M + \Delta m} \quad \text{where} \quad M = \sum_n m_n \tag{A6.1}$$

is the total mass. Then, for molecules in which the orientation of the principal axes are not changed by substitution, linear molecules and symmetric tops (on axis substitution),

$$|r_\alpha| = [\mu^{-1}(I_\beta^* - I_\beta)]^{1/2} \tag{A6.2}$$

This is the result obtained in Section 6.11. In off-axis substitution in symmetric tops, planar molecules and in some non-planar asymmetric tops the substitution may change the orientation of two axes only, leaving the orientation of one, say the a-axis, unchanged. In symmetric tops

$$|r_b| = \left[\frac{(I_c^* - I_c)(I_b^* - I_c)}{\mu(I_b - I_c)}\right]^{1/2}, \qquad |r_c| = \left[\frac{(I_b^* - I_b)(I_c^* - I_b)}{\mu(I_c - I_b)}\right]^{1/2} \tag{A6.3}$$

One can eliminate I_c, which may not be known, if c is the symmetry axis, using the relation $I_c = I_c^* + I_b^* - I_a^*$ and thus obtain

$$|r_b| = \left[\frac{(I_a^* - I_b^*)(I_a^* - I_c^*)}{\mu(I_a^* - I_b^* - I_c^* + I_b)}\right]^{1/2}, \qquad |r_c| = \left[\frac{(I_b^* - I_b)(I_c^* - I_b)}{\mu(I_b^* + I_c^* - I_a^* - I_b)}\right]^{1/2} \tag{A6.4}$$

For planar asymmetric rotors, if c is the out-of-plane axis, then

$$|r_a| = \left[\frac{(I_b^* - I_b)(I_a^* - I_b)}{\mu(I_a - I_b)}\right]^{1/2}, \qquad |r_b| = \left[\frac{(I_a^* - I_a)(I_b^* - I_a)}{\mu(I_b - I_a)}\right]^{1/2} \tag{A6.5}$$

For the general asymmetric rotor the required equation is

$$|r_\alpha| = \left[\frac{\Delta P_\alpha}{\mu}\left(1 + \frac{\Delta P_\beta}{I_\alpha - I_\beta}\right)\left(1 + \frac{\Delta P_\gamma}{I_\alpha - I_\gamma}\right)\right]^{1/2} \tag{A6.6}$$

where

$$\Delta P_\alpha = \tfrac{1}{2}(-\Delta I_\alpha + \Delta I_\beta + \Delta I_\gamma) \tag{A6.6a}$$

$$\Delta I_\alpha = I_\alpha^* - I_\alpha \tag{A6.6b}$$

A7 Diatomic Molecular Potential Constants

It is convenient to start from the classical equation of motion (1.6) for a harmonic oscillator. The general solution of this differential equation is

$$x = a \cos \bar{\omega} t \tag{A7.1}$$

where $\bar{\omega}$ is the angular frequency and t is the time. If $\bar{\omega} = 2\pi\omega$ then the associated frequency is

$$\omega = \frac{1}{2\pi}\sqrt{\frac{k}{m}} \tag{A7.2}$$

where ω is in Hz (cps), k in dynes per cm and m in grams. We have defined the force field in relations (1.30) and (1.31c) in terms of the coordinate q, which is related to x by the relation $q = \sqrt{m\bar{\omega}}\,x$. In the literature the following potential function is often used[118,119]

$$V(x) = \tfrac{1}{2}F_2x^2 + \tfrac{1}{2}F_3x^3 + \tfrac{1}{2}F_4x^4 + \cdots \tag{A7.3}$$

where the F_ns are in dynes/cm^{n-1} and x is in cm (essentially $r - r_e$). We can transform the potential function which we have found most convenient for calculation (1.30) using $q = \sqrt{m\bar{\omega}}\,x$

$$V(x) = \tfrac{1}{2}m\bar{\omega}^2x^2 + \tfrac{1}{6}\varphi_3(m\bar{\omega})^{3/2}x^3 + \cdots \tag{A7.4}$$

and thus see that

$$F_2 = m\bar{\omega}^2 \tag{A7.5}$$

$$F_3 = \tfrac{1}{3}\varphi_3(m\bar{\omega})^{3/2} \tag{A7.6}$$

Using the table of fundamental quantities, we obtain

$$F_2 = [5{\cdot}891\,80 \times 10^{-2}]m\omega^2 \tag{A7.7}$$

where F_2 is in dynes/cm, m is in amu and ω in cm^{-1}. According to our result (6.36b) we can substitute for φ_3 in (A7.6) and thus obtain the result

$$F_3 = a_1(F_2/r_e) \tag{A7.8}$$

The result for F_4 is

$$F_4 = a_2(F_2/r_e^2) \tag{A7.9}$$

where

$$a_2 = (\tfrac{5}{4})a_1^2 - (\tfrac{2}{3})(\omega_e x_e/B_e) \tag{A7.10}$$

The results are now readily related to tables of force constants such as those given by Herschbach and Laurie.[118] Using the results in Section 6.2

we find that for AlF

$$F_2 = 4{\cdot}230 \times 10^5 \text{ dynes/cm} \qquad (\text{or} \times 10^{-11} \text{ ergs/Å}^2)$$

$$a_1 = -3{\cdot}184$$

$$F_3 = -8{\cdot}1439 \times 10^{13} \text{ dynes/cm}^2 \quad (\text{or} \times 10^{-11} \text{ ergs/Å}^3)$$

and thus the potential function

$$V(x) = 1{\cdot}0647 \times 10^5 x^2 - 2{\cdot}049 \times 10^5 x^3 + \cdots \tag{A7.11}$$

can be generated where x is in Å and $V(x)$ in cm^{-1}. This function was used to plot Figure 6.3. Note that F_3 will almost always be negative for any realistic potential surface which must extrapolate to dissociation as x increases. (Note 1 erg ≡ 1 dyne cm ≡ $5{\cdot}034015 \times 10^{15}$ cm^{-1}.)

A8 The Van Vleck Transformation

This transformation procedure introduced by Van Vleck has been discussed by several authors.[309,146,150,343] It is essentially a type of perturbation theory whose systematic procedure is useful when dealing with a large number of terms of varying orders of magnitude.

Consider the Hamiltonian

$$H = H_0 + \lambda H_1 + \lambda^2 H_2 \tag{A8.1}$$

where H_0 defines the basis representation which contains two sets of states $|m\rangle, |m'\rangle, \ldots$ and $|n\rangle, |n'\rangle, \ldots$ etc. Let us separate the m and n states into separate blocks. The Van Vleck transformation allows us to diagonalize the blocks separately and still take account of off-diagonal terms linking the blocks. It essentially transforms the matrix H to $\tilde{H}$, where $\tilde{H}$ is block diagonal. The matrix elements of $\tilde{H}$ are given *to second order* by

$$\langle m|\tilde{H}|m'\rangle = \langle m|H_0|m'\rangle + \lambda\langle m|H_1|m'\rangle + \lambda^2\langle m|H_2|m'\rangle + \langle m|\Delta H|m'\rangle \tag{A8.2}$$

where

$$\langle m|\Delta H|m'\rangle = \frac{\lambda^2}{2}\sum_n\left[\frac{\langle m|H_1|n\rangle\langle n|H_1|m'\rangle}{E_m^0 - E_n^0} + \frac{\langle m|H_1|n\rangle\langle n|H_1|m'\rangle}{E_{m'}^0 - E_n^0}\right]$$
$$+\lambda^3\sum_n\left[\frac{\langle m|H_1|n\rangle\langle n|H_2|m'\rangle}{E_m^0 - E_n^0} + \frac{\langle m|H_2|n\rangle\langle n|H_1|m'\rangle}{E_{m'}^0 - E_n^0}\right] \tag{A8.2a}$$

The technique is particularly powerful when the energies separating the m states are much smaller than those separating the states $|m\rangle$ from $|n\rangle$. When this is the case we see that $E_m^0 - E_n^0 \sim E_{m'}^0 - E_n^0$ etc. and the first term in (A8.2a) simplifies to

$$\langle m|\Delta H|m'\rangle = \lambda^2 \sum_n \langle m|H_1|n\rangle\langle n|H_1|m'\rangle[\Delta E_{mn}]^{-1} \tag{A8.3}$$

A worked example of the use of this transformation is given in Section 10.5c where the Λ-doubling splitting is calculated.

Table A1 $e^{-\beta\omega}$ as a Function of $\omega(\text{cm}^{-1})$ at −78°, 25° and 100°C

$\omega(\text{cm}^{-1})$	−78°C 195°K	25°C 298°K	100°C 373°K
100	0·478	0·617	0·680
200	0·229	0·381	0·462
300	0·109	0·235	0·314
400	0·052	0·145	0·214
500	0·025	0·089	0·145
600	0·012	0·055	0·099
700	0·0057	0·034	0·067
800	0·0027	0·021	0·046
900	0·0013	0·0129	0·031
1000	0·0006	0·0080	0·021
1500	$0{\cdot}2 \times 10^{-4}$	$0{\cdot}7 \times 10^{-3}$	$0{\cdot}3 \times 10^{-2}$
2000	$0{\cdot}4 \times 10^{-6}$	$0{\cdot}6 \times 10^{-4}$	$0{\cdot}4 \times 10^{-3}$
3000	$0{\cdot}2 \times 10^{-9}$	$0{\cdot}5 \times 10^{-6}$	$0{\cdot}9 \times 10^{-5}$

Table A2 Fundamental Physical Constants and Conversion Factors[256]

Velocity of light	c	2·997 9250	30[a]	10^{8} m sec^{-1}	10^{10} cm sec^{-1}
Fine structure constant	α	7·297 351	33	10^{-3}	10^{-3}
	α^{-1}	137·036 0	6		
Electron charge	e	1·602 192	21	10^{-19} C	10^{-20} emu
		4·803 25	6		10^{-10} esu
Planck's constant	h	6·626 20	15	10^{-34} J sec	10^{-27} erg sec
	$\hbar = h/2\pi$	1·054 592	24	10^{-34} J sec	10^{-27} erg sec
Avogadro's number	N	6·022 17	12	10^{26} kmole^{-1}	10^{23} mole^{-1}
Atomic mass unit	amu	1·660 531	33	10^{-27} kg	10^{-24} g
Electron rest mass	m_e	9·109 56	16	10^{-31} kg	10^{-28} g
	m_e^*	5·485 93	10	10^{-4} amu	10^{-4} amu
Proton rest mass	M_p	1·672 614	33	10^{-27} kg	10^{-24} g
	M_p^*	1·007 276 61	24	amu	amu
Neutron rest mass	M_n	1·674 920	33	10^{-27} kg	10^{-24} g
	M_n^*	1·008 665 20	30	amu	amu
Rydberg constant	R_∞	1·097 373 12	33	10^{7} m^{-1}	10^{5} cm^{-1}
Electron magnetic moment in Bohr magnetons	μ_e/μ_B	1·001 159 638 9	9		
Bohr Magneton	μ_B	9·274 10	19	10^{-24} J T^{-1}	10^{-21} erg G^{-1}
Gas constant	R_0	8·314 3	10	10^{3} J kmole^{-1} K^{-1}	10^{7} erg mole^{-1} K^{-1}
Boltzmann's constant R_0/N	k	1·380 62	18	10^{-23} J K^{-1}	10^{-16} erg K^{-1}

[a] Estimated error based on 3 standard deviations and applies to the last two digits in the preceding column.

Table A3 Conversion Factors

Units	cm^{-1}	Hz	erg	cal $mole^{-1}$	eV
1 cm^{-1}	1	$2{\cdot}997\,925 \times 10^{10}$	$1{\cdot}986\,486 \times 10^{-16}$	$2{\cdot}859\,22$	$1{\cdot}239\,855 \times 10^{-4}$
1 Hz	$3{\cdot}335\,64 \times 10^{-11}$	1	$6{\cdot}626\,20 \times 10^{-27}$	$9{\cdot}537\,31 \times 10^{-11}$	$4{\cdot}135\,707 \times 10^{-15}$
1 erg	$5{\cdot}034\,015 \times 10^{15}$	$1{\cdot}509\,161 \times 10^{26}$	1	$1{\cdot}439\,334 \times 10^{16}$	$6{\cdot}241\,45 \times 10^{11}$
1 cal $mole^{-1}$	$0{\cdot}349\,746$	$1{\cdot}048\,513 \times 10^{10}$	$6{\cdot}947\,66 \times 10^{-17}$	1	$4{\cdot}336\,34 \times 10^{-5}$
1 eV	$8065{\cdot}46$	$2{\cdot}417\,966 \times 10^{14}$	$1{\cdot}602\,192 \times 10^{-12}$	$23\,060{\cdot}9$	1

$$h/(8\pi^2 c) = 27{\cdot}9933 \times 10^{-40}\ \text{g cm}^2 . \text{cm}^{-1} = I\ \text{g cm}^2 \times B\ \text{cm}^{-1}$$
$$= 16{\cdot}8580\ \text{amu Å}^2 . \text{cm}^{-1} = I\ \text{amu Å}^2 \times B\ \text{cm}^{-1}$$
$$h/(8\pi^2 \times 10^6) = 8{\cdot}39218 \times 10^{-35}\ \text{g cm}^2 . \text{MHz} = I\ \text{g cm}^2 \times B\ \text{MHz}$$
$$= 5{\cdot}05391 \times 10^5\ \text{amu Å}^2 . \text{MHz} = I\ \text{amu Å}^2 \times B\ \text{MHz}$$

References

1. Adel, A., and Dennison, D. M., *Phys. Rev.*, **44**, 99 (1933). — Vibration–rotation theory.
2. Allen, H. C., and Cross, P. C., *Molecular Vib-rotors*, Wiley (1963).
3. Amano, T., Hirota, E., and Morino, Y., *J. Phys. Soc. Jap.*, **22**, 399 (1967). — m-wave SO
4. Amano, T., Saito, S., Hirota, E., Morino, Y:, Johnson, D. R., and Powell, F. X., *J. Mol. Spectrosc.*, **30**, 275 (1969). — m-wave ClO
5. Amat, G., and Henry, L., *Cah. Phys.*, **12**, 273 (1958). — Vibration–rotation theory
6. Amat, G., and Henry, L., *Journal de Phys. et le Rad.*, **21**, 728 (1960). — Vibration–rotation theory, C_{3v}, C_{4v}
7. Anderson, P. W., *Phys. Rev.*, **76**, 647, 471A (1949). — Pressure broadening
8. Arfken, G., *Mathematical Methods for Physicists*, Academic Press, New York (1966).
9. Atkins, P. W., *Molecular Quantum Mechanics, I, II, III*, Oxford University Press (1970).
10. Bak, B., Bang, O., Nicolaisen, F., and Rump, O., *Spectrochim Acta*, **27A**, 1865 (1971). — IR NCN_3
11. Baker, J. G., in (192). — mm-wave techniques
12. Bardeen, J., and Townes, C. H., *Phys. Rev.*, **73**, 97 (1948). — Quadrupole interaction theory
13, Bardeen, J., and Townes, C. H., *Phys. Rev.*, **73**, 627, 1204 (1948) — Quadrupole interaction, second-order theory
14. Bell, R. P., *Proc. Roy. Soc. A*, **183**, 328 (1945). — Ring puckering
15. Bell, S., *J. Phys. B*, **2**, 1001 (1969). — 1, 2, 3 dim. oscillator matrix elements
16. Benedict, W. S., and Plyler, E. K., *Can. J. Phys.*, **35**, 1235 (1957). — IR NH_3
17. Bennewitz, H. G., Paul, W., and Schlier, C., *Z. Physik*, **141**, 6 (1955). — Molecular beam fields
18. Benz, H. P., Bauder, A., and Gunthard, H. H., *J. Mol. Spectrosc.*, **21**, 156 (1966). — Quadrupole interaction matrix elements
19. Beringer, R., and Castle, J. G., *Phys. Rev.*, **78**, 581 (1950). — EPR NO
20. Beringer, R., and Rawson, E. B., *Phys. Rev.*, **86** 607A (1952). — EPR NO Λ-doubling

21. Beringer, R., Rawson, E. B., and Henry, A. F., *Phys. Rev.*, **94**, 343 (1954).	EPR NO
22. Blackman, G. L., Bolton, K., Brown, R. D., Burden, F. R., and Mishra, A., *J. Mol. Spectrosc.*, **47**, 457 (1973).	m-wave NCN_3, quadrupole interaction with 4 nuclei
23, Born, M., and Huang, K., *Dynamical Theory of Crystal Lattices*, Oxford University Press (1954).	Born–Oppenheimer approximation, in English
24. Born, M., and Oppenheimer, R., *Ann. Physik*, **84**, 457 (1927).	See (23)
25. Boyd, D. R. J., and Longuet-Higgins, H. C., *Proc. Roy. Soc. A.*, **213**, 55 (1952).	Coriolis theory
25a. Bradley, R. H., Brier, P. N., and Whittle, M. J., *J. Mol. Spectrosc.*, **44**, 536 (1972).	mm-wave BrF_5 C_{4v} molecule
26. Bragg, J. K., *Phys. Rev.*, **74**, 533 (1948).	Asymmetric rotor quadrupole interaction
27. Brown, R. L., and Radford, H. E., *Phys. Rev.*, **147**, 6 (1966).	EPR NO
28. Bunker, P. R., *J. Chem. Phys.*, **47**, 718 (1967).	Vibration–rotation theory, $CH_3C{\equiv}CCH_3$
29. Burden, F. R., and Millen, D. J., *J. Chem. Soc. A*, 1212 (1967).	Equilibrium structure, OCS
30. Burkhard, D. G., and Dennison, D. M., *Phys. Rev.*, **84** 408 (1951).	Internal rotation CH_3OH, IAM
31. Burrus, C. A., and Gordy, W., *Phys. Rev.*, **92** 1437 (1953).	mm-wave NO
32. Careless, A. J., Green, M. C., and Kroto, H. W., *Chem. Phys. Letts.*, **16**, 414 (1972).	m-wave $(CH_3)_3SiNCO$
33. Careless, A. J., and Kroto, H. W. To be published.	m-wave SiH_3CN
34. Careless, A. J., Kroto, H. W., and Landsberg, B. M., *Chem. Phys.*, **1**, 371 (1973).	m-wave F_2CS
35. Carrington, A., in (256).	Review of radicals
36. Carrington, A., Dyer, P. N., and Levy, D. H., *J. Chem. Phys.*, **47**, 1756 (1967).	EPR ClO
37. Carrington, A., Fabris, A. R., Howard, B. J., and Lucas, N. J. D., *Mol. Phys.*, **20**, 961 (1971).	EPR NCO, NCS
38. Carrington, A., and Levy, D. H., *J. Phys. Chem.*, **71**, 2 (1967).	Review EPR
39. Carrington, A., Levy, D. H., and Miller, T. A., *Proc. Roy. Soc. A*, **293**, 108 (1966).	EPR SO $^1\Delta$
40. Carrington, A., Levy, D. H., and Miller, T. A., in (244).	Review EPR
41. Carrington, A., and McLachlan, A. D., *Introduction to Magnetic Resonance*, Harper and Row (1969).	
42. Chan, M. Y., Wilardjo, L., and Parker, P. M., *J. Mol. Spectrosc.*, **40**, 473 (1971).	Sextic centrifugal distortion coefficients
43. Chan, S. I., Borgers, T. R., Russell, J. W., Strauss, H. L., and Gwinn, W. D., *J. Chem. Phys.*, **44**, 1103 (1966).	Ring puckering III trimethylene oxide
44. Chan, S. I., and Stelman, D., *J. Chem. Phys.*, **39**, 545 (1963).	Gaussian humps

45. Chan, S. I., Stelman, D., and Thompson, L. E., *J. Chem. Phys.*, **41**, 2828 (1964). — Quartic oscillator basis sets
46. Chan, S. I., Zinn, J., Fernandez, J., and Gwinn, W. D., *J. Chem. Phys.*, **33**, 1643 (1960). — Ring puckering I trimethylene oxide
47. Chan, S. I., Zinn, J., and Gwinn, W. D., *J. Chem. Phys.*, **34**, 1319 (1961). — Ring puckering II trimethylene oxide
47a. Chu, F. Y., and Oka, T. To be published. — $\Delta K = \pm 3$ transitions in PH_3
48. Cleeton, C. E., and Williams, N. H., *Phys. Rev.*, **45**, 234 (1934). — Ye original m-wave experiment
49. Condon, E. U., and Shortley, G. H., *The Theory of Atomic Spectra*, Cambridge University Press (1967).
50. Connes, J., *Rev. Opt. Theor. Instrum.*, **40**, 45, 116, 171, 231 (1961). — Interferometers
51. Costain, C. C., *Phys. Rev.*, **82**, 108 (1951). — NH_3 inversion potential
52. Costain, C. C., *J. Chem. Phys.*, **29**, 864 (1958). — Substitution method
53. Costain, C. C., *Trans. of Am. Crystallographic Assoc.*, **2**, 157 (1966). — Accuracy of substitution method
54. Costain C. C., *Can. J. Phys.*, **47**, 2431 (1969). — m-wave Lamb dip
55. Costain, C. C., and Dowling, J. M., *J. Chem. Phys.*, **32**, 158 (1960). — m-wave NH_2CHO
56. Costain, C. C., and Kroto, H. W., *Can. J. Phys.*, **50**, 1453 (1972). — m-wave NCN_3
56a. Costain, C. C., and Srivastava, G. P., *J. Chem. Phys.*, **41**, 1620 (1964). — m-wave dimers $CF_3COOH.HCOOH$
57. Cox, A. P., Brittain, A. H., and Whittle, M. J., *J. Mol. Spectrosc.*, **35**, 49 (1970). — m-wave Ni(Cpd)NO
58. Cox, A. P., Flynn, G. W., and Wilson, E. B., *J. Chem. Phys.*, **42**, 3094 (1965). — Double resonance
59. Cross, P. C., Hainer, R. M., and King, G. W., *J. Chem. Phys.*, **12**, 210 (1944). — Asymmetric rotor intensities
60. Curl, R. F., and Oka, T., *J. Chem. Phys.*, **58**, 4908 (1973). — IR m-wave double resonance CH_4
60a. Curl, R. F., Oka, T., and Smith, D. S., *J. Mol. Spectrosc.*, **46**, 518 (1973). — Pure rotational transition in CH_4
61. Daniels, J. M., and Dorain, P. B., *J. Chem. Phys.*, **45**, 26. (1966). — EPR SO
62. Darling, B. T., and Dennison, D. M., *Phys. Rev.*, **57**, 128 (1940). — Vibration–rotation Hamiltonian, Δ
63. Davies, P. B., Neumann, R. M., Wofsy, S. C., and Klemperer, W., *J. Chem. Phys.*, **55**, 3564 (1971). — Electric resonance, PH_3
64. Dennison, D. M., *Rev. Mod. Phys.*, **12**, 175 (1940). — 2nd ed. of (1)
65. Dicke, R. H., and Wittke, J. P., *Introduction to Quantum Mechanics*, Addison-Wesley, Reading, Mass. (1961).
66. Dickinson, R. G., Dillon, R. T., and Rasetti, F., *Phys. Rev.*, **34**, 583 (1929). — Raman, NH_3
67. Dirac, P. A. M., *The Principles of Quantum Mechanics*, Oxford University Press, (1947).
68. Dixon, R. N., *Trans. Faraday Soc.*, **60**, 1363 (1964). — Quasi-linearity

69. Dorman, F., and Lin, C. C., *J. Mol. Spectrosc.*, **12**, 119 (1964). — Vibration–rotation theory of linear molecules
70. Dorney, A. T., and Watson, J. K. G., *J. Mol. Spectrosc.*, **42**, 135 (1972). — Centrifugal distortion induced spectra
71. Dowling, J. M., *J. Mol. Spectrosc.*, **6**, 550 (1961). — Planar molecules centrifugal distortion
72. Dowling, J. M., *J. Mol. Spectrosc.*, **27**, 527 (1968). — Far IR NH_3
73. Dowling, J. M., *Investigations in Far I.R. with Lamellar Grating Interferometer*, Airforce Rept. SSD-TR-67-30. Aerospace Rept. TR-1001 (9260-01)-7 (1967). — Includes spectrum of CO
74. Dowling, J. M., and Hall, R. T., *J. de Physique Colloque*, **28**, C2-156 (1967). — Interferometer instrumental
75. Dreizler, H., in (256). — Flexible mols. with two degrees of freedom
76. Dreizler, H., and Dendl, G., *Z. Naturforsch.*, **20a**, 30 (1965). — Centrifugal distortion Me_2O
77. Dreizler, H., and Rudolph, H. D., *Z. Naturforsch.*, **20a**, 749 (1965). — Centrifugal distortion Me_2S
78. Dunham, J. L., *Phys. Rev.*, **41**, 721 (1932). — Diatomic molecule theory
79. Duxbury, G., and Jones, R. G., *Mol. Phys.*, **20**, 721 (1971). — Stark spectroscopy
80. Dyke, T. R., Howard, B. J., and Klemperer, W., *J. Chem. Phys.*, **56**, 2442 (1972). — $(HF)_2$ quasi-molecule
81. Dymanus, A., Dijkerman, H. A., and Zijderveld, G. R. D., *J. Chem. Phys.*, **32**, 717 (1960). — m-wave OCS line width and intensities
82. Engerholm, G., Luntz, A. C., Gwinn, W. D., and Harris, D. O., *J. Chem. Phys.*, **50**, 2446 (1969). — 5 mem. rings. pseudorotation
83. Eyring, H., Walter, J., and Kimball, G. E., *Quantum Chemistry*, Wiley (1958).
84. Falick, A. M., Mahan, B. H., and Myers, R. J., *J. Chem. Phys.*, **42**, 1837 (1965). — EPR $O_2(^1\Delta)$
85. Fellgett, P., Thesis, Cambridge. (1951). — Multiplex gain
86. Fermi, E., *Z. Physik*, **71**, 250 (1931). — Fermi resonance
87. Flygare, W. H., *Ann. Rev. Phys. Chem.*, **18**, 325 (1967). — Review
88. Flygare, W. H., and Benson, R. C., *Mol. Phys.*, **20**, 225 (1971). — Zeeman effect in diamag. molecules
89. Frosch, R. A., and Foley, H. M., *Phys. Rev.*, **88**, 1337 (1952). — Hyperfine structure EPR, NO
90. Gallagher, J. T., Bedard, F. D., and Johnson, C. M., *Phys. Rev.*, **93**, 729 (1954). — m-wave NO
91. Gebbie, H. A., Burrows, W. J., and Bird, G. R., *Proc. Roy. Soc. A*, **310**, 579 (1969). — Far IR O_2
91a. Georgiou, K., Kroto, H. W., and Landsberg, B. M., *J. Chem. Soc. Chem. Comm.*, 739 (1974). — m-wave H_2CCS thioketene
92. German, K., and Zare, R., *Phys. Rev. Letts.*, **23**, 1207 (1969). — Optical m-wave double resonance
93. Gerry, M. C. L., Thompson, J. C., and Sugden, T. M., *Nature*, **211**, 846 (1966). — m-wave SiH_3NCO

94. Glidewell, C., Robiette, A. C., and Sheldrick, G. M., *Chem. Phys. Letts.*, **16**, 526 (1972). — Electron diffraction SiH_3NCO

95. Golden, S., and Wilson, E. B., *J. Chem. Phys.*, **16**, 669 (1948). — Stark effect asymmetric rotors

96. Gordon, H. R., and McCubbin, T. K., *J. Mol. Spectrosc.*, **18**, 73 (1965). — IR CO_2

97. Gordon, H. R., and McCubbin, T. K., *J. Mol. Spectrosc.*, **19**, 137 (1966). — IR CO_2

98. Gordon, J. P., Zeiger, H. J., and Townes, C. H., *Phys. Rev.*, **99**, 1264 (1955). — THE MASER

99. Gordy, W., *Pure Appl. Chem.*, **11**, 403 (1965). — mm-waves

100. Gordy, W., and Cook, R. L., *Microwave Molecular Spectra*, Interscience (1970).

101. Gordy, W., Smith, W. V., and Trambarulo, R. F., *Microwave Spectroscopy*, Dover, N.Y. (1953).

102. Grenier-Besson, M. L., *J. Phys. Rad.*, **21**, 555 (1960). — *l*-doubling theory

103. Grenier-Besson, M. L., *J. Phys. Rad.*, **25**, 757 (1964). — *l*-doubling theory

104. Grenier-Besson, M. L., and Amat, G., *J. Mol. Spectrosc.*, **8**, 22 (1962). — Vibration–rotation analysis *l*-doubling

105. Guarnieri, A., and Favero, P., *Microwave Gas Spectroscopy Bibliography* [1954–1967], Inst. Chimico G. Ciamician, Univ. di Bologna (1968). — Review

106. Gwinn, W. D., and Luntz, A., *Transactions of Am. Crystallographic Assoc.*, **2**, 90 (1966). — Structure of flexible molecules

107. Hall, R. T., and Dowling, J. M., *Chem. Phys.*, **45**, 1899 (1966). — Far IR $NO(^2\Pi)$

108. Hall, R. T., Vrabec, D., and Dowling, J. M., *App. Opt.*, **5**, 1147 (1966). — Interferometer instrumental

109. Harrington, H. W., *J. Chem. Phys.*, **49**, 3023 (1968). — m-wave intensities

110. Harris, D. O., Engerholm, G. G., Tolman, C. A., Luntz, A. C., Keller, R. A., Kim, H., and Gwinn, W. D., *J. Chem. Phys.*, **50** 2438 (1969). — 5 mem. rings pseudo-rotation

111. Harris, D. O., Harrington, H. W., Luntz, A. C., and Gwinn, W. D., *J. Chem. Phys.*, **44**, 3467 (1966). — Puckering

112. Hayashi, M., and Pierce, L., *J. Chem. Phys.*, **35**, 1148 (1961). — Internal rotation tables

112a. Heath, G. A., Thomas, L. F., Sherrard, E. I., and Sheridan, J., *Disc. Faraday Soc.*, **19**, 38 (1955). — m-wave $CH_3(C{\equiv}C)_2H$

113. Heitler, W., *The Quantum Theory of Radiation*, Oxford University Press (1954).

114. Helminger, P., and Gordy, W., *Bull. Am. Phys. Soc.*, **12**, 543 (1967). — mm-wave NH_3

115. Henry, L., and Amat, G., *Cah. Phys.*, **14**, 231 (1960). — Vibration–rotation theory

116. Herschbach, D. R., *J. Chem. Phys.*, **27**, 975 (1957). — Internal rotation tables

117. Herschbach, D. R., *J. Chem. Phys.*, **31**, 91 (1959). — Internal rotation theory

118. Herschbach, D. R., and Laurie, V. W., *Tables of Vibrational Force Constants*, UCRL 9694, Univ. of Calif. Rad. Lab. Berkeley (1960).

119. Herschbach, D. R., and Laurie, V. W., *J. Chem. Phys.*, **35**, 458 (1961).	Force constants of diatomic mols.
120. Herschbach, D. R., and Laurie, V. W., *J. Chem. Phys.*, **37**, 1668 (1962).	Vibration–rotation theory
121. Herschbach, D. R., and Laurie, V. W., *J. Chem. Phys.*, **40**, 3142 (1964).	Vibration–rotation theory Δ
122. Herzberg, G., *Molecular Spectra and Molecular Structure, Vol.* 1: *Spectra of Diatomic Molecules*, 2nd ed., Van Nostrand (1950).	
123. Herzberg, G., *Molecular Spectra and Molecular Structure, Vol.* 2: *Infrared and Raman Spectra of Polyatomic Molecules*, Van Nostrand (1945).	
124. Herzberg, G., *Molecular Spectra and Molecular Structure, Vol.* 3: *Electronic Spectra of Polyatomic Molecules*, Van Nostrand (1966).	
125. Hill, E. L., and Van Vleck, J. H., *Phys. Rev.*, **32**, 250 (1928).	Hund's case [*a*–*b*]
126. Hirota, E., *J. Mol. Spectrosc.*, **7**, 242 (1961).	m-wave $CH_2(CN)_2$
127. Hirota, E., *J. Chem. Phys.*, **37**, 283 (1962).	Internal rotation $CH_3CH_2CH_2F$
128. Hougen, J. T., *J. Chem. Phys.*, **36**, 519 (1962).	Linear molecule theory
129. Hougen, J. T., *J. Chem. Phys.*, **37**, 1433 (1962).	Classification of symmetric top eigenfunctions
130. Hougen, J. T., *J. Chem. Phys.*, **39**, 358 (1963).	$^4\Sigma$ states
131. Hougen, J. T., *Can. J. Phys.*, **42**, 1920 (1964).	Vibration–rotation theory $CH_3C{\equiv}CCH_3$
132. Hougen, J. T., *Can. J. Phys.*, **43**, 935 (1965).	Vibration–rotation theory $CH_3C{\equiv}CCH_3$
133. Hougen, J. T., *Pure and App. Chem.*, **11**, 481 (1965), Butterworths.	Flexible molecules review
134. Hougen, J. T., *N.B.S. Monograph* 115 (1970), 'The Calculation of Rotational Energy levels and Rotational Line Intensities in Diatomic Molecules'.	
135. Hougen, J. T., Bunker, D. R., and Johns, J. W. C., *J. Mol. Spectrosc.*, **34**, 136 (1970).	Quasi-linear molecules
136. Hughes, H. K., *Phys. Rev.*, **72**, 614 (1947).	Mol. beam electric resonance
137. Hughes, H. K., *Phys. Rev.*, **76**, 1675 (1949).	Stark effect, linear molecules
138. Hughes, R. H., and Wilson, E. B., *Phys. Rev.*, **71**, 562 (1947).	The Stark spectrometer
139. Ishiguro, E., and Kobori, M., *J. Phys. Soc. Japan*, **22**, 263 (1967).	Calculation of spin–orbit coupling constants
140. Jaffé, H. H., *Accounts of Chem. Res.*, **2**, 136 (1969).	Semi-empirical SCF calculations
141. James, T. C., and Thibault, R. J., *J. Chem. Phys.*, **41**, 2806 (1964).	IR NO

142. Johns, J. W. C., *Can. J. Phys.*, **45**, 2639 (1967). — Quasi-linearity Gaussian and Lorentzian humps
143. Johnson, D. R., Powell, F. X., and Kirchhoff, W. H *J. Mol. Spectrosc.*, **39**, 136 (1971). — m-wave H_2CS
144. Jones, L. U., Shoolery, J. N., Shulman, R. G., and Yost, D. M., *J. Chem. Phys.*, **18**, 990 (1950). — m-wave HNCO
145. Jones, W. J., and Stoicheff, B. P., *Phys. Rev. Letts.*, **13**, 657 (1964). — Inverse Raman effect
146. Jordahl, O. M., *Phys. Rev.*, **45**, 87 (1934). — Van Vleck transf.
147. Kaiser, E. W., *J. Chem. Phys.*, **53**, 1686 (1970). — Molecular beams
148. Kakar, R. K., Rinehart, E. A., Quade, C. R., and Kojima, T., *J. Chem. Phys.*, **52**, 3803 (1970). — m-wave C_6H_5CHO benzaldehyde
149. Karplus, R., and Schwinger, J., *Phys. Rev.*, **73**, 1020 (1948). — Power saturation theory
150. Kemble, E. C., *Fundamental Principles of Quantum Mechanics*, Dover (1958).
151. Kemp, J. D., and Pitzer, K. S., *J. Chem. Phys.*, **4**, 749 (1936). — Internal rotation
152. Kewley, R., Murty, K. R., and Sugden, T. M., *Trans. Faraday Soc.*, **56**, 1732 (1960). — m-wave SF_5Cl
153. Kewley, R., Sastry, K. V. L. N., Winnewisser, M., and Gordy, W., *J. Chem. Phys.*, **39**, 2856 (1963). — mm-wave CS, free space cell
154. Kilb, R. W., Lin, C. C., and Wilson, E. B., *J. Chem. Phys.*, **26**, 1695 (1957). — Theory of internal rotation, PAM
155. Kilb, R. W., and Pierce, L., *J. Chem. Phys.*, **27**, 108 (1957). — CH_3SiH_3, barrier
156. Kimmit, M. F., *Far Infrared Techniques*, Pion Limited, London (1970).
157. Kimura, K., Katada, K., and Bauer, S. H., *J. Am. Chem. Soc.*, **88**, 416 (1966). — Electron diffraction $(CH_3)_3SiNCO$
158. King, G. W., *Spectroscopy and Molecular Structure*, Holt, Rinehart and Winston (1964).
159. King, G. W., Hainer, R. M., and Cross, P. C., *J. Chem. Phys.*, **11**, 27 (1943). — Gen. theory of rigid asym. rotor
160. Kirchhoff, W. H., *J. Mol. Spectrosc.*, **41**, 333 (1972). — Centrifugal distortion analyses
161. Kirchhoff, W. H., and Johnson, D. R., *J. Mol. Spectrosc.*, **45**, 159 (1973). — Centrifugal distortion NH_2CHO
162. Kivelson, D., and Wilson, E. B., *J. Chem. Phys.*, **20**, 1575 (1952). — Centrifugal distortion
163. Koehler, J. S., and Dennison, D. M., *Phys. Rev.*, **57**, 1006 (1940). — Internal rotation IAM CH_3OH
164. Koningstein, J. A., *Introduction to the Theory of the Raman Effect*, D. Reidel, Dordrecht (1972).
165. Kraitchman, J., *Am. J. Phys.*, **21**, 17 (1953). — Structure
165a. Kroto, H. W., and Landsberg, B. M. To be published. — m-wave thioacetone
166. Kroto, H. W., Landsberg, B. M., and Suffolk, R. J., *Chem. Phys. Letts.* In press. — m-wave and photo-electron spectra, thioacetaldehyde
166a. Kroto, H. W., and Maier, M., To be published. — Vibrational satellites of methyl diacetylene

167. Krupnov, A. F., *Preprint N24*, Radiophysical Research Inst., Gorky, Russia.	Acoustic detection of mm-wave absorption
168. Lafferty, W. J., *J. Mol. Spectrosc.*, **25**, 359 (1968).	Direct *l*-doubling transitions
169. Landau, L. D., and Lifshitz, E. M., *Quantum Mechanics*, Pergamon (1956).	
170. Landsberg, G., and Mandelstam, L., *Naturwiss.*, **16**, 557, 772 (1928).	Raman effect in quartz
171. Laurie, V. W., *Acc. of Chem. Res.*, **3**, 331 (1970).	Review of dynamics of flexible mols.
172. Laurie, V. W., and Herschbach, D. R., *J. Chem. Phys.*, **37**, 1687 (1962).	Vibration–rotation theory
173. Laurie, V. W., Pence, D. T., and Jackson, R. H., *J. Chem. Phys.*, **37**, 2995 (1962).	m-wave F_2CO
174. Lepard, D. W., *Can. J. Phys.*, **48**, 1664 (1970).	Raman intensities
175. Lett, R. G., and Flygare, W. H., *J. Chem. Phys.*, **47**, 4730 (1967).	m-wave CH_3NCO
176. Lewis, C. M., and Houston, W. V., *Phys. Rev.*, **44**, 903 (1933).	Raman CO_2, C_2H_2, C_2H_4
177. Lide, D. R., *Ann. Rev. Phys. Chem.*, **15**, 225 (1964).	m-wave review
178. Lide, D. R., *Rev. Sci. Instr.*, **35**, 1226 (1964).	High temp. cell
179. Lide, D. R., and Mann, D. E., *J. Chem. Phys.*, **29**, 914 (1958).	Torsion in $(CH_3)_3CH$, etc.
180. Lide, D. R., and Matsumura, C., *J. Chem. Phys.*, **50**, 3080 (1969).	Linear molecules
181. Lindfors, K. R., and Cornwell, C. D., *J. Chem. Phys.*, **42**, 149 (1965).	Dipole moments
182. Lin, C. C., and Mizushima, M., *Phys. Rev.*, **100**, 1726 (1955).	Theory of EPR, NO
183. Lin, C. C., and Swalen, J. D., *Rev. Mod. Phys.*, **31**, 841 (1959).	Review of internal rotation problem
184. Longuet-Higgins, H. C., *Mol. Phys.*, **6**, 445 (1963).	Symmetry classification of flexible molecules
184a. Lovas, F. J., and Lide, D. R., in *Advances in High Temperature Chemistry*, Vol. 3, Academic Press Inc., New York (1971).	High-temperature m-wave spectroscopy
185. Ludwig, G., *Wave Mechanics*, Pergamon (1968).	Original papers on Q.M. in English
186. McKnight, J. S., and Gordy, W., *Phys. Rev. Letts.*, **21**, 1787 (1968).	mm-wave O_2
187. McLennan, J. C., and McLeod, J. H., *Nature*, **123**, 160 (1929).	Raman H_2
188. McWeeny, R., and Sutcliffe, B. T., *Methods in Molecular Quantum Mechanics*, Academic Press (1969).	
189. Maes, S., *Cah. Phys.*, **14**, 125, 164 (1960).	Vibration–rotation theory η terms
190. Maki, A. G., and Johnson, D. R., *J. Mol. Spectrosc.*, **47**, 226 (1973).	m-wave OCS
191. Margenau, H., and Henry, A., *Phys. Rev.*, **78**, 587 (1950).	EPR theory

192. Martin, D. H. (Ed.), *Spectroscopic Techniques for Far I.R. Submillimetre and Millimetre Waves*, North Holland, Amsterdam (1967).

193. Martins, J. F., and Wilson, E. B., *J. Mol. Spectrosc.*, **26**, 410 (1968). — m-wave $XeOF_4$ dipole moment by double resonance

194. Marton, L. (Ed.), *Methods of Experimental Physics, Vol.* 3, Williams, D. (Ed.), 'Molecular Physics', Academic (1962).

195. Meal, J. H., and Polo, S. R., *J. Chem. Phys.*, **24**, 1119 (1956). — Coriolis ζ sum rules

196. Mecke, R., *Z. Physik*, **81**, 313 (1933). — Δ, H_2O

197. Mills, I. M., in (256). — Vibration–rotation review

198. Mills, I. M., Watson, J. K. G., Smith, W. F., *Mol. Phys.*, **16**, 329 (1969). — Vibration induced rotational spectra

199. Mirri, A. M., Scappini, F., Innamorati, L., and Favero, P., *Spectrochimica Acta*, **25A**, 1631 (1969). — mm-wave F_2CO

200. Mizushima, M., and Venkateswarlu, P., *J. Chem. Phys.*, **21**, 705 (1953). — Vibration induced rotational spectra

201. Moffit, W., and Liehr, A. D., *Phys. Rev.*, **106**, 1195 (1957). — Doubly degenerate harmonic oscillator

202. Morino, Y., 'Pure and Applied Chemistry', *Molecular Spectroscopy IX*, Butterworths (1969). — Anharmonic force field

203. Morino, Y., in (256). — Review of microwave research

204. Morino, Y., and Hirota, E., *Ann. Rev. Phys. Chem.*, **20**, 139 (1969). — Review

205. Morino, Y., Kikuchi, Y., Saito, S., and Hirota, E., *J. Mol. Spectrosc.*, **13**, 95 (1964). — Equilibrium structure, SO_2 effect of including γ terms

206. Morino, Y., Kuchitsu, K., and Oka, T., *J. Chem. Phys.*, **36**, 1108 (1962). — Average structures

207. Morino, Y., and Nakagawa, T., *J. Mol. Spectrosc.*, **26**, 496 (1968). — Equilibrium structure OCS

208. Morino, Y., and Saito, S., *J. Mol. Spectrosc.*, **19**, 435 (1966). — Equilibrium structure of F_2O, Fermi resonance

209. Muenter, J. S., *J. Chem. Phys.*, **48**, 4544 (1968). — Dipole moment of OCS $\mu = 0{\cdot}71521 \pm 0{\cdot}00020$ Debye molecular beam electric resonance, parallel plates

210. Murrell, J. N., and Harget, A. J., *Semi-empirical SCF–MO Theory of Molecules*, Wiley–Interscience (1972).

211. Nakagawa, T., and Morino, Y., *J. Mol. Spectrosc.*, **31**, 208 (1969). — Vibration–rotation study in HCN and OCS. *l*-doubling

212. Naylor, R. E., and Wilson, E. B., *J. Chem. Phys.*, **26**, 1057 (1957). — Barrier CH_3BF_2 V_6 terms
213. Naude, S. M., and Hugo, T. J., *Can. J. Phys.*, **35**, 64 (1957). — Electronic spectrum AlF
214. Nielsen, A. H., *J. Chem. Phys.*, **11**, 160 (1943). — Vibration–rotation theory linear XYZ molecules
215. Nielsen, H. H., *Phys. Rev.*, **40**, 445 (1932). — Internal rotation IAM
216. Nielsen, H. H., *Phys. Rev.*, **77**, 130 (1950). — *l*-doubling
217. Nielsen, H. H., *Rev. Mod. Phys.*, **23**, 90 (1951). — Vibration–rotation theory review
218. Nielsen, H. H., in *Handbuch der Physik*, Vol. 37/1, Springer–Verlag, Berlin (1959). — Vibration–rotation theory review
219. Nielsen, H. H., and Dennison, D. M., *Phys. Rev.*, **72**, L86, 1101 (1947). — Interactions in $|K| = 3$ levels of NH_3
220. Oelfke, W. C., and Gordy, W., *J. Chem. Phys.*, **51**, 5336 (1969). — mm-wave H_2O_2
221. Oka, T., *J. Phys. Soc. Japan*, **15**, 2274 (1960). — Average structures
222. Oka, T., *J. Chem. Phys.*, **45**, 752 (1966). — $\Delta J = 3$ transitions
223. Oka, T., *J. Chem. Phys.*, **47**, 13 (1967). — Collision induced transitions
224. Oka, T., *J. Chem. Phys.*, **47**, 5410 (1967). — *l*-doubling theory C_{3v} and C_{4v} mols.
225. Oka, T., *Can. J. Phys.*, **47**, 2343 (1969). — Double res.
226. Oka, T., *Mémoires Société Royale des Sciences de Liège*, 6*e* serie, tome III, 37 (1972). — Interstellar spectra
227. Oka, T., and Morino, Y., *J. Mol. Spectrosc.*, **6**, 472 (1961). — Δ, inertial defect
228. Oka, T., and Morino, Y., *J. Mol. Spectrosc.*, **8**, 9 (1962). — Δ, inertial defect
229. Oka, T., and Morino, Y., *J. Mol. Spectrosc.*, **8**, 300 (1962). — H_2Se average structure
230. Oka, T., and Morino, Y., *J. Mol. Spectrosc.*, **11**, 349 (1963). — Δ, inertial defect
231. Oka, T., Shimizu, F. O., Shimizu, T., and Watson, J. K. G., *Astrophys. J. Letts.*, **165**, 15L (1971). — $\Delta K = 3$ transitions
232. Ozier, I., *Phys. Rev. Letts.*, **27**, 1329 (1971). — CH_4, μ
233. Ozier, I., Ho, W., and Birnbaum, G., *J. Chem. Phys.*, **51**, 4872 (1969). — Far IR and dipole moment of CH_3D
234. Pariseau, M. A., Suzuki, I., and Overend, J., *J. Chem. Phys.*, **42**, 2335 (1965). — Anharmonic force field of CO_2
235. Parkin, J. E., *Ann. Rep. Progr. Chem.*, **64**, 181 (1967). — Review
236. Parkin, J. E., *Ann. Rep. Progr. Chem.*, **65**, 111 (1968). — Review
237. Pierce, L., *J. Chem. Phys.*, **29**, 383 (1958). — m-wave CH_3SiH_2F barrier
238. Pierce, L., Di Cianni, N., and Jackson, R. H., *J. Chem. Phys.*, **38**, 730 (1963). — Force field from centrifugal distortion data
239. Pierce, L., and Peterson, D. H., *J. Chem. Phys.*, **33**, 907 (1960). — $(CH_3)_3SiH$

240. Placzek, G., *Handbuch der Radiologie*, 6 (2), 205 (1934). Akademische Verlagsgesellschaft mb H Leipzig. — Theory of Raman effect
240a. UCRL. 526(L) (1962), Clearinghouse, USAEC. — English trans.
241. Placzek, G., and Teller, E., *Z. Phys.*, **81**, 209 (1933). — Raman intensities
242. Plíva, J., *Collection Czech. Chem. Commun.*, **23**, 777, 1846 (1958) — Vibrational coordinates
243. Powell, F. X., and Lide, D. R., *J. Chem. Phys.*, **41**, 1413 (1964). — $^3\Sigma$ state of SO
244. Prigogine, I., and Rice, S. A. (Ed.), *Advances in Chemical Physics Vol. XVIII*, Interscience (1970).
245. Quade, C. R., *J. Chem. Phys.*, **48**, 5490 (1968). — C_6H_5OH torsion
246. Quade, C. R., and Lin, C. C., *J. Chem. Phys.*, **38**, 540 (1963). — Internal rotation CH_2DCHO
247. Racah, G., *Phys. Rev.*, **61**, 186 (1942). — Angular momentum theory
248. Racah, G., *Phys. Rev.*, **62**, 438 (1942). — Angular momentum theory
249. Racah, G., *Phys. Rev.*, **63**, 367 (1943). — Angular momentum theory
250. Radford, H. E., *Phys. Rev.*, **122**, 114 (1961). — EPR OH $^2\Pi_{3/2}$
251. Radford, H. E., *Phys. Rev.*, **126**, 1035 (1962). — EPR OH $^2\Pi_{1/2}$
252. Raman, C. V., and Krishnan, K. S., *Nature*, **121**, 501 (1928). — Smekal–Raman–Krishnan–Landsberg–Mandelstam effect
253. Ramsey, N. F., *Nuclear Moments*, Wiley (1953).
254. Ramsey, N. F., *Molecular Beams*, Oxford (1963).
255. Rank, D. M., Townes, C. H., and Welsh, W. J., *Science*, **174**, 1083 (1971). — Review on interstellar spectroscopy
256. Rao, K. N., and Mathews, C. W. (Ed.), *Molecular Spectroscopy Modern Research*, Academic Press (1972).
257. Rasetti, F., *Phys. Rev.*, **34**, 367 (1929). — Raman effect O_2, N_2, H_2
258. Ray, B. S., *Z. Physik*, **78**, 74 (1932). — Energy levels of the asymmetric rotor
259. Rose, M. E., *Elementary Theory of Angular Momentum*, Wiley, N.Y. (1957).
260. Rosenberg, A., and Ozier, I., *J. Chem. Phys.*, **58**, 5168 (1973). — Rotational spectrum GeH_4
261. Rosenberg, A., and Ozier, I., *Chem. Phys. Letts.*, **19**, 400 (1973). — Rotational spectrum SiH_4
262. Rosenberg, A., Ozier, I., and Kudian, A., *J. Chem. Phys.*, **57**, 568 (1972). — Rotational spectrum CH_4
263. Rudolph, H. D., *Ann. Rev. Phys. Chem.*, **21**, 733 (1970).
264. Saegebarth, E., and Wilson, E. B., *J. Chem. Phys.*, **46**, 3088 (1967). — m-wave CH_2FCFO int. rotation
265. Saito, S., *J. Mol. Spectrosc.*, **30**, 1 (1969). — Equilibrium structure SO_2
266. Sanderson, R. B., in (256). — Review of interferometer techniques

267. Schiff, L. I., *Quantum Mechanics*, McGraw-Hill (1955).
268. Schlapp, R., *Phys. Rev.*, **51**, 342 (1937). — $^3\Sigma$ states theory (303)
269. Schnabel, E., Törring, T., and Wilke, W., *Z. Physik*, **188**, 167 (1965). — Inversion spectrum of NH_3
270. Schrödinger, E., *Ann. der Physik*, **79**, 361 (1926). — Solution of $H\psi = E\psi$ for hydrogen. (185) for English translation
271. Schrödinger, E., *Ann. der Physik*, **79**, 734 (1926). — Comparison of matrix and wave mechanics. (185) for English translation
272. Shimizu, F., *J. Chem. Phys.*, **53**, 1149 (1970). — Laser–Stark spectroscopy in NH_3
273. Shimoda, K., and Shimizu, T., 'Non-linear Spectroscopy of Molecules', *Progress in Quantum Electronics, Vol.* 2., Pergamon (1972).
274. Shoolery, J. N., Shulman, R. G., Sheehan, W. F., Schomaker, V., and Yost, D. M., *J. Chem. Phys.*, **19**, 1364 (1951). — (and Schw . . .?) m-wave $CF_3C{\equiv}CH$
275. Shulman, R. G., and Townes, C. H., *Phys. Rev.*, **77**, 500 (1950). — Stark effect in Hughes–Wilson cell
276. Shulman, R. G., and Townes, C. H., *Phys. Rev.*, **78**, 347A (1950). — *l*-doubling transitions HCN
277. Silvers, S. J., Bergeman, T. H., and Klemperer, W., *J. Chem. Phys.*, **52**, 4385 (1970). — Optical m-wave double resonance in $^1\Pi^*$ state CS
278. Slawsky, Z. I., and Dennison, D. M., *J. Chem. Phys.*, **7**, 509 (1939). — Centrifugal distortion in symmetric tops
279. Smekal, A., *Naturwiss.*, **11**, 873 (1923). — Raman effect prediction
280. Smith, D., in (194). — m-wave spectroscopy review
281. Snyder, L. E. in *M.T.P. Int. Rev. of Science, Series one*, **3**, 193 (1972), Ed. D. A. Ramsay, Butterworths, London. — Interstellar spectra review
282. Snyder, L. E., and Buhl, D. To be published — Interstellar HNC?
283. Solomon, P. M., *Physics Today*, **26**, 32 (1973). — Interstellar spectra review
284. Starck, B., 'Molecular constants from microwave spectroscopy', in *Landolt–Bornstein Group II Atomic and Molecular Physics, Vol.* 4, Springer, Berlin (1967).
285. Stoicheff, B. P., in *Adv. in Spectroscopy*, **1**, 91 (1959), Interscience. — Review of gas phase Raman spectra and techniques
286. Stoicheff, B. P., in (194). — Review of Raman spectroscopy
287. Strandberg, M. W. P., *Microwave Spectroscopy*, Methuen (1954). — Rotational spectroscopy in a nutshell

288. Strandberg, M. W. P., Wentink, T., and Kyhl, R. L., *Phys. Rev.*, **75**, 270 (1949). — m-wave and Stark effect in OCS
289. Sugden, T. M., and Kenney, C. N., *Microwave spectroscopy of Gases*, Van Nostrand, London (1965).
290. Suzuki, I., *J. Mol. Spectrosc.*, **25**, 479 (1968). — Force field CO_2
291. Suzuki, I., Pariseau, M. A., and Overend, J., *J. Chem. Phys.*, **44**, 3561 (1966). — Force field CO_2
292. Swalen, J. D., and Costain, C. C., *J. Chem. Phys.*, **31**, 1562 (1959). — m-wave $(CH_3)_2CO$ barrier
293. Swalen, J. D., and Ibers, J. A., *J. Chem. Phys.*, **36**, 1914 (1962). — Inversion barrier in NH_3
294. Takeo, H., Hirota, E., and Morino, Y., *J. Mol. Spectrosc.*, **34**, 370 (1970). — Equilibrium structure SeO_2 isotope independence
295. Tanaka, T., and Morino, Y., *J. Mol. Spectrosc.*, **33**, 538 (1970). — Coriolis resonance O_3
296. Tannenbaum, E., Myers, R. J., and Gwinn, W. D., *J. Chem. Phys.*, **25**, 42 (1956). — Internal rotation CH_3NO_2, V_6
297. Thorson, W., and Nakagawa, I., *J. Chem. Phys.*, **33**, 994 (1960). — Quasi-linear molecules
298. Tinkham, M., *Group Theory and Quantum Mechanics*, McGraw-Hill (1964).
299. Tinkham, M., and Strandberg, M. W. P., *Phys. Rev.*, **97**, 937, 951 (1951). — EPR O_2
300. Tolles, W. M., Handelman, E. T., and Gwinn, W. D., *J. Chem. Phys.*, **43**, 3019 (1965). — m-wave CF_3NO_2 barrier
301. Törring, T., *Z. Physik*, **161**, 179 (1961). — *l*-doubling trans. HCN
302. Townes, C. H., Holden, A. N., and Merritt, F. R., *Phys. Rev.*, **74**, 1113 (1948). — m-wave of linear molecules, structure, quadrupole moments, OCS, ClCN, BrCN and ICN
303. Townes, C. H., and Schawlow, A. L., *Microwave Spectroscopy*, McGraw-Hill (1955).
304. Trischka, J. W., in (194). — Molecular beams review
305. Ueda, T., and Shimanouchi, T., *J. Chem. Phys.*, **47**, 5018 (1967). — Ring puckering potentials
306. Uehara, H., *Bull. Chem. Soc. Japan*, **42**, 886 (1969). — EPR SO
307. Vanasse, G. A., and Sakai, H., *Progr. Opt.*, **6**, 261 (1967). — Interferometers
308. Van der Waerden, B. L., *Sources of Quantum Mechanics*, Dover (1967). — Birth of quantum mechanics; labour pains. Key papers in English
309. Van Vleck, J. H., *Phys. Rev.*, **33**, 467 (1929). — Λ-doubling, Van Vleck transformation
310. Van Vleck, J. H., *Rev. Mod. Phys.*, **23**, 213 (1951). — What practically everybody knows about angular momentum

311. Van Vleck, J. H., and Weisskopf, V. F., *Rev. Mod. Phys.*, **17**, 227 (1945). Pressure broadening
312. Von Neumann, J., *Mathematical Foundations of Quantum Mechanics*, Princeton University Press (1955).
313. Watson, J. K. G., *Can. J. Phys.*, **43**, 1996 (1965). Correlation rule for flexible molecules
314. Watson, J. K. G., *J. Chem. Phys.*, **45**, 1360 (1966). Centrifugal distortion asymmetric rotors
315. Watson, J. K. G., *J. Chem. Phys.*, **46**, 1935 (1967). Centrifugal distortion asymmetric rotors
316. Watson, J. K. G., *J. Chem. Phys.*, **48**, 181 (1968). Centrifugal distortion asymmetric rotors
317. Watson, J. K. G., *J. Chem. Phys.*, **48**, 4517 (1968). Sextic centrifugal distortion coefficients
318. Watson, J. K. G., *Mol. Phys.*, **15**, 479 (1968). Simplification of Wilson–Howard Hamiltonian
319. Watson, J. K. G., *Mol. Phys.*, **19**, 465 (1970). Hamiltonian for linear molecules
320. Watson, J. K. G., *J. Mol. Spectrosc.*, **40**, 536 (1971). Centrifugal distortion in spherical tops
321. Watson, J. K. G., *J. Mol. Spectrosc.*, in press. Structure determination r_e r_s r_0 and r_m
322. Weatherburn, C. E., *Advanced Vector Analysis*, G. Bell and Sons Ltd., London (1962).
323. Weiss, R., *Rev. Sci. Inst.*, **32**, 397 (1961). Molecular beam detector
324. Weiss, S., and Leroi, G. E., *J. Chem. Phys.*, **48**, 962 (1968). IR CH_3CH_3 barrier
325. Welsh, H. L., Crawford, M. F., Thomas, T. R., and Love, G. R., *Can. J. Phys.*, **30**, 577 (1952). Toronto Arc Raman source
326. Welsh, H. L., Stansbury, E. J., Romanko, J., and Feldman, T., *J. Opt. Soc. Am.*, **45**, 338 (1955). Raman techniques
327. Westerkamp, J. F., *Phys. Rev.*, **93**, 716 (1954). *l*-doubling transitions HCN
328. White, J. U., Alpert, N. L., and De Bell, A. G., *J. Opt. Soc. Am.*, **45**, 154 (1955). multiple reflection (White) mirrors
329. White, R. L., *J. Chem. Phys.*, **23**, 249 (1955). Bridge spectrometer
330. Wigner, E. P., *Group Theory and its application to Quantum Mechanics of Atomic Spectra*, Academic Press, N.Y. (1959).
331. Wilson, E. B., *J. Chem. Phys.*, **3**, 276 (1935). Statistical weights
332. Wilson, E. B., *J. Chem. Phys.*, **27**, 986 (1957). Symmetric top centrifugal distortion constants and force field
333. Wilson, E. B., *Adv. Chem. Phys. II*, Interscience, N.Y. (1959). Review of internal rotation

334. Wilson, E. B., *Science*, **162**, 59 (1968). — Review

335. Wilson, E. B., Decius, J. C., and Cross, P. C., *Molecular Vibrations*, McGraw-Hill, N.Y. (1955). — Vibration–rotation Hamiltonian in Chapter 11.

336. Wilson, E. B., and Howard, J. B., *J. Chem. Phys.*, **4**, 260 (1936). — The vibration–rotation Hamiltonian, see also (318) and (62)

337. Wilson, E. B., Lin, C. C., and Lide, D. R., *J. Chem. Phys.*, **23**, 136 (1955). — Theory of internal rotation CH_3NO_2

338. Winnewisser, G., *J. Chem. Phys.*, **56**, 2944 (1972). — mm-wave HSSH, and DSSD where $\kappa = -0{\cdot}999\,999\,34$!

338a. Winnewisser, G., Winnewisser, M., and Winnewisser, B. P., in *M.T.P. Int. Rev. of Science, Series one*, **3**, (1972), Ed. D. A. Ramsay, Butterworths, London. — mm-wave review

339. Winnewisser, M., and Winnewisser, B. P., *J. Mol. Spectrosc.*, **41**, 143 (1972). — mm-wave HCNO

340. Winton, R. S., and Gordy, W., *Phys. Rev. Letts. A*, **32**, 219 (1970). — mm-wave Lamb dip

341. Winton, R. S., and Winnewisser, G. To be published. — Internal rotation splitting, mm-wave Lamb dip (338a)

342. Wofsy, S. C., Muenter, J. S., and Klemperer, W., *J. Chem. Phys.*, **53**, 4005 (1970). — Electric resonance CH_3D

343. Wollrab, J. E., *Rotational Spectra and Molecular Structure*, Academic Press (1967).

344. Wyse, F. C., Gordy, W., and Pearson, E. F., *J. Chem. Phys.*, **52**, 3887 (1970). — mm-wave AlF

Tables

345. *Tables Relating to Mathieu Functions*, Columbia University Press, N.Y. (1951). — Two-fold barrier and *A* states of three-fold barrier only

346. *Microwave Spectral Tables*, Nat. Bur. Stand. U.S. Monograph 70. — Monograph 70
   Vol. I, *Diatomic Molecules* (1964).
   Vol. II, *Line Strengths of Asymmetric Rotors* (1964).
   Vol. III, *Polyatomic Molecules with Internal Rotation* (1969).
   Vol. IV, *Polyatomic Molecules without Internal Rotation* (1968).
   Vol. V, *Spectral Line Listings* (1968).

Major texts on microwave spectroscopy: 100, 101, 287, 289, 303, 343.
General review papers on aspects of rotational spectroscopy: 35, 105, 106, 133, 171, 177, 184a, 197, 202, 203, 204, 235, 236, 280, 285, 286, 304, 339a.
Tables: 112(343), 284, 345, 346.

Author Index

Subject Index

A CATALOG OF SELECTED

DOVER BOOKS

IN SCIENCE AND MATHEMATICS

A CATALOG OF SELECTED

DOVER BOOKS

IN SCIENCE AND MATHEMATICS

CATALOG OF DOVER BOOKS

NUMERICAL METHODS FOR SCIENTISTS AND ENGINEERS, Richard Hamming. Classic text stresses frequency approach in coverage of algorithms, polynomial approximation, Fourier approximation, exponential approximation, other topics. Revised and enlarged 2nd edition. 721pp. 5⅜ × 8½.
65241-6 Pa. $14.95

THEORETICAL SOLID STATE PHYSICS, Vol. I: Perfect Lattices in Equilibrium; Vol. II: Non-Equilibrium and Disorder, William Jones and Norman H. March. Monumental reference work covers fundamental theory of equilibrium properties of perfect crystalline solids, non-equilibrium properties, defects and disordered systems. Appendices. Problems. Preface. Diagrams. Index. Bibliography. Total of 1,301pp. 5⅜ × 8½. Two volumes. Vol. I 65015-4 Pa. $12.95
Vol. II 65016-2 Pa. $12.95

OPTIMIZATION THEORY WITH APPLICATIONS, Donald A. Pierre. Broad-spectrum approach to important topic. Classical theory of minima and maxima, calculus of variations, simplex technique and linear programming, more. Many problems, examples. 640pp. 5⅜ × 8½. 65205-X Pa. $13.95

THE MODERN THEORY OF SOLIDS, Frederick Seitz. First inexpensive edition of classic work on theory of ionic crystals, free-electron theory of metals and semiconductors, molecular binding, much more. 736pp. 5⅜ × 8½.
65482-6 Pa. $15.95

ESSAYS ON THE THEORY OF NUMBERS, Richard Dedekind. Two classic essays by great German mathematician: on the theory of irrational numbers; and on transfinite numbers and properties of natural numbers. 115pp. 5⅜ × 8½.
21010-3 Pa. $4.95

THE FUNCTIONS OF MATHEMATICAL PHYSICS, Harry Hochstadt. Comprehensive treatment of orthogonal polynomials, hypergeometric functions, Hill's equation, much more. Bibliography. Index. 322pp. 5⅜ × 8½. 65214-9 Pa. $9.95

NUMBER THEORY AND ITS HISTORY, Oystein Ore. Unusually clear, accessible introduction covers counting, properties of numbers, prime numbers, much more. Bibliography. 380pp. 5⅜ × 8½. 65620-9 Pa. $8.95

THE VARIATIONAL PRINCIPLES OF MECHANICS, Cornelius Lanczos. Graduate level coverage of calculus of variations, equations of motion, relativistic mechanics, more. First inexpensive paperbound edition of classic treatise. Index. Bibliography. 418pp. 5⅜ × 8½. 65067-7 Pa. $10.95

MATHEMATICAL TABLES AND FORMULAS, Robert D. Carmichael and Edwin R. Smith. Logarithms, sines, tangents, trig functions, powers, roots, reciprocals, exponential and hyperbolic functions, formulas and theorems. 269pp. 5⅜ × 8½. 60111-0 Pa. $5.95

THEORETICAL PHYSICS, Georg Joos, with Ira M. Freeman. Classic overview covers essential math, mechanics, electromagnetic theory, thermodynamics, quantum mechanics, nuclear physics, other topics. First paperback edition. xxiii + 885pp. 5⅜ × 8½. 65227-0 Pa. $18.95